Mechanical Behavior of Concrete

Mechanical Behavior of Concrete

Edited by
Jean-Michel Torrenti
Gilles Pijaudier-Cabot
Jean-Marie Reynouard

iSTE

WILEY

First published 2010 in Great Britain and the United States by ISTE Ltd and John Wiley & Sons, Inc.

ISTE Ltd
27-37 St George's Road
London SW19 4EU
UK

www.iste.co.uk

John Wiley & Sons, Inc.
111 River Street
Hoboken, NJ 07030
USA

www.wiley.com

Library of Congress Cataloging-in-Publication Data

Comportement du béton au jeune âge and comportement mécanique du béton . English
Mechanical behavior of concrete / edited by Jean-Michel Torrenti, Gilles Pijaudier-Cabot, Jean-Marie Reynouard.
p. cm.
Includes bibliographical references and index.
ISBN 978-1-84821-178-0
1. Concrete--Analysis. 2. Concrete--Curing. I. Torrenti, Jean-Michel. II. Reynouard, Jean-Marie. III. Pijaudier-Cabot, Gilles. IV. Title.
TP882.3.C6613 2010
624.1'834--dc22

2009049963

British Library Cataloguing-in-Publication Data
A CIP record for this book is available from the British Library
ISBN 978-1-84821-178-0

Printed and bound in Great Britain by CPI Antony Rowe, Chippenham and Eastbourne

FSC
Mixed Sources
Product group from well-managed forests and other controlled sources
Cert no. SGS-COC-002953
www.fsc.org
© 1996 Forest Stewardship Council

Table of Contents

Foreword

Because it is a versatile, durable, and cost-effective building material, concrete is still the most used construction material today: we cast an average of 1 ton of concrete per year per inhabitant in the world. Therefore, it is important to understand the mechanical response of this material and to model it as accurately as possible. Depending on the required use and geographic location, concrete structures are subjected to different loading histories on different time scales. The aim of this book is to discuss the behavior of concrete under these different conditions. Its content is directed toward masters students, doctoral students, engineers, and researchers who wish to have a concise overview of modern modeling of concrete and concrete structures.

The first part of the book concerns instantaneous or time-independent models. Chapter 1 is a comprehensive presentation of the advanced experimental techniques and experimental characterization that is required before modeling is undertaken. Chapter 2 details two classical types of constitutive models for concrete, elastic plastic models and continuum damage-based models, and their combinations. The implementation of these models in the framework of the finite element method calls for robust numerical techniques to integrate non-linear constitutive laws step by step. Therefore a section in Chapter 2 is dedicated to this aspect of modeling, which is so often neglected. The subsequent chapter deals with an original aspect of the behavior of concrete: structural size effect. Concrete is a quasi-brittle material. When cracking occurs, a very large microcracked area appears at the tip of the crack. The interaction between this area, denoted the "fracture process zone", and the rest of the structure is the source of a pronounced size effect. The fractal nature of the cracks could also explain this effect. Whatever the physical explanation of the structural size effect, experimental results show that the nominal stress at rupture of a structural element could be decreased by 25% when the size of the structure is

---

Written by Jean-Michel TORRENTI, Gilles PIJAUDIER-CABOT, Jean-Marie REYNOUARD.

increased four times. A more accurate model and reduced safety coefficients in the design codes imply a consistent appraisal of this effect, which is still frequently concealed in safety factors today.

In the second part of the book, cyclic and dynamic loadings are considered. In the case of exceptional loads, such as earthquakes, complex models have to be implemented in order to describe the response of concrete. Chapter 4 is devoted to the cyclic behavior of concrete and shows how, in a structural calculation, complex aspects such as closure of cracks, and the existence of a hysteresis during loading-unloading cycles may be captured. Chapter 5 concerns the response of concrete to fatigue loads. The process of fatigue of concrete is directly related to the propagation of microcracks inside this material. Microcracking due to fatigue constitutes an irreversible phenomenon that contributes to the slow, but progressive, deterioration of the inner structure of the material. In the case of repeated loadings these effects must be taken into account and a review of existing models for the fatigue behavior of concrete is presented. Chapter 6 deals with the rate-dependant behavior of concrete and models for transient analyses. Rate dependency and the influence of a high level of hydrostatic pressure are examined regarding experimental, theoretical, and modeling.

Because concrete is a continually developing material, the last part is devoted to the evolution of the properties of concrete with time and to delayed strains. Concrete undergoes many phenomena during the first dozen hours that follow batching and setup when fresh, and it progressively hardens due to hydration. Concrete may be subjected to different thermal (thermal curing, ambient temperature), hydric (drying, wet curing), or mechanical stresses. In this context, its behavior will strongly depend on its composition and on the environmental loads during its early age. The aim of Chapter 7 is to describe these two groups of influential parameters. Chapter 8 presents an example of the modeling of concrete at early age. The modeling element is completed by the experimental determination of the parameters of the behavior of concrete at early age. Finally, Chapter 9 deals with two important phenomena: shrinkage and creep. Delayed strain is still subject of intensive research, from the structural scale down to the nanoscale. These efforts contribute then to delimit the range of validity of existing simplified models and to elaborate more advanced constitutive relations, e.g. based on multiple-scale analyses of concrete viewed as a cementitious composite. It is in this spirit that outstanding and future issues are outlined in the concluding remarks of this book.

PART 1

Instantaneous or Time-Independent Models for Concrete

Chapter 1

Test Techniques and Experimental Characterization

1.1. Introduction

Detailed experimental analysis of concrete material has always been a complex task. This complexity is due to the variability of the material, which depends on its composition, its conditioning, and on the mechanical, thermal, or chemical loads it is undergoing. For precise experimental characterization it is necessary to know the composition and "history" of the material. Accuracy and rigor, as always, are required for an experimental study on this type of material.

Moreover, the complexity of concrete characterization has amplified the need for experimental studies. The goal is to understand, explain, and model the behavior of concrete. Experimental studies of concrete will always have at least one of the following aims: *characterization*, *modeling,* or *validation*. Indeed it is necessary to characterize the behavior of material under different loadings experimentally. An understanding of the physical phenomena involved is then possible. Phenomenological or micro-mechanical modeling is improved using the observations made during tests, and the obtained models have to be validated via an experiment on a specimen or a structure.

In order to be concise, we will focus here on the experimental characterization of concrete under mutli-axial mechanical loading. Moreover, we will limit ourselves to the study of quasi-static loads, that is for deformation rates ranging from 10^{-5} to

---

Chapter written by Nicolas BURLION.

$10^{-2}\ s^{-1}$. We can therefore suppose that long-term effects (mainly shrinkage and creep) and dynamic effects are negligible. We will consider the conditions to be isothermal and at room temperature. Indeed, thermal strains on concrete material leads to various physical and physicochemical processes (internal pressure due to water evaporation, aggregates setting off, phase changing, Calcium Silicate Hydrate CSH dehydration, etc.) which change the microstructure of the material, and therefore, its mechanical behavior (more information can be found in the following papers: [BAZ 96; HEI 98; NEV 00; NEC 00]). In addition, concretes undergoing chemical attack will not be considered in this study as microstructural modifications influence their mechanical behavior (see as examples [BAR 92; CAR 96; GER 96; TOR 99a]). The concretes and mortars studied herein are archetypal; specifically, their composition, assembly, and conservation modes are common in civil engineering, which allows us to show the general characteristics of their mechanical behaviors.

The initial focus of the chapter will be ways to consider the specificities of the materials towards a reliable and adequate experimental analysis. The interpretation of an experimental result of a concrete or a mortar cannot be performed out of this study's context. Boundary conditions, curing conditions, or the maturation of the material have to be specified. In the second part, classical extensometry will be described. The influence of the boundary conditions will also be related to extensometry which can only be performed on the external surfaces of the specimen. Moreover, we will see that it is not always possible to obtain a "material" test, i.e. where all the characteristics and the applied stresses are homogenous. Depending on the observation scale, concrete can be considered as either a homogenous material (which we will assume in most of the interpretations) or as a composite or heterogenous material (implying an additional difficulty regarding the analysis of the physical phenomena explaining these types of behavior). In this chapter, we will consider the material as homogenous, even if microstructural observations cannot support our reasoning. The last section of this chapter will focus on the study of the main test techniques used to test a concrete material under multiaxial stresses. The results obtained for mortar or concrete will be given in parallel. Of course, these different techniques can be used to identify a mechanical behavior or validate a model.

1.2. Experimental specificities related to concrete material

1.2.1. *Composition and variability of the material*

"Classical" concretes (see [BAR 96a; DRE 98] for more information) that will be studied are made of a granular skeleton of a sand-lime type, with constant grain size distribution and with common cleanness [BER 96]. The binding agent is a commercially available cement without any particular properties [BAR 96b]. Workability is standard and concrete implementation is done as usual. We will not cover high-performance concretes (more information can be found in the following papers: [MAL 92; DEL 92; LAP 93; BAR 94; BEL 96; AIT 01]). Once mixed, concrete hardens and its mechanical performances increase with time. It then becomes a porous material, made of aggregates and of hydrated cement, which contains a fluid interlayer. The influences of this liquid interlayer can be delayed: it can modify the short-term behavior of concrete via evaporation; for example, leading to desiccation cracking, and therefore, the existence of defects located at the surface. Interactions between the microstructure, the fluid interlayer and the environment can also modify the mechanical behavior of concrete. The action of the environment can become a major factor and sometimes leads to a change of the poromechanical properties of concrete [COU 95].

Moreover, the aggregates and cement used to make the material are often of local origin, which implies a significant range from one production site to another. It should be noted that the experimental results obtained for a particular concrete or mortar are a function of all ingredients entering into the mix. Therefore, the comparison of two results must be done carefully. Nevertheless, it is possible to provide general characteristics of concrete behavior as long as it is assumed that concrete, even if made of sand, gravel, cement, and water, is a homogenous material at the considered mechanical test scale. In addition, a reliable comparison of various results can be performed by using a similar cement and Leucate's normalized sand (complying with the EN 196-1 and ISO 679 rules), in the case of a mortar. This choice of cement and similar sand whatever the test, under constant conditions, allows the variability to be reduced. In the case of a concrete, it is commonly accepted that the mineralogical and geometrical nature of the aggregates used are kept the same.

In the same way, the behavioral variability of one specimen to another has to be considered. Even if the specimen is of the right size, concrete material is still a heterogenous material. This implies that the mechanical characteristics of the different components are not exactly identical. Under the action of mechanical loads, the stress and deformation fields within the specimen are also heterogenous [TOR 87], and vary from one specimen to another as the granular arrangement is not the same for a series of specimens assumed to be identical. A variation of the mechanical properties of the material (elasticity modulus, uniaxial tensile strength, uniaxial compressive strength, etc.) is then observed. These are more pronounced as the tensile stresses or the tensile deformations evolve in the specimen. Locally, those tensile stresses lead to microcracking, by exceeding the strength threshold, which is not constant. As a result, a dispersion of the mechanical characteristics appears when the applied stresses allow some extension (in the case of uniaxial tension or uniaxial compression). Nevertheless, when those extensions are not possible or are limited, experimental dispersion is decreased (in the case of a triaxial compression test). To conclude, at least three identical tests are performed to correctly quantify the mechanical behavior of mortar and concrete materials.

1.2.2. *Specimen preparation*

1.2.2.1. *Specimen characteristics*

Regarding all the mechanical tests we are about to present, it is necessary to use, or in some cases to design, a specimen that allows a number of rules to be respected: first, this specimen (therefore, its shape and its dimensions) have to respect the basic condition of the test of a material, i.e. it must be a representative elementary volume (usually called REV). This REV must comply with the assumption of homogeneity of the concrete; therefore, it should be large with respect to the heterogeneities within the material and must not be too large in order to limit the structural effects. Moreover, a size effect exists regarding concrete (see section 1.2.3), leading to the constant use of samples with the same dimensions. Finally, it must allow the application of mechanical effects.

Two types of samples are usually used: cylindrical specimens for uniaxial and triaxial tests; cubic specimens for biaxial and triaxial real tests.

1.2.2.2. *How to obtain specimens*

There are two ways in which specimens can be obtained. The first is molding. In France, the measurement of simple compressive strength is done on specimens cast using steel, cardboard, or polypropylene cylindrical molds with a height to diameter ratio of 2. Their diameter is chosen as a function of the diameter of the largest aggregate. The diameter of the specimen has to be at least three-times longer than the largest aggregate's diameter in order to minimize the wall effects. These effects are due to an over-concentration of the fine elements in the mix near the mold's wall. For specimens of large dimension with respect to the aggregate size, this effect is negligible but has to be considered for the smallest ones. The dimensions of the molds are normalized (NF P 18-400), the most commonly used have a 160 mm diameter and a 320 mm height. In the central zone of the specimen (effective zone) compressive stress may be considered as uniaxial and uniform (see section 1.3.2).

The second way of obtaining specimens is saw-cutting or core drilling (NF P 18-405 codes). This consists of drilling the specimens from a large block of the material. With this technique, any specimen geometry can be obtained and wall effects are avoided. Drilling can lead to a superficial cracking of the material, which is an intrinsic bias of this technique. Figure 1.1 shows three specimens obtained from molding, core drilling and saw-cutting. Regarding molding (picture on the left side), the obtained surface state does not permit the aggregates to be seen. It should be noted that bubbles can be hidden by the cement paste at the surface and be a source of error if the strain is measured nearby. Therefore, it is necessary to prepare the surface before gluing strain gages in order to avoid this potential problem (see section 1.3.1) Conversely, the saw-cut and drilled specimens show the aggregates. The deformation gages stuck to their surfaces have to be large enough to "average" the differential deformations between the cement paste and aggregates. Indeed, those two constituents do not usually have the same elasticity modulus, which implies that on the scale of an aggregate, there is a heterogenous deformation.

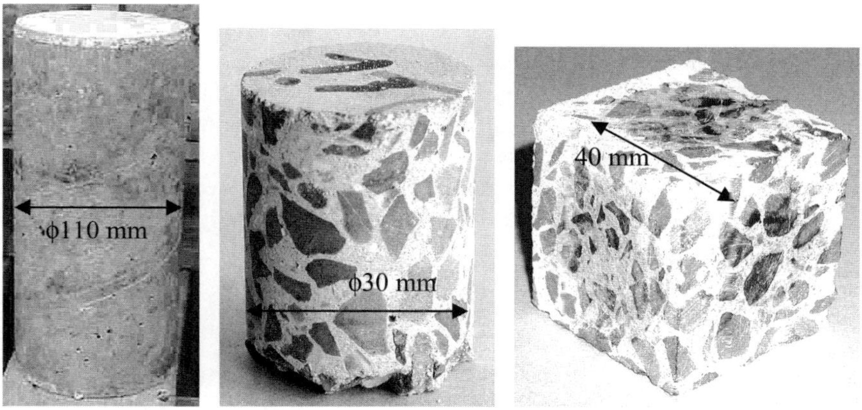

Figure 1.1. *Concrete specimens obtained using molding, core drilling and saw-cutting*

1.2.2.3. *Initial anisotropy*

Mechanical behavior of concrete material is quite complex as it can be influenced by an initial anisotropy due to different phenomena, mainly related to its formation and its mode of implementation. In particular, the direction of casting has a non-negligible importance. Under the action of gravity, compaction of the granular skeleton occurs in a privileged way. Moreover, during the set up of the material into its mold, the rise of air bubbles dragged along the formworks and inside the material (Figure 1.2) can be limited to the presence of aggregates that block this entrapped air. Initial and oriented defects will then be present "under" the aggregates. Those initial defects will then influence the initiation and propagation of microcracking.

The initial anisotropy on the uniaxial tensile behavior is characterized by a decrease of the peak strength when the tension direction is parallel to the casting direction, compared with the one measured in the perpendicular direction [HUG 70]. Similar results have been obtained in uniaxial compression [HUG 70; MIE 84]:

Figure 1.3 shows deformation curves for a compressive stress parallel to casting direction and along a perpendicular direction, the maximal stresses are logically the same, whereas peak deformation is more important than when loading is parallel to casting direction. The elastic properties, measured in both directions, are also different. It is necessary to consider this phenomenon in the characterization of concrete material, especially when the water to cement ratio is high (due to the excess of water and significant porosity).

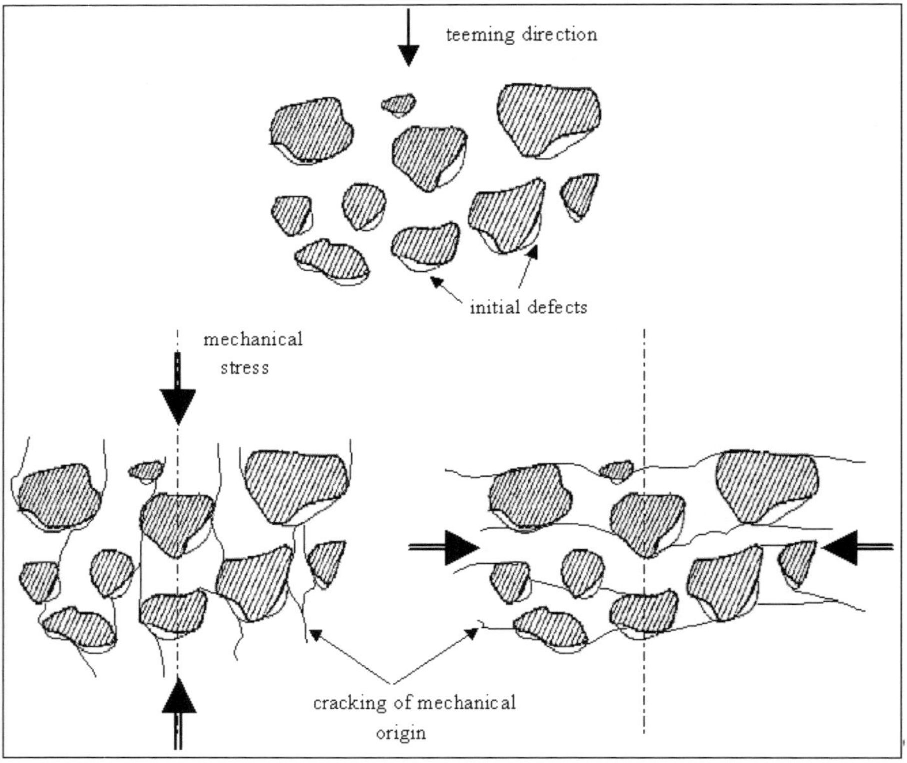

Figure 1.2. *Origin of the initial anisotropy induced by teeming direction [MIE 84]*

Other sources of initial anisotropy can be identified: the hydric gradient existing between the surroundings and the core of a specimen; the effects of hydration heat of the cements (leading to differential dilatations); or also the segregation of the aggregates. As an example, in Figure 1.4 the deformations measured on three perpendicular sides of a concrete cube (40 mm side) under hydrostatic pressure are presented. This concrete cube has been obtained from sawing the central part of a prism of 40×40×160 mm^3 dimensions. Regarding hydric gradient, the cube was placed for a month in water (in order to avoid desiccation) before any mechanical tests, and then stored under a relative humidity of 60% for a year, in order to enable the drying and desiccation shrinkage of the concrete to evolve. This physical phenomenon leads to microcracking of the material, oriented along the main hydric direction of the gradients, i.e. perpendicularly to the longitudinal direction of the prism. After sawing, four faces of the cube underwent drying, and therefore, significant microcracking. As the behavior of concrete is elastoplastic, there is an initial anisotropy due to drying, the elastic properties at the start of loading being different. In addition, this initial anisotropy influences the non-linear behavior, in particular, the irreversible deformations that evolve differently depending on their measurement on dried faces or faces prevented from drying.

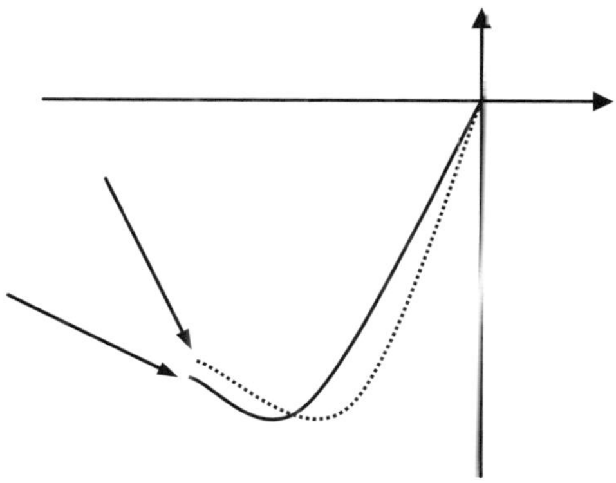

Figure 1.3. *Influence of initial anisotropy induced by the direction of casting on the uniaxial compressive behavior [MIE 84]*

1.2.3. *Preservation and curing conditions*

Concrete is a "living" material, or more exactly, a material whose physicochemical and mechanical properties evolve with time as a function of its initial composition, its implementation, the environment in which it is being preserved, and the applied mechanical stresses (for example, creep of concretes will be the more important if loading is applied directly after the setting). Among them, the main factors modifying the mechanical behavior of the material in time are the temperature and the surrounding humidity. We can list other factors that will not be considered here, for example, the action of chemicals (leaching [CAR 96; GER 96; LEB 01; HEU 01], attack by sulfates and chlorides [BAR 92; NEV 00], etc.) or the effects of delayed deformations (shrinkage and creep, for a general picture see [BAZ 82; ULM 99; BAR 00; ULM 01]).

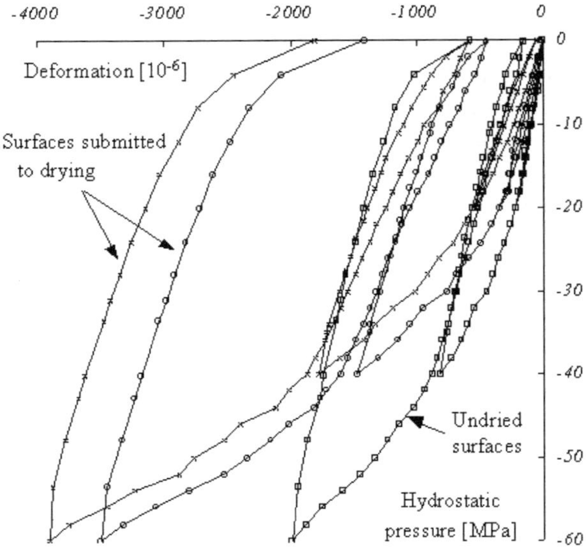

Figure 1.4. *Influence of the initial anisotropy induced by the drying of concrete on the hydrostatic compressive behavior: hydrostatic pressure/deformations curves [BUR 01a]*

The effects of temperature, resulting from the heat release during the hydration reaction of cement, on the maturation of concrete have been widely studied (see [NEV 00]), both from a kinetics standpoint, resulting in increasing strength, and from a thermal shrinkage standpoint. It should be noted that these effects do not modify the characteristics of the material. In the same way, the surrounding humidity in which the specimens are preserved has an influence on the maturation of the material and its delayed deformations [BAZ 82; ULM 99; NEV 00].

It is important to define "stable" curing and conservation conditions in order to precisely characterize the mechanical behavior of a concrete or a mortar. Usually, a 28 day humid curing (a shorter duration is obviously possible to investigate the variations of the mechanical behavior during this period), then conservation at room temperature equal to 20 ± 1°C and relative humidity of 60 ± 5% are commonly used conditions in research laboratories. Ideally, specimens have to be preserved in water or under a hygrometry higher than 90% until the test is carried out. Unfortunately, due to experimental reasons (gage sticking, surface preparation, etc.), this condition is barely observed in the literature. Due to the dependence of the mechanical behavior of concrete on its maturation conditions, for every mechanical characterization test, the following points (proposed by the TC14-CPC technical committee of The International Union of Laboratories and Experts in Construction

Materials, Systems and Structures (RILEM) [RIL 72; 73]) have to be stated as a minimum:

– type and dimensions of the specimen;

– composition of concrete;

– how to implement concrete;

– how to obtain specimens;

– curing conditions;

– conservation conditions;

– number of identical tests performed or experimental scattering of the results.

These points contribute to the quality and traceability of the published results.

1.2.4. *Size effect*

A particular phenomenon involving materials with a cementitious matrix is the size effect; specifically, the measured mechanical behavior, in particular the maximal strength, depends on the size of the specimen used during the test (see [BAZ 98; VLI 00] for more information). Overall, uniaxial tensile strength and uniaxial compressive strength decrease when the size of the specimen increases. In addition, this relation is not linear.

This size effect, which will be described in more detail in Chapter 3, can be explained e.g. using the theory of the weakest link: the bigger the volume of the specimen, the more important the probability of finding a defect is (i.e. the more defects there are), and the more the probability of the decrease of the maximal strength increases. This size effect is less sensitive when there is a decreased possibility to get tractions in the specimen, which is the case of triaxial compression with a strong confinement. More generally, the size effect decreases with the increase of hydrostatic pressure [ROS 97]. Depending on the tests, this phenomenon has to be considered, especially when comparing different experimental results between each other.

Another important factor influencing the mechanical behavior of concrete is that, regarding temperature and hygrometry, the specimen is actually a "structure" whose two parameters evolve with time as a function of the conservation conditions and the sample size. This induces some size effect due to the ratio of the size of the specimen to the characteristic length/times involved in thermal and hydric analyses. For example, the maximal stress in uniaxial compression increases with the drying of concrete (all the other properties being constant), and the elasticity modulus decreases with regard to the microcracking induced during desiccation [TER 80; BUR 00; YUR 03]. The hydric and thermal state of the material will be a function of the date the test was performed, but also of the size of the specimen (hydric and thermal equilibriums will be reached more or less rapidly). We then have to be aware of the current state of the material that is undergoing the mechanical test.

1.3. Extensometers and experimental conditions

Before characterizing the behavior of concrete under multiaxial stresses, we will detail some of the classical techniques of strain and displacement measurement and the boundary conditions on a concrete specimen. It is necessary in order to provide some consistent modeling of the experiments, and thus consistent interpretation of test data. The experimental issues are as follows:

– stress measurement is not possible, as it is a theoretical concept: it can only be deduced from the surface strain measurement or from a pressure measurement of a specimen. Stress determination is still the result of a calculation from other measured values or reverse methods, which initially postulate a constitutive law;

– deformation measurement in the core of a specimen is difficult and the global deformation field is usually extrapolated from surface measurements;

– metrology must not modify the deformation or stress field where the measurements are being carried out;

– boundary conditions in the experiment have to match, almost exactly, the desired distribution of stress or strain, for instance homogenous distributions if the constitutive law is to be directly calibrated.

1.3.1. *Classical extensometry on a concrete specimen*

Regarding concrete, the physical quantities usually measured for a multiaxial quasi-static test are local displacements of the sample, deformations, and applied forces (force, couple, or pressure). It should be noted that measurements of temperature, relative humidity, or acoustic emissions are also performed, but will not be dealt with in this chapter.

Sensors can be the cause of *errors*, i.e. there is an intrinsic difference between the measured value and the value given by the sensor. The *sensitivity* of the sensor is the ratio between the variation of physical effects (displacement, deformation, temperature variation, etc.) and the value (usually electrical) variation given by the sensor. A sensor works on a *measurement range,* which has to be considered. When a sensor is used it is then necessary to consider the following: *accuracy* describing the aptitude of the sensor in giving results close to the real value and the value required: the more sensitive a sensor is, the more accurate it is; *fidelity* describing the reproducibility of the performed measurement. Accuracy and fidelity both make a sensor ideal. To keep this accuracy, calibration is necessary on a regular basis.

Three main groups of sensors are used in civil engineering: "active" sensors which turn intrinsically physical effects into electronic data that can be used directly: a piezo-electrical sensor directly turns its applied stress into electricity; "passive" sensors whose parameters are sensitive to the physical value to be measured indirectly: i.e. dynamometric rings that are deformed (this is what is being measured) under a mechanical load (this is what we want to know); or strain gages where what is really being measured is the electrical resistance of a wire stuck to the specimen, caused by its deformation.

Figure 1.5. *Photograph of a Vishay-Micromesures® gage stuck to a regular mortar*

1.3.1.1. *Strain gages*

The principle of strain gages stuck on the surface of a specimen undergoing deformation is to measure the variation of electrical resistance of a thin wire whose cross-section diminishes upon stretching (due to Poisson's effect). After multiplying by the *gage factor*, we obtain the local deformation of the specimen. The data acquisition to be processed is relatively simple [AVR 84]. The most common ones have a resistance of 120 Ω, those used to get more precision are of about 350 Ω. Due to the reduced size of gages, deformation measurement is performed locally in the direction of the grid. Rosettes (made of three gages at 45° or 60°) allow the whole deformation tensor to be locally determined. The size and type of gage is a function of the size of the specimen and of the aggregate size. The dimension of the gage has to be large enough so that it is not affected by local heterogenities. The drawbacks of gages are their fragility and often complex implementation (in the case of humid or damaged surfaces). The choice of the adequate glue can also be difficult, in particular, for really porous materials.

The measurable strain range, with an accuracy of 0.1%, goes from 5×10^{-6} to 10^{-1}. For significant deformations (from 1 to 10%), one should be extremely careful with the choice of the glue, which has to accommodate those deformations. Numerous factors, such as temperature, humidity, test duration, and strain amplitude during the test, have to be considered when the "gage+glue" complex is to be determined [AVR 84]. Some particular gages are used to detect and monitor cracking propagation (gage with brittle wire), or pressure measurement (known as pressure gages, which is used for dynamic tests). Finally, it should be noted that gages are sensitive to temperature and other factors, such as hydrostatic pressure or transversal deformation of the grid. Appropriate corrections have to be made as a function of those different parameters.

1.3.1.2. *Displacement sensors*

Regarding displacement measurement, the tools usually used are "LVDT" (Figure 1.6b) and gage displacement sensors. A LVDT (linear variable differential transducer) is a linear displacement sensor based on variations of the magnetic properties of a ferromagnetic bar moving in a winding, which can easily be turned into electrical data. This sensor measures the displacement of a point at the surface of the specimen in the desired direction; it needs specific electronic conditioning, and the contact between the sensor and specimen must always be maintained (using glue, for example). Displacement sensors with gages have the advantage of using the same electronic data acquisition as is used by a single gage. Turning displacement into electrical data is done via measuring the 350 Ω strain gages stuck to a string's blade, which get deformed when the measuring rod goes into the sensor's body. It is worth noting that a small effort is then applied to the specimen (lower than 4.4 N in the case of a Vishay® sensor), which insures a glue-free constant contact.

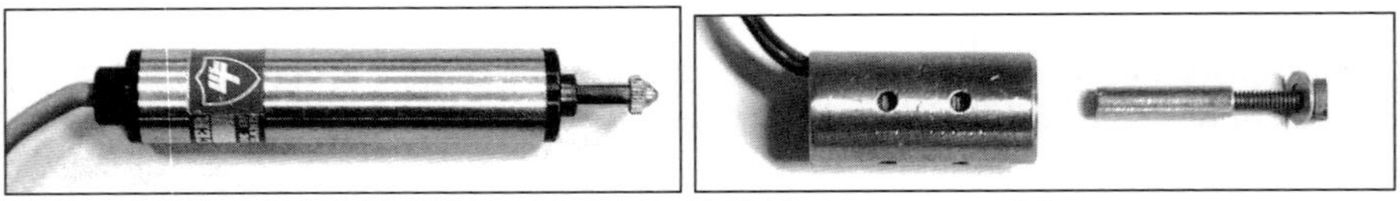

a) b)

Figure 1.6. *Photograph of a) a gage displacement sensor, and b) a LVDT*

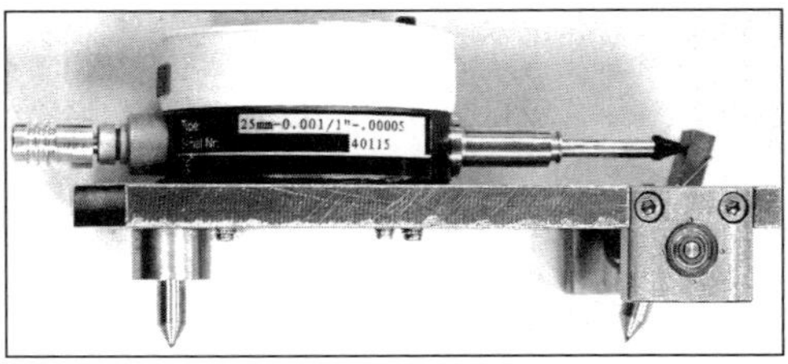

Figure 1.7. *Photograph of a cone deformeter (Controlab®)*

The displacement measurements performed with these sensors are global. Analyses of specimen dilation or shrinkage, during bending, for example, can also be done with displacement sensors. For instance, a cone deformeter is presented in Figure 1.7, which is used to measure a deformation. The idea is to use a displacement sensor to measure the variation of the distance between two cones stuck on the surface of the specimen, the cones being present to ensure a perfect relative position between the deformeter and the measured point.

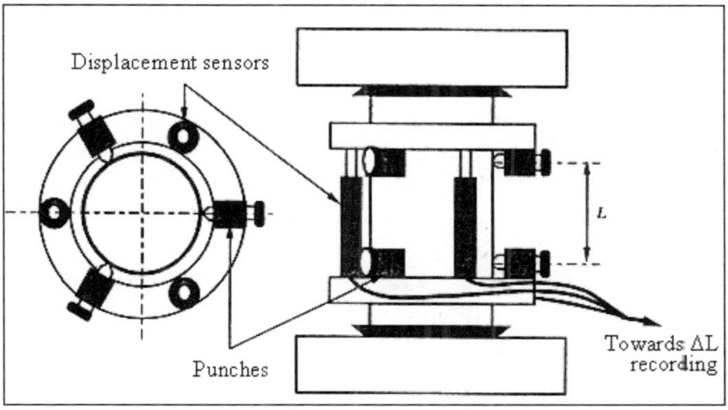

Figure 1.8. *Compression extensometer [DUP 95]*

Other displacement sensors such as *vibrating strings* are used in civil engineering. The vibration frequency of a string submitted to excitation depends on its tension. By fixing the two extreme points of a metallic string to a specimen, we can monitor its frequency variation, and thus the variation of length between two points of the specimen.

The uniaxial compression test is common on concrete (see section 1.4.5), and axial deformation measurement is easily carried out. For this reason a specific compression extensometer has been developed [BOU 80]. Its principle is similar to a cone deformeter as the measured displacement corresponds to the connection of two circular rings fixed on a concrete specimen using some punches (Figure 1.8). With the help of three displacement sensors located at 120°, and data acquisition, this sensor directly provides the average axial deformation of the specimen during loading. Its handling is simple and the accuracy of the obtained results is excellent [BOU 99].

To obtain the lateral deformation of the specimen during a uniaxial compression test, the use of a chain sensor can be considered (Figure 1.9). The principle is based on the measurement of the relative displacement of the two units of the roll chain which monitor the inflation of the specimen, and gives a global measure of the lateral deformation.

Finally, other types of measurements are possible, such as acoustical or optical measurements. In the case of acoustical measurements, the celerity of elastic waves in concrete are used to determine the elasticity modulus of the specimen. Optical measurements present the advantage that there is no contact between the measuring device and the sample. This can be interesting in the case of a specimen at high temperature or during the localization of deformation, when strain field is very inhomogenous. With image analysis, the relative displacements of a grid or a speckle pattern printed on the specimen before the test are measured optically. Recent studies on concrete undergoing bi-traction have produced interesting results [AST 01]. This principle corresponds to field measurements: some laser or image analysis based devices are real "equivalent" displacement sensors and present similar characteristics to classical sensors but without any contact.

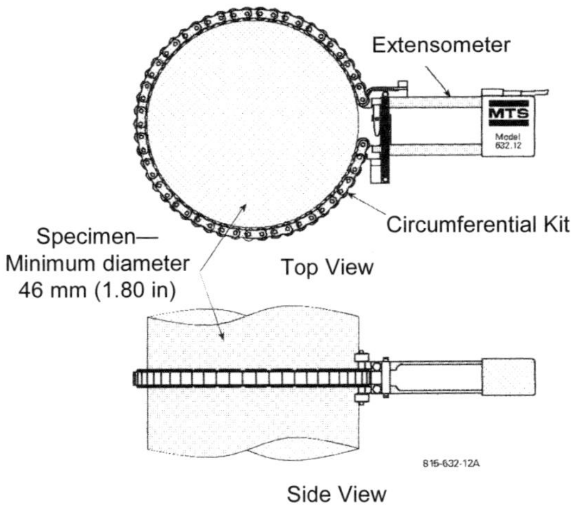

Figure 1.9. *Lateral deformation chain sensor (MTS®)*

1.3.1.3. *Force sensors*

A force sensor is chosen by default and belongs to the test machine. Nevertheless, the precision of this sensor, its measurement range, and its rigidity should be carefully considered. The working principle of a force sensor is to measure the deformation of a probe body (Figure 1.10), precisely known as a function of the applied force (determined in a factory using calibration). Depending on the boundary conditions imposed to the specimen during the test, the rigidity of the sensor has to be analyzed.

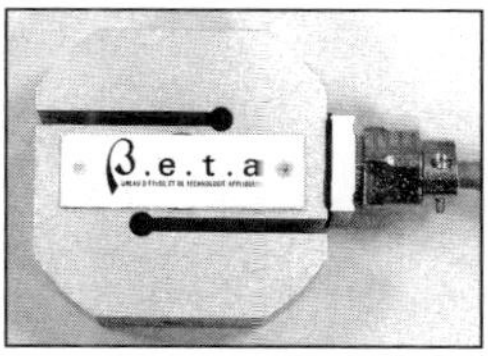

Figure 1.10. *Photograph of a force sensor (Béta®)*

1.3.2. *Boundary conditions and experimental set-up*

Application of the desired boundary conditions is always done using a testing machine and supports on which the specimen is placed. Even if those supports can seem to be ideal, the specimen-machine interface influences the experimental result, and the testing frame as well. For example, during a compression loading test on a high-strength concrete, a brittle failure is observed (usually with the explosion of the specimen), mainly due to the elastic unloading of the test frame during the post-peak loading phase. The stiffer the frame is, the weaker its elastic deformation and the weaker the "explosion" of the specimen will be. Concerning the plates or the supports of the specimen, the ideal case would be that for contacts between the specimen and the test machine, the experimental apparatus exactly follows the deformations of the specimen (under the Poisson effect, for example). This is difficult to obtain, but it is possible to get close to it. Friction between the specimen and the plates of the machine can lead to experimental bias such as spurious bracing.

Therefore, preparation of the surfaces of the specimen should be done extremely cautiously. It is essential that that force transmission between the test machine (assumed to be as perfect as possible) and the specimen is done without any imperfections. In order to guarantee this, surfaces in contact with the plates of the machine have to be perfectly flat (general case) or, in general, that the plate-specimen contact is homogenous (in the case of brush platen supports, Figure 1.14). In addition, parallelism between the outermost surfaces of a uniaxial or triaxial compressive specimen has to be as perfect as possible in order not to introduce a bending moment, which could alter the homogeneity of the compressive stress.

Along the same line, perpendicularity between the axis of the specimen (cylinder or cube) and the support surfaces has to be guaranteed. To overcome these two defects, a ball joint and fixed support system can be used (see Figure 1.11 for a uniaxial compression test). The fixed support enables the reference frame of the specimen to be aligned with that of the machine, whereas the ball joint mostly erases the perpendicularity and parallelism defects. Other types of boundary conditions can be used as a function of the relative importance of those defects regarding the mechanical behavior of the material, for example in the case of uniaxial deformation where the homogeneity of the strain field within the specimen is extremely sensitive to the boundary conditions. Figure 1.12 shows an example of a behavior obtained in uniaxial compression in terms of axial stress as a function of the axial deformations measured by two gages located at both ends of the specimen in its center. When the parallelism of the faces of the cylinder, and also the perpendicularity of its generator are not perfect regarding the machine, there might initially be some induced tensile strain in the specimen due to the bending moment, which obviously can lead to spurious damage of the material. Moreover, the obtained behaviors present total and irreversible different deformations, whereas a real material test should give an identical response. Figure 1.12 shows the stress versus (axial and radial) deformation curves obtained on a specimen similar to the previous one, with ball joint to support boundary conditions, which gives much better results. Finally, the choice of ball joint is a function of the size of the specimen so as to minimize the friction forces (within the ball joint) induced by the specimen regarding the ball joint effects.

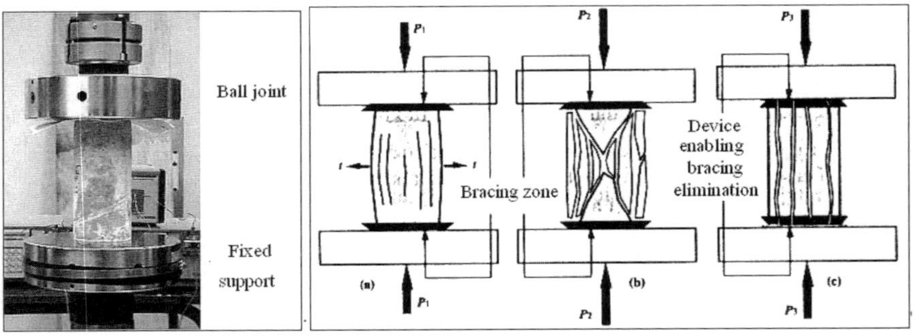

Figure 1.11. *Classical supports of a test machine: ball joint and fixed. Fracture mode of concrete under uniaxial compression [DUP 95]*

Figure 1.13 shows the consequence of boundary conditions and the size of the specimens on the uniaxial tensile behavior of sandstone [MIE 96] (whose tensile behavior is similar to the behavior of a mortar). In the case of blocked supports in rotation (Figure 1.13a), the obtained behavior depends on the size of the specimen (different maximal value, different ductility). When the supports are free to rotate (Figure 1.13b), the obtained experimental responses are independent of the size of the specimen. This difference can be explained by the fact that in the case of fixed rotating supports, right at the initiation of a microcracking on the edge of the specimen, a bending moment occurs which is not avoided by the experimental set up. This moment is more important as the diameter of the specimen increases, which leads to lower apparent strengths for larger sizes (without correction for this bending moment). If the supports are ball jointed, this bending moment is balanced using the rotation supports.

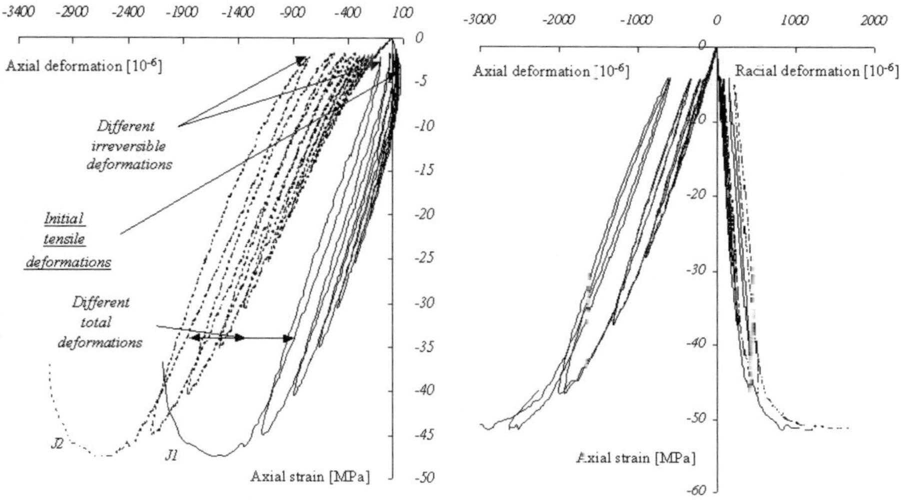

a) b)

Figure 1.12. *Stress-strain curves on regular mortar (water:cement = 0.5): a) in case of a poor capping and rigid support; b) in case of perfectly ball jointed [YUR 03]*

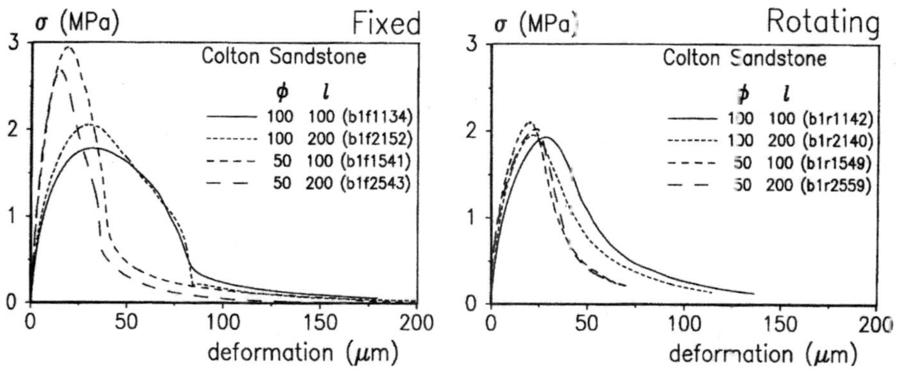

a) b)

Figure 1.13. *Effect of the supports on the mechanical behavior of sandstone in traction: a) fixed rotating supports; b) ball jointed supports [MIE 96]*

Friction between the specimen and the test machine is often unavoidable, and leads to the bracing of the edges of the specimen (Figure 1.11). In the case of uniaxial compression, bracing cones appear if friction is not avoided, resulting from a triaxial compression state at the ends of the specimen. The obtained cracking facies is then dependant on friction (Figure 1.11b). If there is very little friction between the plates of the machine and the specimen (Figure 1.11c), the compressive stress is uniform within the specimen and leads to vertical cracks. Reduction of friction is provided by anti-bracing systems, for example, grease, brush platens (Figure 1.14a), sand boxes (Figure 1.15), deformable supports, aluminum foil covered with talc, or a Teflon® plate. In general, those systems follow the specimen during its deformations perpendicular to the stress direction (usually due to the Poisson effect within the pane of the supports), while ensuring that no spurious force appears. Regarding uniaxial compressive tests, one of the first ideas was to lubricate the plate-specimen contact using grease. This works well for small applied loads, but for higher applied loads grease is driven away. Talc and stacked aluminum foils can also be placed between the specimen and the machine. It is also possible to use the plates of a material with a low friction coefficient between steel and concrete, such as Teflon®. Finally, a sand box can also be used [BOU 89] (Figure 1.15), whereby the sand grains act like small marbles allowing the lateral motion of the specimen surface under the Poisson effect.

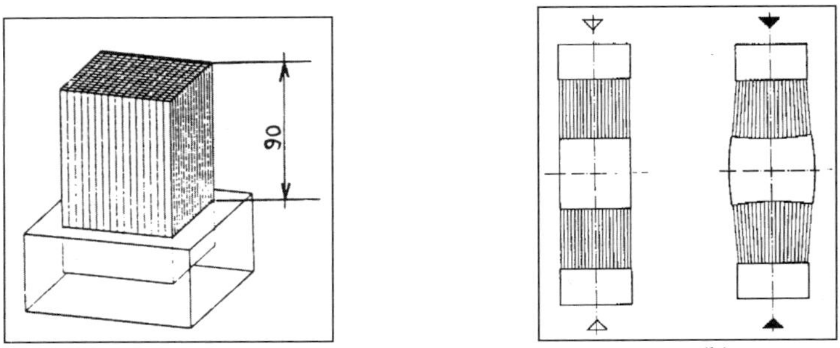

(a) (b)

Figure 1.14. *a) Example of a brush platen; b) brush rotation effects [MIE 84]*

These devices are suitable for uniaxial compression. In the case of multiaxial stresses, brush platens are more suitable (Figure 1.14a). The idea is to have a brush applying a compressive force on the specimen. The bending stiffness of the brush is as small as possible in order to avoid bracing effects (Figure 1.14b). The brushes are steel barrettes maintained together. The design of those special supports is based on the buckling loads of the brushes.

In general, the specific type of support used in the experiments induces some consequences on the experimental results, after the peak load is reached especially [KOT 83; MIE 84; TOR 87; VON 89; MIE 91; TOR 91; VLI 96; MIE 97a]. We can sum them up regarding uniaxial compression as follows [TOR 96]:

– elastic behavior is virtually the same, whatever the boundary conditions are;

– post-peak behavior of the specimens is strongly dependant on the boundary conditions of the specimen;

– localization of the deformations is observed (generally occurring just before the peak), and its appearance is maximal at the stress peak, leading to a macro-cracking being parallel to the compression axis (Figure 1.11c);

– this localization depends on the boundary conditions;

– friction effects are more important when the specimen is small, a test without any anti-friction system leads to an overestimated ultimate strength;

– characterization of the post-peak behavior of the material is very difficult and problematic (see section 1.4.5.3).

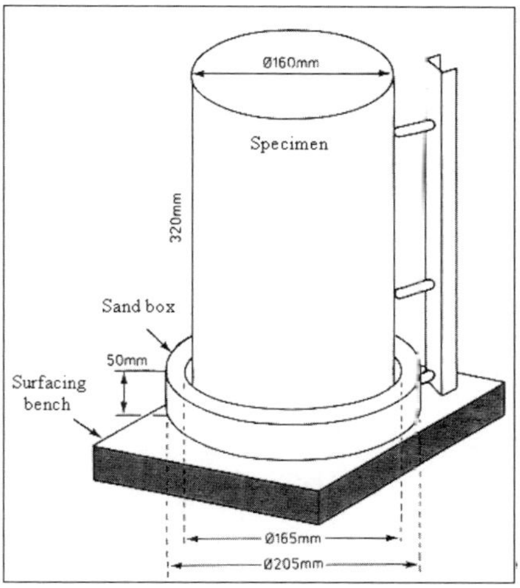

Figure 1.15. *Sand box [BOU 89]*

An alternative to the use of specific supports is the inter-position of material interfaces between the specimen and the test machine, which bear transverse deformations similar to the deformation of the concrete tested: the ratios of the Poisson coefficient with Young's modulus have to be close for the specimen and the interface. Aluminum [TOU 95] or reactive powder concretes [TOR 99a] may be used as interface materials.

1.4. Behavior of concrete under uniaxial stress: classical tests

In the first section of this chapter, we introduced most of specificities of experimental analysis on concrete. With what follows, we will describe classical tests. The first part will deal in more details with the uniaxial tests that are the most used, as most of the stress states in civil engineering are uniaxial. In the second part, multiaxial tests, necessary for particular applications, will be presented.

1.4.1. *Direct uniaxial tension*

During this test, a homogenous tensile stress is applied on the specimen using steel or aluminum heads stuck to the ends of the specimen (Figure 1.16). A double ball jointing of the ends of the specimen is used to make sure the tensile strain is homogenous within the specimen (see section 1.3.2). In general, the test is carried out at an imposed rate. This is one of the hardest test to perform as the brittle behavior of concrete under this type of stress only provides the first part of the curve, that is to say until the loading peak. At that point, local rupture of the material occurs causing the propagation of a through crack within the specimen. Localization of deformation comes along with cracking and causes an elastic unloading of the other parts of the specimen. Subsequent deformations are not homogenous within the specimen, the zone where they monotonically increase is called the "zone of deformation localization". Elastic unloading may cause the loss of stability of the specimen and brutal failure. It is then difficult to obtain the behavior of the softening phase. In order to obtain the post peak response of the specimen, it is often necessary to implement some local displacement control during the test (local relative displacement measured on the specimen) [TER 80, GUO 87, VIS 94].

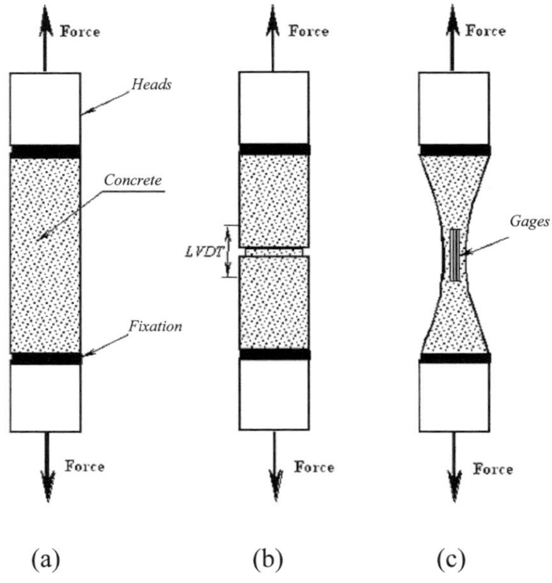

Figure 1.16. *Principle of direct traction test*

In order to better control post-peak behavior, in other words, to control the localization of the deformations, some authors rely on notched specimens (Figure 1.16b; [MIE 93]) or "diabolo"-shaped specimens (Figure 1.16c; [BEL 96]) allowing the localization zone to be known and then equipped with displacement transducers. Typical responses are shown in Figure 1.17a: initially, the material is almost elastic and linear until the peak, for stresses of about 2 to 5 MPa regarding usual concretes, and a softening phase follows the loading peak. During softening the material becomes damaged, and its elasticity modulus drops sharply, due to significant cracking. Some irreversible deformations are then observed if the material is unloaded. Visser and van Mier have shown that the saturation degree of concrete has a small influence on tensile behavior (Figure 1.17b; [VIS 94]). Nevertheless, the ratio between the peak stress of a dry mortar and of a humid mortar (saturated in water) in direct traction is always higher or equal to 1, and varies as a function of the nature of the aggregates [TER 80]. Starting at the onset of localization of the deformations and after, the curve $\sigma - \varepsilon$ has no real intrinsic meaning if the strain ε is not the strain measured inside the localization zone (where the strain is assumed to remain almost constant).

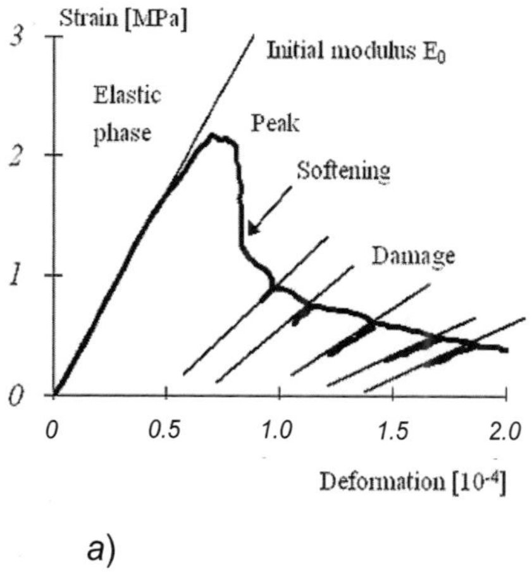

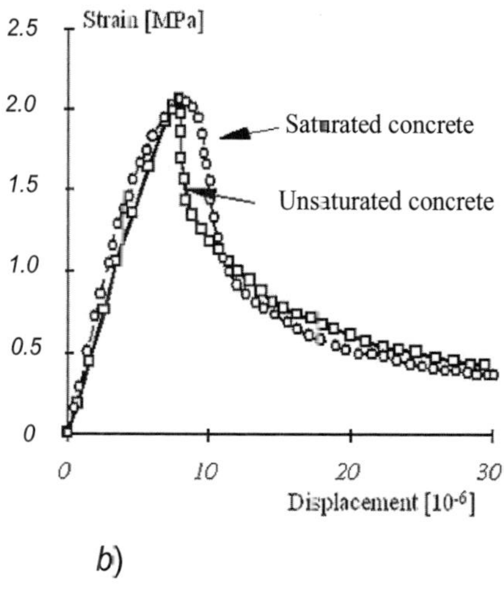

Figure 1.17. *Behavior of a strained concrete in direct tension: a) case of loading-unloading [TER 80]; b) influence of the water content [VIS 94]*

1.4.2. *Indirect uniaxial tension*

Due to the experimental difficulty of carrying out direct traction tests, different tests are more commonly used, relying on the dissymmetry of the compressive and tensile strengths of the concretes. In some specific experimental situations, it is possible to obtain locally a tensile fracture of the specimen being loaded in compression. The most commonly used test is called the "Brazilian" or splitting test which is the diametral compression of a disk (Figure 1.18; NF P 18-408 rule). The distribution of the normal stress is sketched in Figure 1.18b. A quasi-uniform tensile stress develops along the diameter of the specimen. This tensile stress is proportional to the applied compressive force. Only the maximal force is measured, this test is only useful to determine the maximal tensile strength. It is not useful to determine the tensile response of concrete. The stress distribution is far from being homogenous over the specimen and the determination of the tensile response would require some structural analysis and then be determined indirectly.

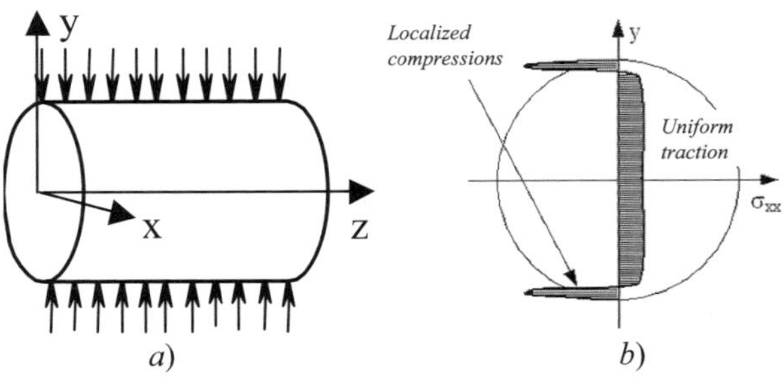

Figure 1.18. *Indirect traction tests with splitting and stress distribution*

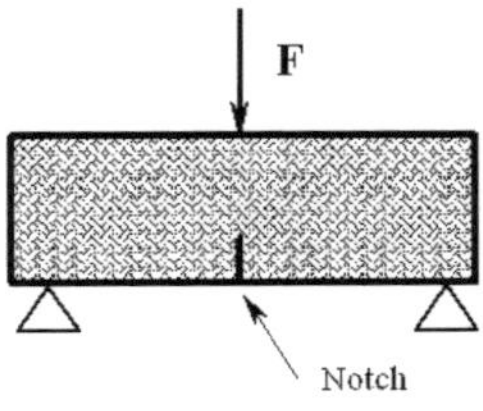

Figure 1.19. *Indirect traction tests through three-point bending*

Another type of indirect traction test is the three-point bending test on a concrete specimen that may or may not be notched (Figure 1.19, RILEM recommendation [RIL 73]) or the four-point bending test. The principle is to develop a moment within the beam, and therefore, to call upon the tensile lower fibers, the higher fibers being elastic due to dissymmetry behavior. The boundary conditions used are roll supports at the ends of the beam, to enable shrinkage during loading. The applied force and the deflection at the center of the beam are measured. It is possible to stick deformation gages on the beam in order to get local information, or cracking recording gages above the notch. The test control is achieved through an imposed displacement or an imposed force.

Because the tensile strain is not homogenous within the body of the specimen, same as in the splitting tests, the interpretation of the obtained results is difficult [SAO 88]. Nevertheless, with a three-point bending test it is possible to determine the tensile response of the material using a reverse analysis because of the simplicity of the distribution of the normal stress within the median section of the beam. Moreover, a notch is generally placed in this median cross section in order to introduce a defect minimizing the resistant section, and therefore to enable better control of the cracking and collapse process. It is then possible to control the test as a function of the crack propagation, which allows the tensile softening response of concrete to be determined because loss of stability of the loading process is avoided. This test is most often used for the determination of rupture parameters, or the characterization of the efficiency of reinforcement in concrete (reinforced concrete, fiber-reinforced concrete, adhesive composite material supports).

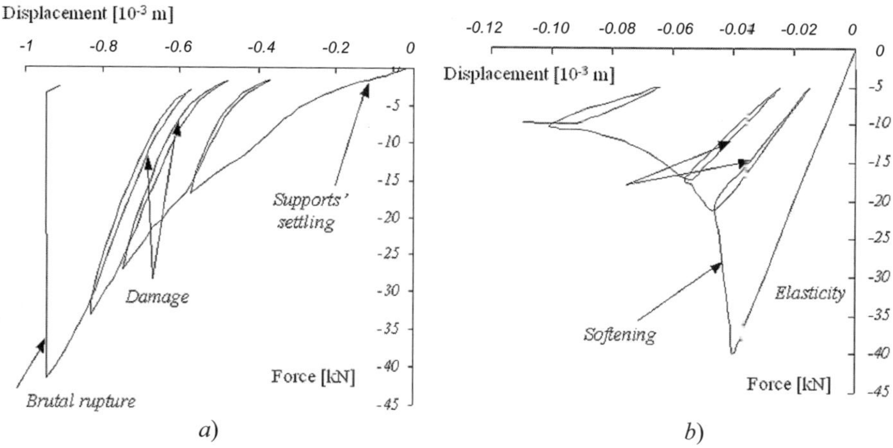

Figure 1.20. *Typical force: deflection result with three-point bending (without any initial notches): a) without any particular precautions and b) with precautions*

The three-point bending test requires some caution. Crushing of concrete at the supports often occurs (Figure 1.30a). This leads to a variation of the absolute position of the center fiber: therefore, the reference of the deflection has to be independent from possible crushing. This is performed by associating the concrete beam with a bar supported by two points belonging to the center fiber and using it as a measurement base. If those precautions are taken, we obtain a reliable representative response, from which it is possible to back calculate the tensile response of the material.

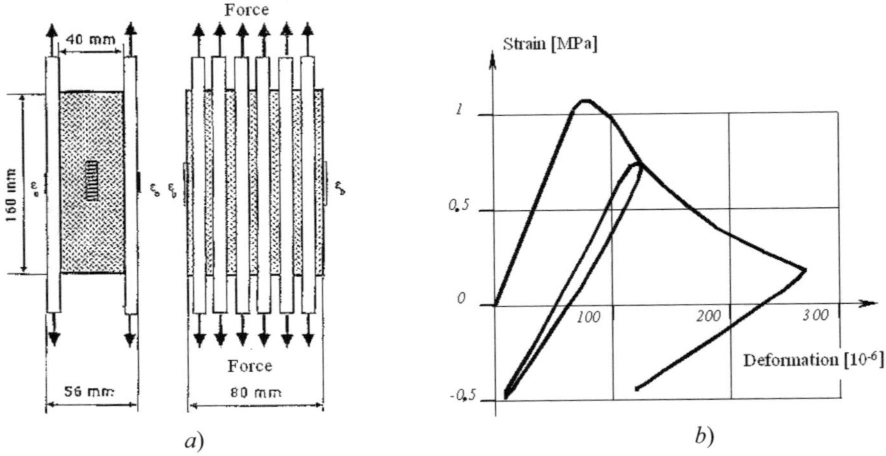

Figure 1.21. *a) Principle of the PIED test; b) axial stress-axial deformation curve of the obtained concrete [RAM 90]*

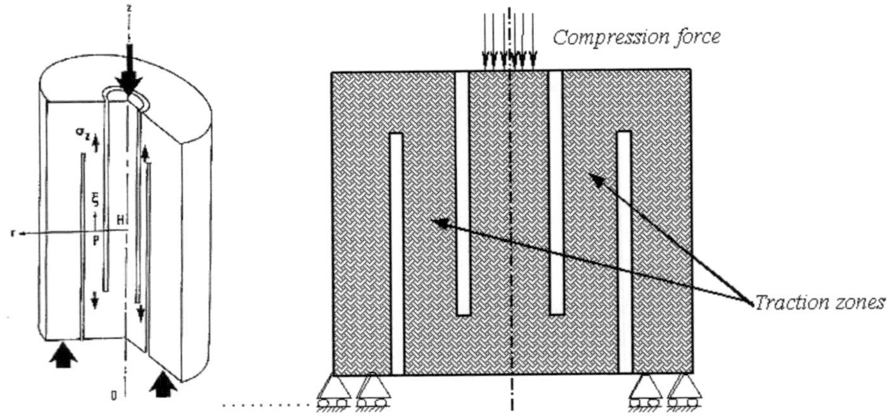

Figure 1.22. *Principle of bi-tube traction test [LUO 86]*

1.4.3. *Diffuse tension test*

This diffuse tension test called PIED (acronym in French for "Pour Identifier l'Endommagement Diffus", Figure 1.21a) has been developed to "stabilize" the test, by controlling the appearance of one or several micro-cracks [PIJ 87; BAZ 89; MAZ 89; RAM 90; BER 91].

The idea is to apply the tensile force on concrete using aluminum rods stuck to the specimen. Those rods are elastically deformed, and impose a homogenously distributed deformation on concrete. Analysis of the results is done by uncoupling the behaviors of aluminum (the elastic behavior is well known) and the behavior of the concrete, same as in reinforced concrete. The results are similar to those obtained for a direct tension test. The process has been improved in order to achieve cyclic loads [RAM 90] (results are presented in Figure 1.21b) or to allow the material permeability to be measured upon microcracking (BIPEDE test [GER 96]).

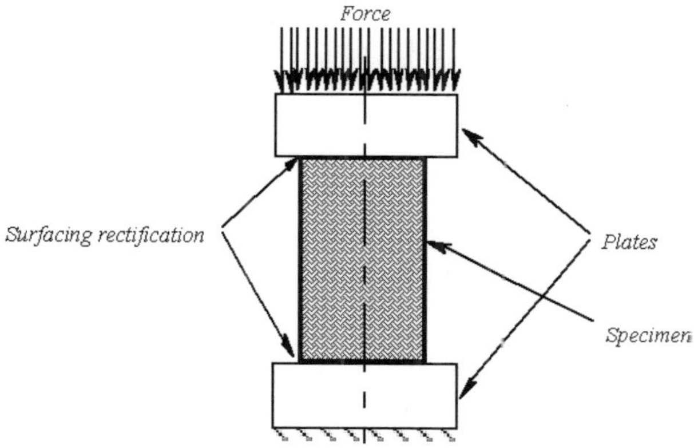

Figure 1.23. *Principle of simple compressive test*

1.4.4. *Bi-tube tension test*

The principle behind this tension test is again to use the tensile/compressive dissymmetric behavior of the material. Luong had the idea to design a "specimen-structure" subjected to compression in order to induce tension in the material (Figure 1.22; [LUO 86]). The test consists of compressing the center part of the specimen whilst simply supporting the outer part. The intermediate part is then subjected to tesion, on a quasi-homogenous way. Both the center and outer parts of the specimen remain elastic during compression. The advantage of this method is that setting up the test is as simple as a regular simple compression test (see the following section).

1.4.5. *Uniaxial compression*

1.4.5.1. *Behavior and rupture*

This is the most commonly test (Figure 1.23). It is carried out on cylinders or cubes of concrete. In general, the normalized test is controlled at an imposed stress rate, but an imposed displacement allows the post-peak regime of the response to be obtained. Regular concretes are slightly more ductile than rocks. Their maximal strengths vary from 20 to 800 MPa (from ordinary concretes to reactive powder concrete now called Ductal®).

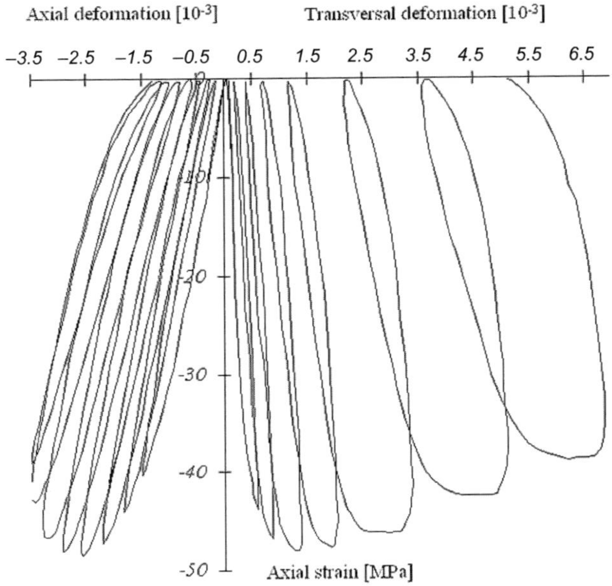

Figure 1.24. *Typical results of simple compression test on concrete [RAM 90]: axial (ε_1) and radial (ε_2) deformations as a function of the axial stress*

During loading, deformations perpendicular to the principal compressive stress appear, creating micro-cracks as the tensile deformation threshold is being exceeded. Micro-cracks coalescence leads to the collapse of the specimen. Moreover the elastic characteristics of the material evolve; the elasticity modulus decreases during the loading whereby the material becomes damaged due to micro-cracking. Some irreversible deformations appear. As we have already seen, the boundary conditions of the specimen play an important role on the characterization of the behavior of the material during simple compression. Due to friction, bracing cones appear at failure. Just the central part of the specimen is subjected to a uniaxial compression stress. Various "anti-bracing" devices have been developed (see section 1.3.2). Figure 1.24 shows a typical compressive uniaxial response of concrete. After the peak load, the Poisson coefficient suddenly increases. In the same way, damage growth is occurring more rapidly. The various stages of damage during a uniaxial compression test are presented in Figure 1.25 [MAZ 84].

In addition, it can be seen on Figure 1.24 that the material is slightly visco-elastic during compression. This is illustrated by the curved shape of the unloading curves as well as the hysteresis during loading-unloading cycles. Visco-elasticity is a function of the water content of concrete and decreases with the drying of the material.

PHASES	X-RAY AND OPTIC MICROSCOPY	VELOCITY OF ULTRASONIC WAVES	EXTENSOMETRY	COMMENTS
$0 < \sigma < 0.4 fc$	–Micro-crack occurs before loading especially localized under the biggest grains	– Almost no evolution	– Elasticity modulus E and constant Poisson coefficient $(E = E_0 . v = v_0)$ Longitudinal ε_i^* and ε_1^2 transversal permanent deformations	– Little damage: quasi-elastic behavior under a short-term loading
$0.4 < \sigma < 0.8 fc$	– Micro-cracking propagation to the grains' periphery, from bottom to top (regarding the direction, the stress axis being vertical)	– Low decrease (of a few %)	$E \searrow$, $\varepsilon_i^* \nearrow$ } Low speed evolutions $\boxed{v = v_0}$ $\varepsilon_i^* \neq 0$	–Micro-cracking with constant Poisson coefficient
$0.8 < \sigma < fc$	– Micro-cracks get oriented vertically and reach the paste	– Slighter decrease (about a dozen %)	$E \searrow$, $\varepsilon_i^* \nearrow$, $\boxed{v \nearrow}$, $\varepsilon_i^* \nearrow$ } Stronger evolutions	–Anisotropic micro-cracking with constant Poisson coefficient
Beyond *fc* (until rupture)	– Creation of vertical cracks	– Strong decrease (up to 100%)	$E \searrow$ up to $E_\sigma / 2$ $\varepsilon_i^* \nearrow$ up to 0.5 10^{-3} $v \nearrow$ up to 0 5, $\varepsilon_i^* \nearrow$ up to 10^{-3} } Rap. evol.	– Strongly anisotropic micro-cracking with volume increase: cracks appear

Figure 1.25. *Results comparing different analyses of concrete deterioration under uniaxial compression [MAZ 84]*

1.4.5.1.1. Water content influence

Interlayer water within the material influences its visco-elastic behavior. Several other phenomena are related to the water content and the presence of hard inclusions (aggregates) within a cementitious matrix, which retracts as the water leaves. Concrete and mortar strengths increase when free water leaves [TER 80; POP 86; BAL 94; BUR 00; YUR 04], because of the capillary effect and increased suction within the partially saturated porous medium. At the same time, micro-cracking induced by desiccation shrinkage leads to a decrease of the initial elasticity modulus of the material [TER 80; TOU 95; BUR 00; YUR 04], thus a decrease of Young's modulus [YUR 04]. This micro-cracking is due to the structural effect induced by the shrinkage difference between the core and the surface, and also to the effect of rigid inclusions within a retracting matrix, which leads to micro-cracking. The latter effect is the more important if the size of the aggregate is small [BIS 01], it will then be more sensitive within concretes than within mortars.

1.4.5.2. *Measurements of the elastic constants of the material*

Concrete can be considered as an initially isotropic material. The elastic parameters of the material are Young's modulus (E) and Poisson coefficient (ν). Regarding common concretes, those usual values of the parameters are *30,000 MPa* and *0.2*, respectively, and are used in numerous constitutive laws and numerical calculations for concrete structures, as well as for the determination of the delayed (time dependent) deformations of concrete. Therefore it is important that the measurement of these parameters be accurate [BOU 99; TOU 99] following the rules of RILEM [RIL 72] or specific procedures such as the one proposed by the laboratoire central des ponts et chaussées [TOR 99b], in order to avoid any errors and misinterpretations due to the visco-elastic behavior of concrete. Moreover, it is also necessary to use an adequate extensometer [BOU 99] or deformation gages stuck at the center of the specimen. Figure 1.36 provides an example. After three cycles of loading-unloading, which can reach 9 MPa (about 25% of the compression strength), the elasticity modulus (E_0) is measured as the slope of the line passing by the vertexes of the third hysteresis loop (if this loop is really occurring, otherwise, regular measurement is being carried out). The Poisson coefficient is equal to the ratio of the transversal modulus (E_t) and the longitudinal modulus (E_0), multiplied by (*–1*). Concretes present an elastic behavior up to 60% of their peak strength.

1.4.5.3. *Characterization of post-peak behavior*

The peak corresponds to the maximal value reached by the compression stress. In general, at this state, we observe the formation of macro-cracking parallel to the direction of compression [TOR 93a]. Experimentally, it is difficult to obtain the softening response because redistribution inside the specimen occurs and the strain distribution is no longer homogenous over the specimen. Moreover, the testing frame will also unload. The induced stress drop can be unstable with respect to the control mode of the test. In the case of imposed force, instability occurs at the force peak. In the case of imposed displacement, instability will occur when acceptable displacement increments are negative. We then talk about *snap-back*.

The post-peak response of concrete is necessary, for example, in studies related to the durability of the material as transport properties are very sensitive to the cracking of the material (see [BAZ 82; ACK 88; BAR 92] for general information). Displacement control of the plates of a test machine, which is very rigid, provides the post-peak response of a surfaced concrete specimen experimentally. Figure 1.27 shows an example axial stress-strain curve. Stress is calculated from the force of the machine and strain is global, that is to say computed from the variation of distance between the supports of the specimen. This type of curve cannot be used for the absolute measurement of elastic parameters or behavior, but allows comparisons (peak stress, peak deformation, softening, ductility, damage evolution, irreversible deformation evolution, etc.) between the tests performed under the same conditions.

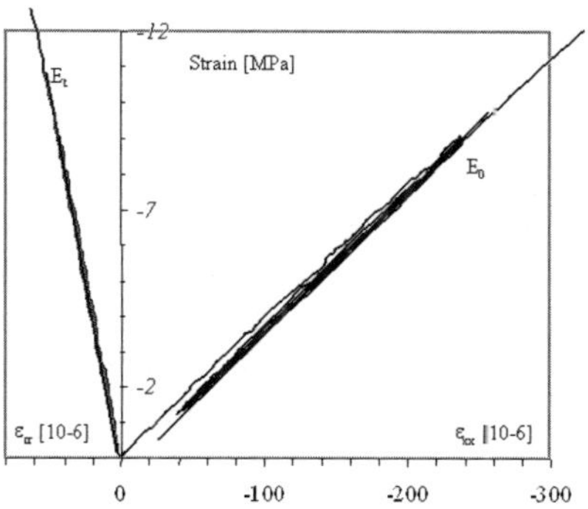

Figure 1.26. *Characteristics of post-peak behavior*

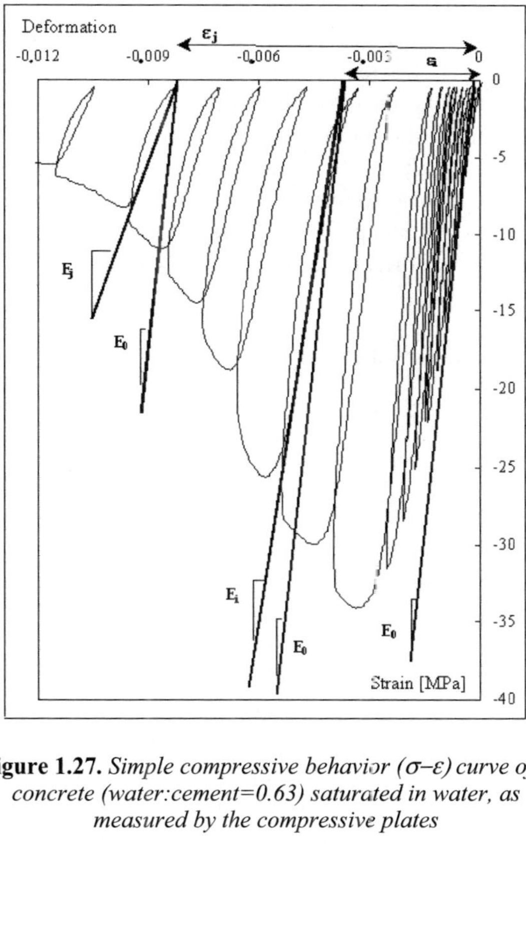

Figure 1.27. *Simple compressive behavior (σ–ε) curve of a concrete (water:cement=0.63) saturated in water, as measured by the compressive plates*

Another possibility to obtain the complete curve of the uniaxial compressive mechanical behavior of concrete is to control the test as a function of an increasing parameter, for instance, the circumferential lateral deformation of the specimen [SHA1 87; GET 96; TOR1 99]. Indeed, this quantity always increases, even during the softening phase. It is also possible to "turn" the classical plane σ-ε, in order to control the test within a plane for which "turned deformation" will always be increasing: the "turned stress" axis is a tangent to the curve of the behavior of concrete at the origin [MIE2 97; JAN 97], the rotation angle of these axis being that between the stress axis and the elastic part of the behavior of the material.

1.4.6. *Uniaxial torsion*

In a similar fashion to direct tension, the torsion test is not often carried out for two reasons: the test itself is difficult to perform and torsion is not often present in civil engineering, and does not, generally, cause to structures to collapse, except in very particular cases. Nevertheless, its study is necessary as it implies crack opening in mixed mode I and II, whereas most of the tests favor mode I. Torsion as a function of the applied rotation is similar to the uniaxial tensile stress-strain curve [LIL 01]: it presents an elastic phase, and then once the peak is reached, significant softening is observed. Cracking is oriented at about 45° to the twisting axis of the cylinder. In the case of a test performed on a notched cylinder, cracking is located within the notch plane; if a compression stress is applied during torsion, a circular-based cone is formed and is oriented towards the twisting axis [LIL 01].

1.5. Concrete under multiaxial stresses

In most real structures, stresses are multiaxial. In most cases, knowledge of the uniaxial behavior of concrete is enough to solve the issue. The knowledge of the multiaxial response is necessary for specific structures such as confinement vessels. In order to better characterize the response of concrete under multiaxial stresses, as well as to investigate the possibility of checking the calibration of models performed on simple tests, an inventory (inevitably non-exhaustive) of the various responses for various states of stress is proposed in the next sections.

Two main test families will be presented in order to show the evolution of the behavior of concrete. First of all, we will describe biaxial and triaxial tests. In each case, we will try to find out the mechanisms and main parameters influencing the material response. The role of the loading path is important during multiaxial loads, as anisotropy is induced by the creation of microcracks: for conciseness, this aspect will not be fully developed in this section (for more information, refer to [MAZ 84; MIE 84; PIJ 85; TOR 87; RAM 90; TOR 93b]).

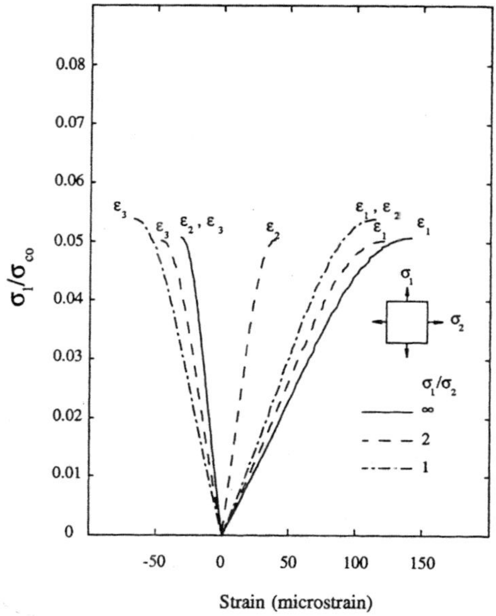

Figure 1.28. *Biaxial tensile behavior [HUS 00]*

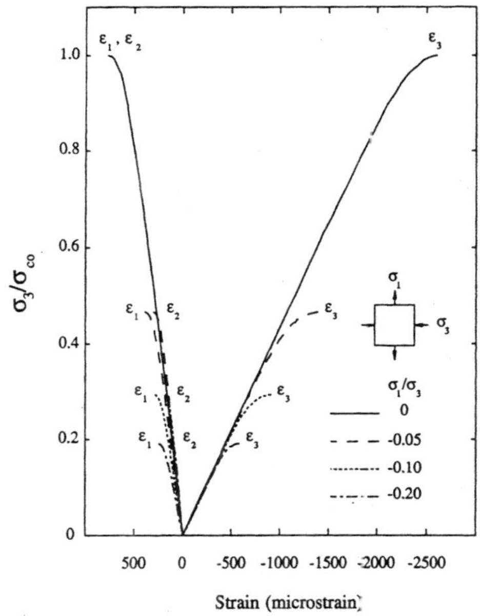

Figure 1.29. *Confined tensile behavior [HUS 00]*

1.5.1. *Multiaxial stresses with tension*

1.5.1.1. *Biaxial tension*

Biaxial tension tests, same as simple tension tests, are really delicate to be performed. The material behaves in the same way as during simple tension, that is to say elastic-brittle with an ultimate strength almost equal to that measured in simple tension. Failure planes are perpendicular to the main direction of traction [KUP 73].

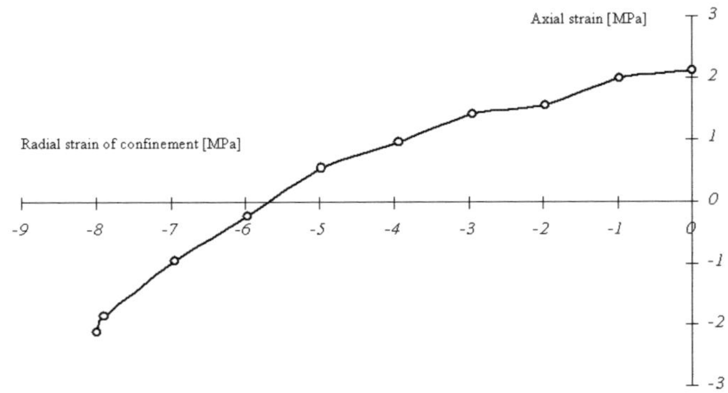

Figure 1.30. *Rupture threshold as a function of confinement applied to the non-saturated concrete specimen [VIS 94]*

An example of biaxial tensile behavior is given in Figure 1.28, in terms of normalized axial stress (stress divided by the uniaxial compressive strength), as a function of the deformations and the ratio between the values of two tensile stresses. We can observe just a few differences compared to uniaxial tension. Deformations only are modified, which seems logical if we consider that the tension "level" controls the behavior of concrete. The triaxial tensile behavior of the material is very close to the response measured in uniaxial and biaxial tension (see [BUR 97]).

1.5.1.2. *Confined tension*

This type of state of stress is rare in civil engineering structures. Nevertheless, with particular conditions, such as pre-stress of concrete in two perpendicular directions, it is interesting to quantify the threshold of the tensile rupture of confined concrete, i.e. when the material is stressed towards a direction using traction, while it is under compression in the other direction. Tests (Figure 1.29) show the evolution of compressive stress (σ_3, normalized regarding uniaxial compressive strength) as a function of the tensile stress applied in a perpendicular direction (normalized σ_1).

We can observe that the compressive strength significantly decreases as soon as tension is increased in the perpendicular direction to the compression axis.

In addition, Visser and van Mier studied the threshold of tensile rupture as a function of the confinement of the specimen in a triaxial cell (with the same idea as in Figure 1.35), adapted to tension-compression-confinement stresses [VIS 94]. This threshold is presented in Figure 1.30. When the intensity of the confinement stress increases (negative values), the maximal axial stress necessary to break the specimen decreases and becomes negative after 5.5 MPa of confinement. Confinement induces a positive deformation in the axial direction. If this extension is above a threshold, cracking occurs. When this confinement becomes important, the specimen has to be axially compressed in order to avoid cracking. This illustrates the fact that extension is the driving force inducing the non linear response, e.g. during biaxial compression or triaxial compression.

1.5.2. *Bi-axial compression*

Numerous authors have performed biaxial compression tests [KUP 69; BAS 74a; MIE 84; COL 85; NAH 86; TOR 87]. They have tested either cubes or prisms on four of the six sides, or confined cylinders. Figure 1.31 presents the different rupture thresholds under biaxial stresses obtained by different authors. The observed differences mainly come from the boundary conditions of each specimen. The more they are equipped with an efficient anti-bracing system, i.e. they are submitted to a plane state of uniform stress, the less important the threshold [MAZ 84]. The behavior of the specimens is more or less ductile as a function the mode of the test and rupture occurs perpendicularly to the maximal extension plane. The mode of rupture is a combination of mode I and mode II.

1.5.3. *Triaxial compression*

Triaxial compression has been studied using three different test types: compression tests with passive confinement, tests with active confinement, and "true" triaxial tests. In the following section, we shall discuss such tests.

The triaxial compressive response of concrete is influenced by its composition, which is itself influenced by the initial porosity of the material. One of the most influencing parameters on the porosity of the material is the water to cement ratio of the formulation of concrete. For a constant granular skeleton, at higher water to cement ratios, the concrete will be more porous, and at lower ratios the porosity will be lower and the mechanical characteristics (compressive strength, tensile strength, etc.) will be higher [REI 94]. This water to cement ratio is then an important variable of the characterization of concrete or mortar.

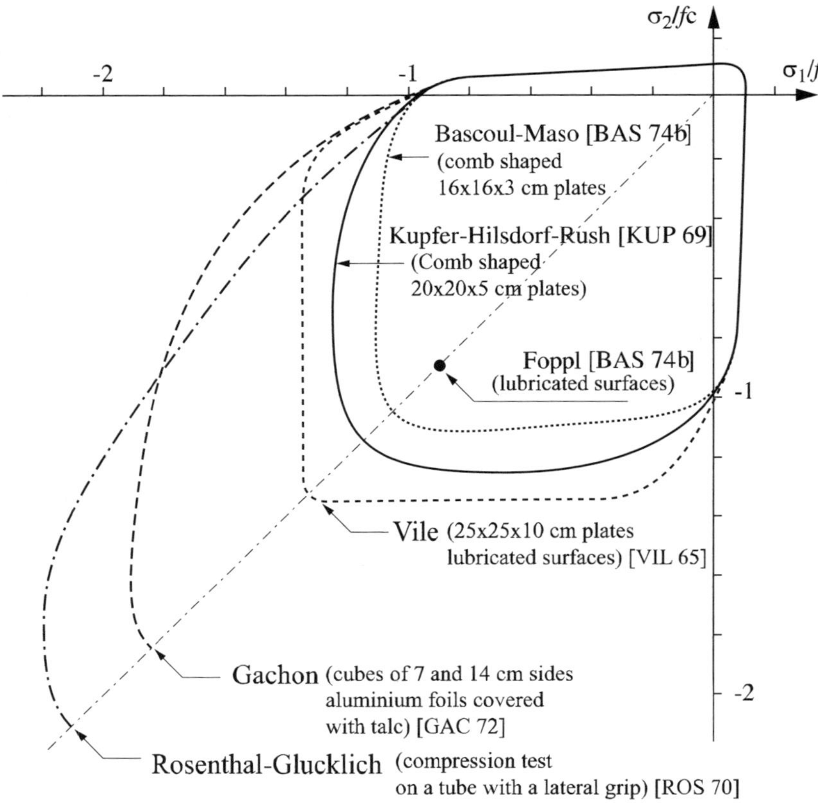

Figure 1.31. *Rupture threshold during bi-axial stress experimentally obtained by different authors for different boundary conditions [MAZ 84]*

Excess of gravel   Optimal density   Excess of sand

Figure 1.32. *Influence of the composition on density and porosity type*

Another factor influencing the porosity of concrete is granular density. The skeleton is usually made of two or three types of aggregates. In general, sand with a diameter of 0 to 5 mm and gravel with a diameter of 5 to 25 or 40 mm are chosen. The skeleton is obtained using the combination of these two constituents in order for the mix to achieve the highest density, that is to say the minimal porosity. Figure 1.32 illustrates the effect of an excess of sand or coarse aggregate. In both cases, the porosity increases. In the case of excess sand, the porosity will have a lower "average diameter" than in the case of an excess of gravel.

1.5.3.1. *Tri-axial compression on cylindrical specimens*

A typical test set-up is depicted in Figure 1.35. Transverse (confinement) stresses are applied on the specimen, with a fluid under pressure and an axial force is superposed. The test is denoted as deviatoric, meaning that a constant confinement pressure is being applied to a cylindrical specimen, and that a stress deviator axially compresses the specimen. In general, the test is being carried out until rupture if confinement is not that high.

Tri-axial response of concretes under low confinement is similar to that which occurs with rock, and may be described using different phases:

– a phase of tightening corresponding to the closure of existing cracks (Figure 1.33a);

– a linear elastic response (Figure 1.33b);

– a phase of stable cracking where randomly dispersed micro-cracks appear and grow (Figure 1.33c), this phenomenon has been shown using acoustic emission [MAZ 84];

– a "dilatancy" phase where the transverse deformations of the specimen become significant, this is close to peak load (Figure 1.33d);

– a post-peak phase once the maximal deviatoric strain has been reached. Strain softening occurs. This phase is not observed when the confinement is significant.

The softening regime is strongly dependent on confinement pressure. The weaker the confinement stress, the more brittle the concrete. The more significant the confinement, the more ductile concrete. Figure 1.34 shows the different observed behaviors. The first graph (*Brittle*) shows a sudden failure that can be observed when confinement is low or zero (uniaxial compression of HPC). The second graph (*Softening*) shows a softening phase, which occurs for a confinement pressure of a few MPa. The third graph (*Ductile*) shows a ductile behavior of the concrete when confinement represents pressures of several tens of MPa. Finally, the last type of response (*Compaction and hardening*) occurs under really high confinement pressures or under hydrostatic stress.

1.5.3.1.1. Tri-axial test with active confinement on concrete

Many authors were inspired by the triaxial test on soils and rocks (Figure 1.35) to perform tests on concrete (see [MIE 84; JAM 84; PIJ 85; COL 85; NAH 86; RAM 90] for general information). Axial stress is applied using a testing frame and confinement is applied using a fluid (generally oil). Figure 1.36 illustrates the brittle-ductile transition depending on the applied confinement on concrete. For confinements of 0.3, 10, and 25 MPa, the material undergoes a loading peak that increases with confinement, but the post-peak phase remains a softening regime, For confinements of 50 and 100 MPa, softening ceases and positive hardening occurs.

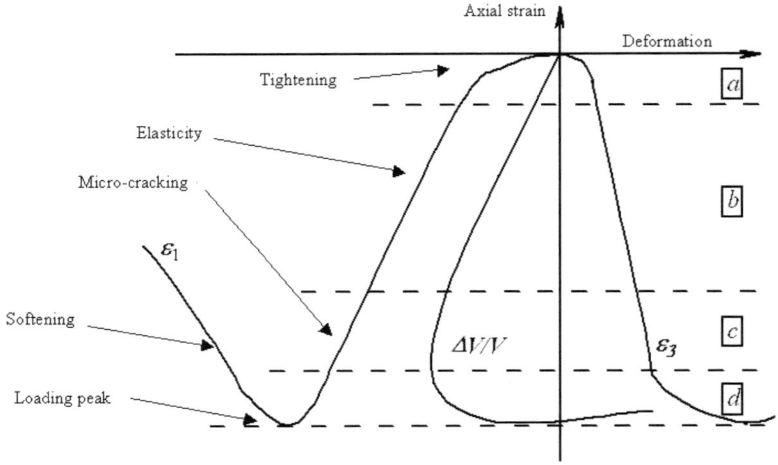

Figure 1.33. *Tri-axial behavior of concrete*

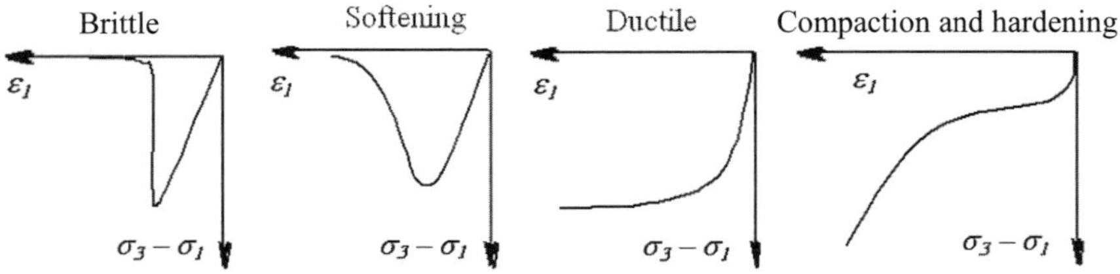

Figure 1.34. *Brittle-ductile transition under increasing confinement*

For low confinement (lower than one-tenth of the deviator applied), van Mier observes the apparition of shear bands within the specimens [MIE 84]. Those bands result from the coupling of mode I and mode II failure, but also from boundary conditions. Indeed, if anti-bracing devices are not used, bracing cones appear, same as in uni-axial compression. Kotsovos showed that the angle between the shear band and the axial axis of the specimen was decreased when anti-bracing devices are used [KOT 79]. When friction becomes zero, this angle is then also zero, leading to the opening of the cracks parallel to the maximal extension direction; thus, mode I.

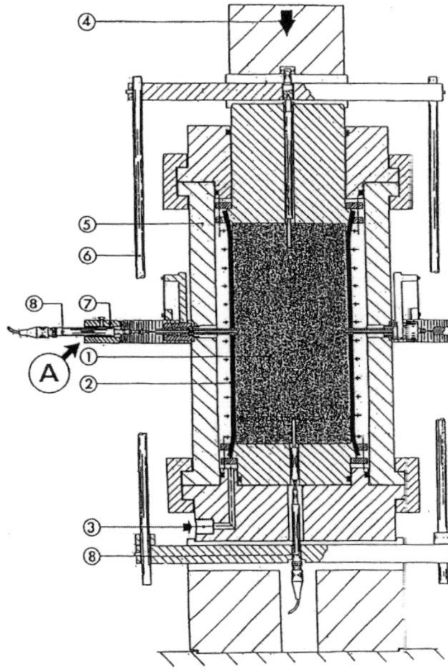

Figure 1.35. *Principle of triaxial test on cylindrical specimens [BEL 91]*

1.5.3.1.2. Tri-axial test with passive confinement

In order to "simplify" the implementation of the triaxial test previously described, numerous studies regarding concrete with passive confinement have been carried out. The first experimental strategy was to confine concrete using a metallic tube with a variable thickness. The second was to insert a specimen within a very rigid cell (very low lateral deformation) and to apply axial compression. In both cases, the confinement is due to the Poisson's effect, the transverse deformations of the specimen are prevented or limited.

Tests on confined concrete using thin tubes [KOT 86; LAL 93]

The aim of these tests is to perform a simple compression test on a composite specimen. A metallic tube is placed around concrete. In general, this tube is being used as a mold during casting of concrete. The force is either applied on concrete itself, or at the same time on both concrete and steel. The second method does not give reliable results as steel has a Poisson coefficient of *0.3,* whereas regular concretes have a coefficient of *0.2*, leading to a more important transverse deformation of the metallic envelope compared to the deformation of concrete. There is no confinement in the elastic phase of the response of the composite specimen, leading to the initiation of damage before confinement is applied by the steel tube.

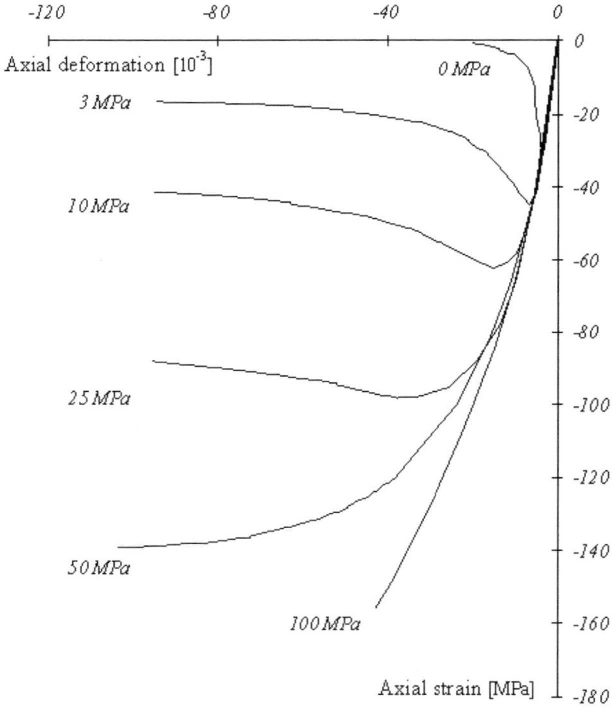

Figure 1.36. *Compression of confined concrete cylinders for different confinement pressures [JAM 84; NAH 86]*

When the compression stress is applied on concrete only, the envelope confines the concrete. Figure 1.37 shows the principle of this test as well as its instrumentation. The test is being carried out same as a simple compression test. The axial deformation of the specimen is measured as a function of the applied force. A series of rosettes placed on the metallic envelope provide the strain state in the steel tube. Figure 1.37b shows the variation of the axial strain applied to the concrete specimen as a function of the axial and transverse deformations measured on the steel envelope at the center of the specimen. The confinement evolves with the thickness of the steel tube. When the tube has a thickness of 2.16 mm, the response of concrete is ductile with an elastic threshold of over 150 MPa.

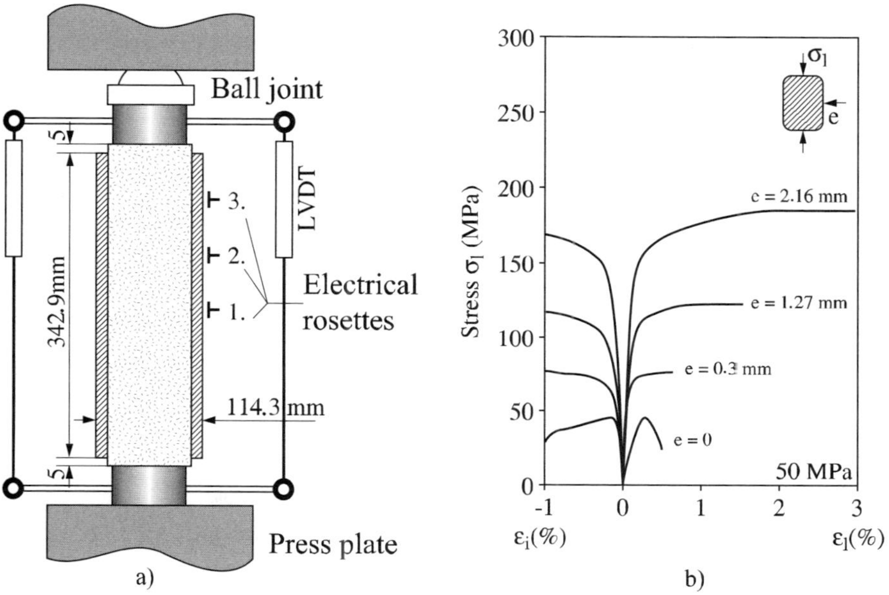

Figure 1.37. *Principle and behavior of a concrete confined using a metallic tube [LAL 93]*

The difficulties encountered in this type of test can be split in two groups: the first is the influence of shrinkage of concrete, the second is the influence of friction between concrete and steel. The specimen is cast into the metallic envelope, its dimensions after hardening are lower than the initial ones because of the different thermal, endogenous, and desiccation shrinkages. Endogenous shrinkage (the most annoying in this case) of concretes or mortars is strongly related to their compositions and increases with the performance of the concrete. Because of shrinkage, concrete is not in contact with steel and the appearance of confinement is delayed upon loading. Figure 1.38 shows the influence of endogenous shrinkage on the confinement of a high-performance concrete [LAL 93]. The curve shows the evolution of a transverse deformation placed on the metallic tube at mid-height (Figure 1.37). We can observe that is necessary to impose an axial displacement of 2 mm to start the confinement of the specimen. This corresponds to an axial deformation of *0.6%,* which is beyond the loading peak during simple compression (about *0.2%*). This phenomenon, important for high-performance concretes, also occurs in regular concretes. This leads to uncontrolled pre-damaging of the concrete.

The problem of steel/concrete friction has been considered by Kotsovos and Perry. These authors recommend that the metallic envelope be greased in order to reduce friction [KOT 86]. Note that the confinement applied to the specimen is always moderate.

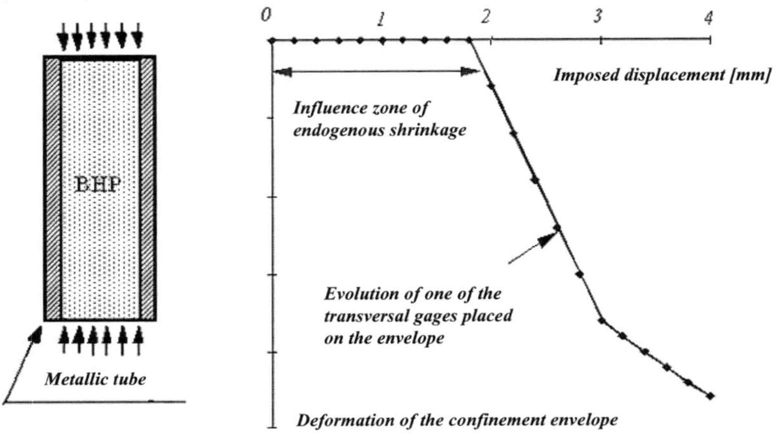

Figure 1.38. *Test principle and influence of endogenous shrinkage on high performance concrete confined with a metallic tube [LAL 93]*

Tests on mortars and cement pates confined with a rigid frame [BAZ 86]

Another type of test using passive confinement has been developed by Bažant based on an experiment developed by geologists. The principle of this test is described in Figure 1.39 and consists of compressing a specimen whose lateral faces are maintained fixed with a rigid cell. Specimens are cylinders of small dimensions with small size aggregate. The loading path is oedometric (see section 1.5.3.2). Specimens are core-drilled and adjusted to be placed within the cell. The edges of the specimen are oiled to minimize friction.

Figure 1.40 shows the evolution of axial stress as a function of the axial strain for micro-concrete and mortar. The tests were carried out up to 2,068 MPa axial stress. The first phase (stress lower than 100 MPa) is elastic, while the second phase is plastic. The material hardens once the porosity of the specimen is almost closed. This hardening occurs when 10% of the axial deformation of cement paste is reached, whereas the hardening phase of micro-concrete begins at a deformation of about 6% and is more progressive. This whole behavior can be called "compaction" as it induces a decrease of porosity. Up to now, for all the tests described in the previous sections, collapse of the specimens occurred simultaneously with a sudden increase in porosity, when micro- and then macro-cracks appeared. In the case of compaction, collapse of the specimen is never observed. Only the capabilities of the test set-up can stop the test. The influence of the initial porosity is shown in Figure 1.40. Cement paste turns out to be more porous than micro-concrete. In addition, it can noted that more porous samples have an elastoplastic response that occur at lower stresses, i.e. lower than an axial stress of 600 MPa in the case of cement paste. Conversely, the low porosity of micro-concrete provides it with a strong hardening response.

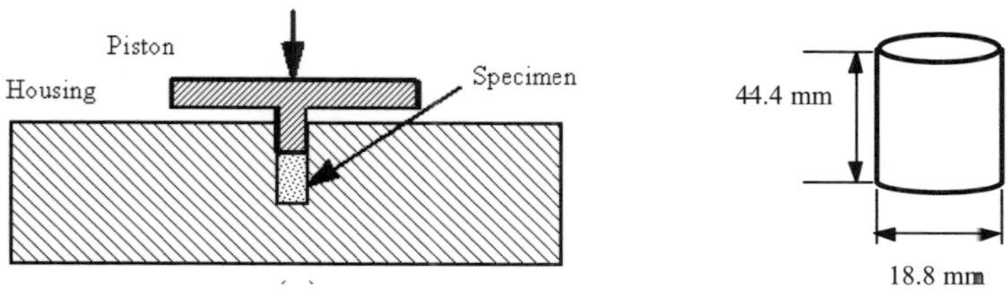

a) *b)*

Figure 1.39. *Principle of passive confinement test using a rigid cell [BAZ 86]*

Axial deformation is equivalent to the volumetric deformation of the specimen because of the stiffness of the cell is very high. Consequently, it is not possible to measure the confinement applied diametrically to the specimen. Nevertheless, it can be estimated if we assume that the Poisson's coefficient of the specimen does not evolve during the deformation. At the end of loading and upon unloading, the material is elastic and has reached a constant irreversible volumetric deformation equal to the initial porosity of the specimen. It is possible to estimate the initial porosity of micro-concrete at about 8% and the porosity of cement paste at 12%. Initially, when the applied stress remains relatively low, only the capillary porosity is reduced. Then, the hydrates porosity is closed for high level axial stresses.

Three quantities need to be defined: *the compaction point*, *elastic hardening*, and *plastic hardening*. It is important to differentiate *elastic hardening* from *plastic hardening* [BUR 97]. *Elastic hardening* corresponds to the stiffening of the material due to the closure of porosity, which corresponds to an increase of the elastic modulus of concrete experimentally. Figure 1.41 illustrates this observation. Conversely, plastic hardening corresponds to the difference between the linear hardening plastic response, which can be encountered when compaction starts, and the non-linear hardening plastic response leading to an increase of the hardening modulus of the material. There is an inflexion point on the material response when linear hardening behavior turns into a nonlinear hardening behavior. This point is called the *compaction point* (Figure 1.41). Elastic and plastic hardenings occur "at the same time" for those concretes. Those two phenomena are strongly related. Some geomaterials do not undergo elastic hardening, but plastic hardening only (in the case of chalks, for example [SHA2 87]).

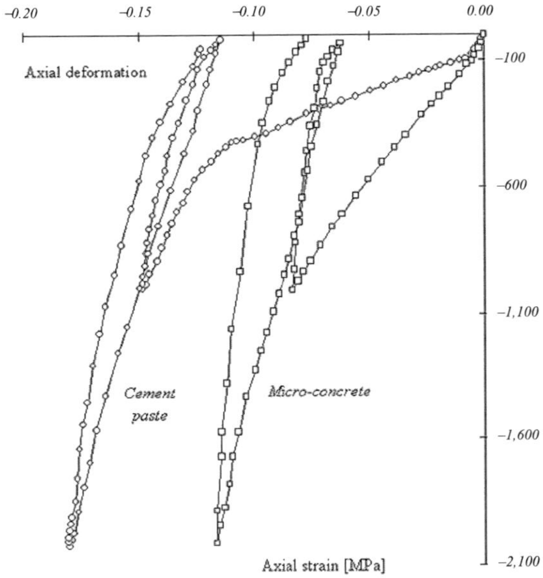

Figure 1.40. *Oedometric test on a cement paste and micro-concrete [BAZ 86]*

Confined uniaxial test [BUR 97; BUR 01a]

This is a quasi-oedometric test in the sense that the specimen is not placed in a rigid cell but rather in a deformable casing. The confinement is provided by the casing around the specimen. As a result, the homogeneity of the confinement and the set up of the specimen have to be carefully considered. Different problems, must be considered:

– the set up of the specimen within the testing cell and its adjustment to the casing;

– the steel-concrete friction under significant strains;

– the uniform and homogenous application of confinement;

– the experimental measurements (confinement pressure, volume variations of the specimen).

Concerning the interface, a special epoxy resin is used to adjust the specimen in the deformable casing/cell (C 6120 Chrysor® type). In order to avoid the resin to adhere on the cell, Teflon© is sprayed on the metallic cell before the test. The experimental set-up comprises (Figure 1.42) a confinement casing/cell where the specimen will be placed, a support the cell will rest on, and a piston needed in order to apply the axial force.

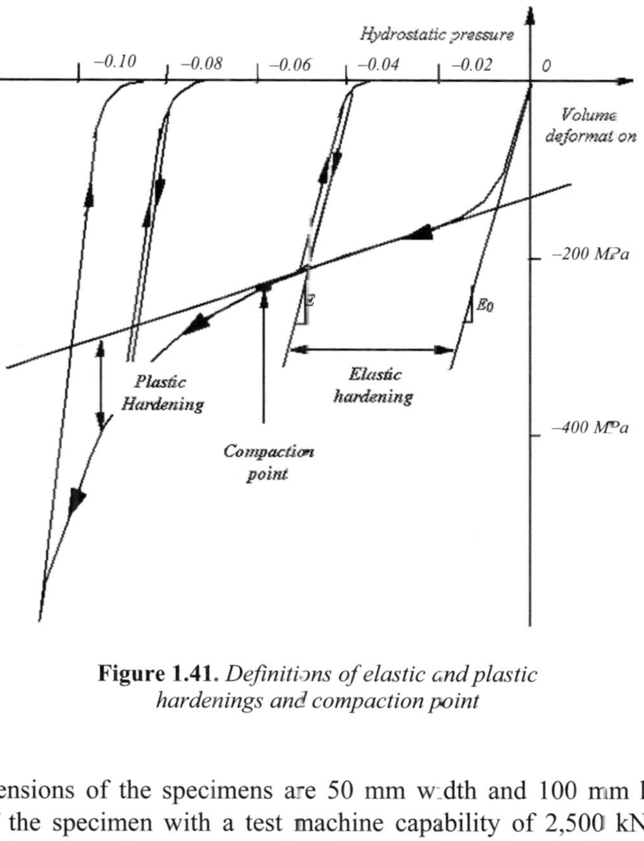

Figure 1.41. *Definitions of elastic and plastic hardenings and compaction point*

The dimensions of the specimens are 50 mm width and 100 mm height. This geometry of the specimen with a test machine capability of 2,500 kN enables to reach an axial compressive stress equal to 1,275 MPa, and a confinement of about 250 MPa (if we assume the Poisson's coefficient does not vary substantially during the test). The displacement of the piston is measured using a linear displacement sensor (Vishay-Micro-measures® type, D S/N 10490); the axial force is measured using the machine force sensor, the deformation of the confinement cell provides information on the applied confinement (provided the casing/cell remain elastic). This deformation is measured on the cell height using the five Vishay-Micro-measures® gage pairs of CEA-06-250UN-120 type stuck with the VM AE-15 glue (1 hour at 100°C). Gages are diametrically opposed (in pairs) and set up to form complete bridges with compensation gages. Confinement homogeneity along the height of the specimen is checked with these strain gages.

In Figure 1.43 is shown a result obtained with the confined uniaxial test in terms of axial stress as a function of the axial deformation of several mortar specimens presenting the same granular skeleton, but different water to cement ratios. The specimen response is compacting and elastoplastic. Materials with a lower water to cement ratio have a lower initial porosity and thus plastic hardening is observed sooner. Figure 1.44 shows the evolution of the elastic modulus of a concrete as a function of the axial deformation in a confined uniaxial test. This modulus progressively increases when the porosity of the material closes. Because of the predominance of axial stress on lateral confinement, there is a reorganization of the material during the test, and the specimen becomes stratified perpendicularly to the maximal stress [BUR 01].

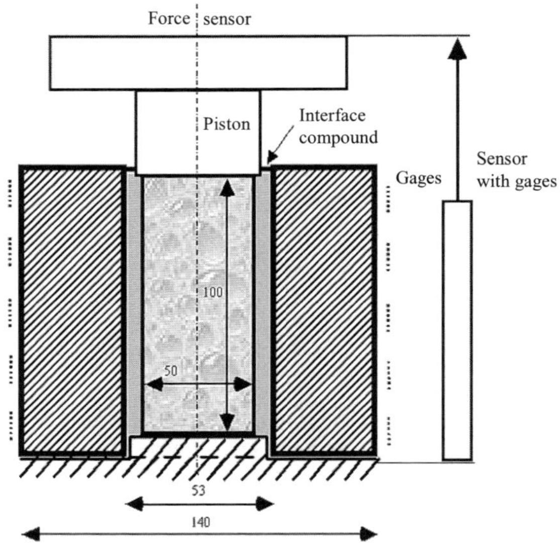

Figure 1.42. *Confined uniaxial test principle [BUR 01]*

1.5.3.2. *"True" triaxial compression tests*

"True" triaxial compression tests can be split into two categories: hydrostatic tests and triaxial tests on cubic specimens. The hydrostatic test is actually a particular case of the triaxial test on cylinders (previously described), but demands significant pressures in order to characterize the compacting elasto-plastic response of the material. True triaxial test relies on complex testing machines, involving six hydraulic jacks with relative displacement control.

1.5.3.2.1. Hydrostatic test

Relatively few authors have carried out purely hydrostatic tests on concrete. This can be explained by the fact that the pressures required in order to characterize the behavior of the concrete are extremely high (several hundreds of MPa).

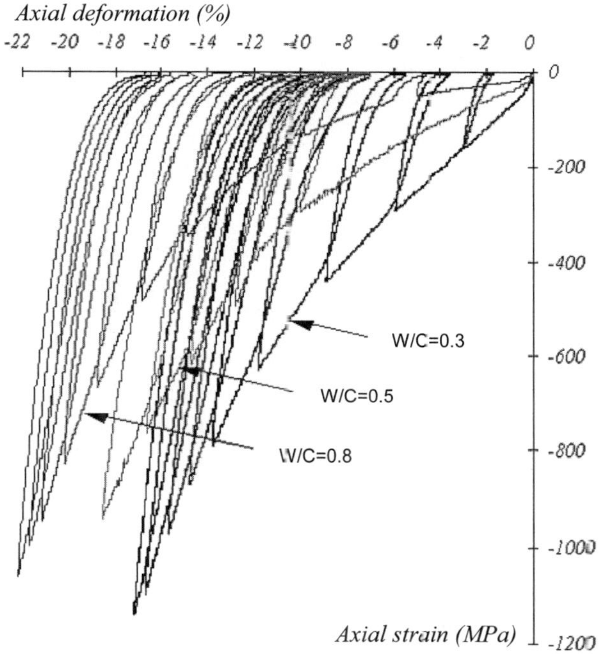

Figure 1.43. *Evolution of the axial strain as a function of the axial deformation during an uniaxial test confined for different mortars (w/c (water/cement ratio)=0.3; 0.5; 0.8) [BUR 01]*

This test is similar to the triaxial test on cylindrical specimens, with an axial stress equal to the confinement pressure; the maximal pressure being here equal to 400 MPa in the reported experimental data [DAH 91; BUR 97; BUR 01a]. Dimensions of the specimens are very small (cylinders with a 17 mm diameter and a 18 mm height). Figure 1.45 shows the high-pressure cell. Pressure measurement is being performed with a 0-400 MPa digital manometer placed within the fluid return circuit. The confinement fluid is a hydrocarbon: pentane. This fluid has been chosen for its exceptionally low viscosity at low and high pressures and its low compressibility properties. Leak-tightness is either static (shielding windows) or dynamic (sleeve bricks) and is provided using "simple and double Bridgman" systems. The deformations of the specimen are measured in situ with strain gages.

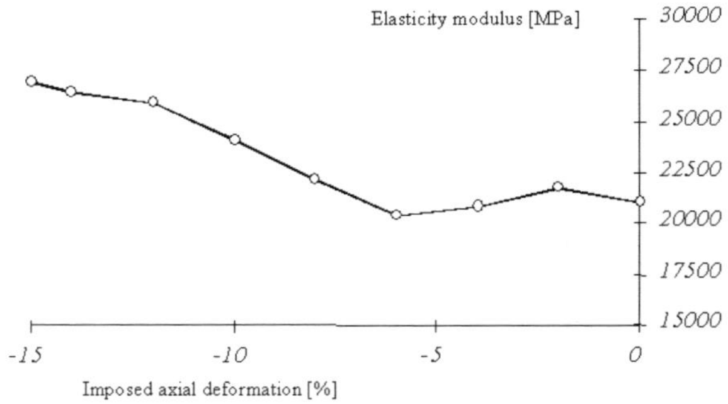

Figure 1.44. *Evolution of the elastic modulus (water:cement = 0.5) of concrete as a function of the total axial deformation imposed during the confined uniaxial test [BUR 01]*

Two aluminum alloy AU4G heads stuck to the sides of the cylinder and a rubber membrane placed over the specimen insures its tightness (Figure 1.46). It prevents the pentane penetrating the pores of the sample and also prevents contact with the glue of the gages, as pentane is a powerful solvent. Note that the heads are made of aluminum to minimize bracing at both ends of the specimen. Vishay-Micro-mesures® deformation gages used in this test (of CEA-06-250UN-120 type, stuck with VM AE-10 glue) provide a deformation range up to 10%. As is shown in Figure 1.46, a gage is axially placed and the other one transversally to allow the volumetric deformations of the specimen to be measured. All the necessary corrections (drift of the strain gage, hydrostatic correction) are performed.

The pressure of the fluid is increased to load the sample. Same as in the confined uniaxial test, the behavior of the material is elastoplastic. This can be seen in Figure 1.47 where the evolution of the hydrostatic pressure is shown as a function of the volume deformation of the specimen, up 400 MPa. The evolution of the elastic modulus of the material can also be determined using this figure (assuming the Poisson's ratio is kept constant). This modulus increases when the porosity of the material is closed (at about 45%). Once the test is performed, and as opposed to the confined uniaxial test, the specimen does not show any visible cracking.

The hydrostatic test directly provides the state law (hydrostatic pressure – volumetric deformation). Here it is illustrated for a regular mortar (water to cement ratio=0.5). A comparison is shown in Figure 1.48 where the evolution of the hydrostatic pressure is shown as a function of the volumetric deformation ΔV/V for the two loading paths: the hydrostatic path and oedometric path. It is clear on this figure that it is difficult for cementitious materials to uncouple the spherical behavior from deviatoric behavior. This uncoupling is common for metallic materials where constitutive models are a state law (relating the hydrostatic pressure to the volumetric deformation) on one side, superimposed to deviatoric response (connecting the deviatoric stress to the deviatoric strain).

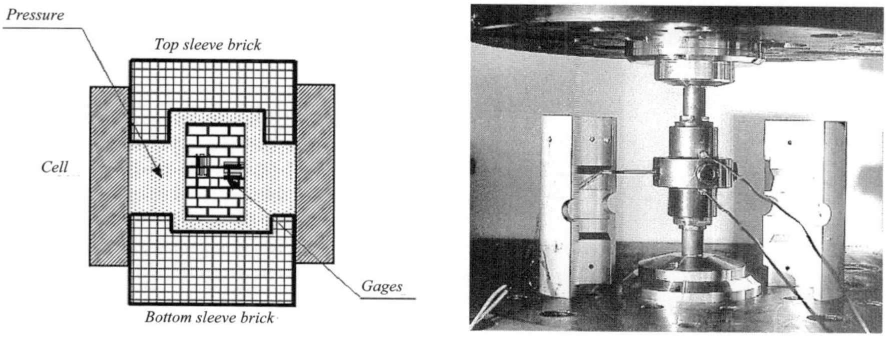

Figure 1.45. *High pressure cell*

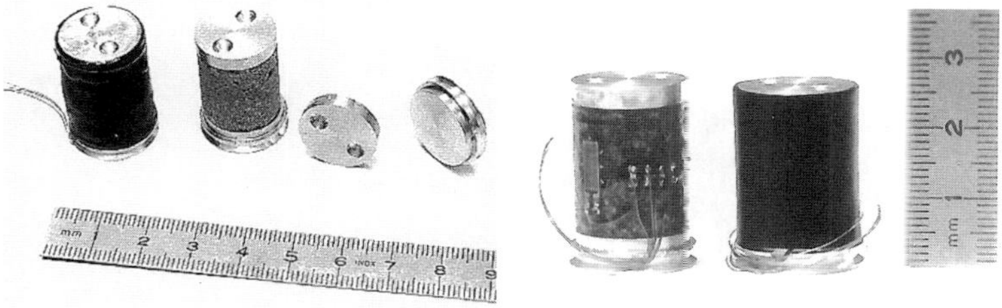

Figure 1.46. *Specimens for the hydrostatic triaxial test*

A few additional observations can be made on the comparison between the state law obtained using an oedometric path and a hydrostatic test. We can note that the elastic threshold is different for both loadings. This seems normal as, for a confined uniaxial test, axial stress is more important than a radial stress. Thus, the plasticity criterion will be reached more rapidly. Regarding the volumetric deformation, it is necessary to develop more significant pressures for the hydrostatic path in order to achieve the same deformation as in the uniaxial confined test. Moreover, we can observe that compressibility modules are slightly different from one path to another for the same deformation. Regarding the hydrostatic path, the consolidation of porosity alone introduces just a few microcracks at this pressure level, whereas in the oedometric case, the movement of the aggregates within the material damages the matrix.

Consequence of pre-confinement

Schickert and Danssman performed hydrostatic tests on mortar and concrete cubes [SCH 84]. The authors have carried out a series of tests to determine the uniaxial compressive strength following a hydrostatic compression of the materials. Figure 1.49 shows the results obtained on two mortars and a concrete. We can observe a decrease in the uniaxial strength as a function of the hydrostatic pressure of the pre-confinement.

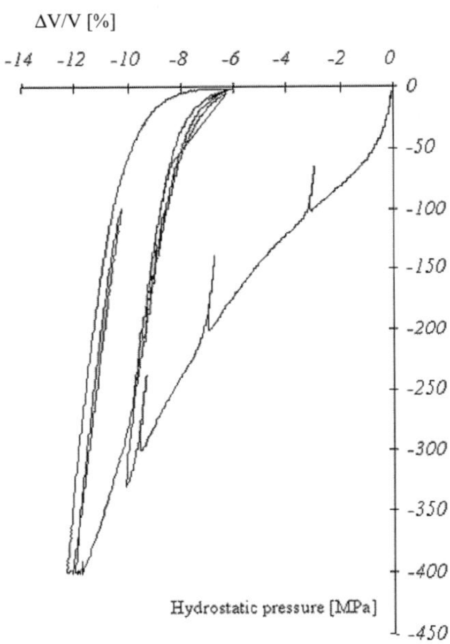

Figure 1.47. *Hydrostatic triaxial test: evolution of the volumetric deformation as a function of the hydrostatic pressure*

This shows that, despite the decrease of the porosity of the material through hydrostatic compaction, damage occurs within the specimen. Intra-matrix microcracks occur due the micromotion of the aggregates, one after another due to the pressure. Similarly, Wong has shown that the density of microcracks increases with confinement pressures for a rock [WON 90], due to the nucleation process.

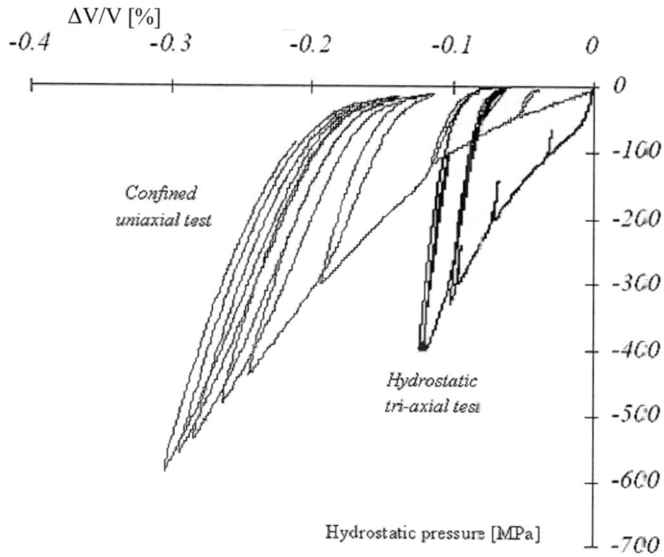

Figure 1.48. *State law for two loading paths (confined uniaxial test and hydrostatic triaxial test) [BUR 01]*

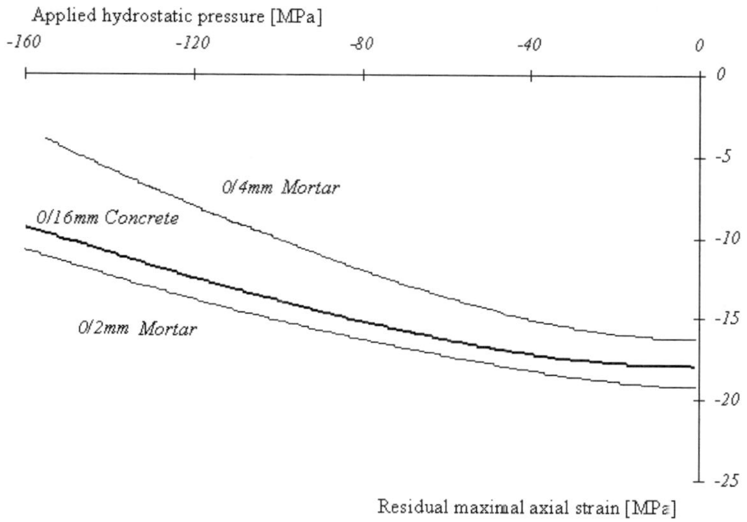

Figure 1.49. *Maximal residual uniaxial strain as a function of the pre-stress hydrostatic pressure [SCH 84]*

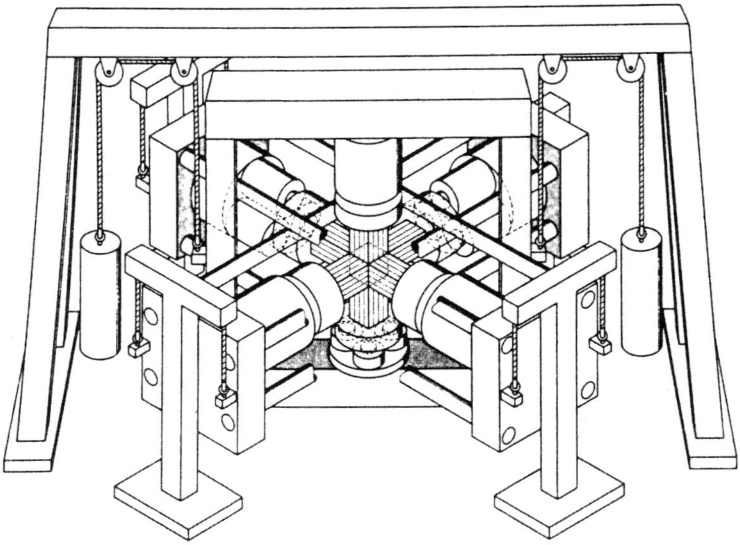

Figure 1.50. *Tri-axial compression machine on concrete [WIN 84]*

Figure 1.51. *Real triaxial machine on a 40×40×40 cm^3 concrete cube*

1.5.3.2.2. True triaxial test

This type of test is performed on concrete cubes. As Figure 1.50 shows, the testing machine has six jacks controlled independently two by two [WIN 84]. It is then possible to investigate the complete principal stress space. Nevertheless, many problems come due to the transmission of the forces on the faces of the cubic specimen. Winkler used brush platens in order to avoid bracing of the specimen. The design of these brushes has to fulfill two conditions: they have to deliver the compressive force without bracing, while opposing a negligible force to lateral loads. The combination of these two conditions is difficult to achieve and it lead Winkler to design triaxial tests in which one of the components is low compared with the other two, thus to make bi-axial compression slightly confined.

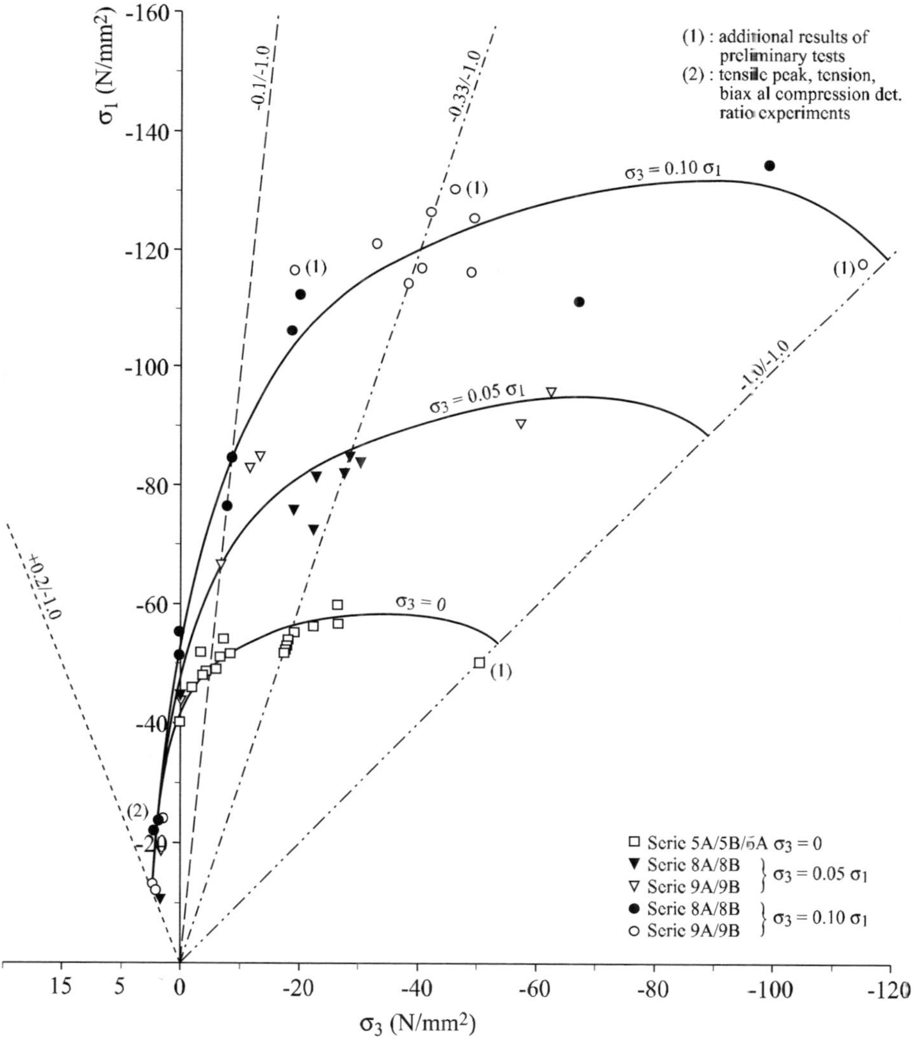

Figure 1.52. *Rupture thresholds obtained under real triaxial compression [MIE 84]*

The picture of a triaxial machine is shown in Figure 1.51. This machine enables cubes of dimensions 40 to 50 cm to be tested [SKO 95]. The anti-bracing system is similar to brush platens and friction is reduced with Teflon® plates. The size of the specimens allows any type of concrete to be tested. Other real triaxial machines have been developed; generally, for use on specimens with a characteristic size of about 10 cm [BAS 74; MIE 84; TOR 87]. Figure 1.52 shows within the σ_1-σ_2 plane, the evolution of the threshold of uniaxial compressive rupture as a function of the increase of the third compression component (σ_3).

1.6. Conclusions regarding the experimental characterization of the multiaxial behavior of concrete

We have analyzed the different tests carried out routinely or not on concrete and highlighted the experimental issues encountered due to the heterogeneity of concrete, to its sensitivity to extensions, to the localization of deformations, to the experimental artifacts that arise during the analysis of the results. Classical uniaxial tests have been detailed. Further studies are required to enable better analysis of the post-peak behavior of concretes and mortars. One should not forget also that concrete is a saturated or partially saturated porous material. This aspect, related to poro-mechanics has not been considered in this chapter.

<table>
<tr><th>Stress type</th><th>Mode of local damage</th><th>Characteristics</th><th>Behavior examples</th></tr>
<tr><td>A:
- uniaxial
- bi-axial
- tri-axial
with traction or low compression in at least one main direction</td><td>Micro-cracking mainly around the grains

Mode I or I + II</td><td>- extension in at least one of the main direction
- more or less "brittle" behavior</td><td>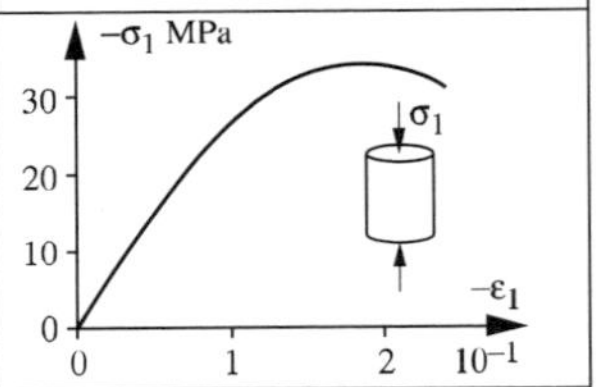
</td></tr>
<tr><td>B: tri-axial adding a significant hydrostatic pressure to a deviator</td><td>Micro-cracking mainly in mode II
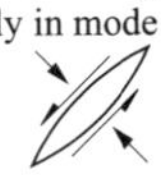</td><td>- no estension
- "ductile" behavior</td><td>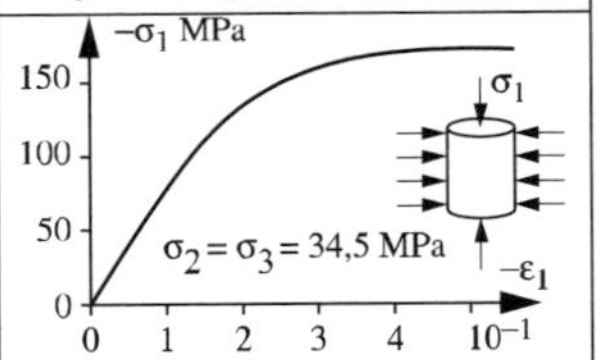
</td></tr>
<tr><td>C: tri-axial mainly hydrostatic pressure</td><td>Crushing of the micro-porous structure
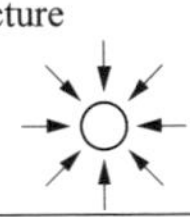</td><td>- compaction (elastic and plastic hardening)</td><td>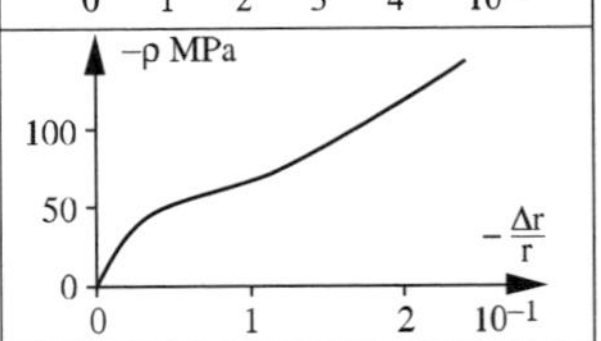
</td></tr>
</table>

Figure 1.53. *Behavior classification as a function of the stress type and the damaging mode [MAZ 84]*

Tri-axial tests performed on concrete show a brittle-ductile transition of the material response due to confinement. This transition is also called the *softening-hardening transition.* It is due to the transition between mode I microcracking (leading to the collapse of the material along with softening and porosity increase), and mode II microcracking (shear bands). For high levels of confinement stresses, crushing of the material is observed and a compaction, with plastic and elastic hardening, occurs. Figure 1.53 provides a useful classification of the main specificities of the mechanical response of concrete [MAZ 84].

1.7. Bibliography

[ACK 88] P. ACKER, Comportement mécanique du béton: apports de l'approche physico-chimique, PhD thesis, Ecole Nationale des Ponts et Chaussées, Paris, 1988.

[AIT 01] P. C. AÏTCIN, *Bétons, Haute Performance*, Eyrolles, Paris, 2001.

[AST 01] E. ASTUDILLO , S. CALLOCH, F. HILD, G. BERNIER, R. BUMEDIJEN, "Etude du comportement biaxial d'un microbéton armé de fibres par intercorrélation d'images", in Y. BERTHAUD *et al.*, *Proc. Colloque Photomécanique 2001, Etude du comportement des matériaux et des structures*, pp. 49-54, 2001.

[AVR 84] J. AVRIL, *Encyclopédie d'analyse des contraintes.* Vishay Micromesures, Malakoff, 1984.

[BAL 94] F. M. BARTLETT, J. G. MACGREGOR, "Effect of moisture condition on concrete core strengths", *ACI Mater. J.*, vol. 91, pp. 227-236, 1994.

[BAR 92] J. BARON, J. P. OLLIVIER, *La durabilité des bétons*, Eyrolles, Paris, 1992.

[BAR 94] V. BAROGHEL-BOUNY, Caractérisation microstructurale et hydrique des pâtes de ciment et des bétons ordinaires et à très hautes performances, PhD thesis, Ecole Nationale des Ponts et Chaussées, Paris, 1994.

[BAR 96a] J. BARON, J. P. OLLIVIER, *Les betons: bases et données pour leur formulation*, Eyrolles, Paris, 1996.

[BAR 96b] J. BARON, J. P. OLLIVIER, J. C. WEISS, "Les ciments courants", in J. BARON and J. P. OLLIVIER, *Les bétons: bases et données pour leur formulation*, Eyrolles, Paris, pp. 1-22, 1996.

[BAR 00] V. BAROGHEL-BOUNY, P. C. AÏTCIN, *Shrinkage of Concrete - Shrinkage 2000*, RILEM Publications PRO 17, Paris, 2000.

[BAS 74a] A. BASCOUL, Etude du comportement mécanique du béton en compression biaxiale, PhD thesis, University of Toulouse, 1974.

[BAS 74b] A. BASCOUL, J. C. MASO, "Influence de la contrainte intermédiaire sur le comportement mécanique du béton en compression biaxiale", *Mater. Struct.*, vol. 42, pp. 411-419, 1974.

[BAZ 82] Z. P. BAŽANT, F. H. WITTMANN, *Creep and Shrinkage in Concrete Structures*, J. Wiley and Sons, Chichester, 1982.

[BAZ 86] Z. P. BAŽANT, F. C. BISHOP, T. P. CHANG, "Confined compression tests of cement paste and concrete up to 300 Ksi.", *ACI J.*, vol. 33, 553-60, 1986.

[BAZ 89] Z. P. BAŽANT, G. P. PIJAUDIER-CABOT, "Measurement of characteristic length of non local continuum", *ASCE J. of Engrg. Mech.*, vol. 115, pp. 755-767, 1989.

[BAZ 96] Z. P. BAŽANT, M. K. KAPLAN, *Concrete at High Temperature: Materials Properties and Mechanical Modeling*, Longman Concrete Design and Construction Series, Longman, London, 1996.

[BAZ 98] Z. P. BAŽANT, J. P. PLANAS, *Fracture and Size Effect in Concrete and Other Quasi-Brittle Materials*, CRC Press, Boca Raton, FL, 1998.

[BEL 91] R. BELLOTTI, P. ROSSI, "Cylinder tests: experimental technique and results. 53-MTC final report", *Mater. Struct.*, vol. 24, pp. 45-51, 1991.

[BEL 96] M. BEHLOUL, Analyse et modélisation du comportement d'un matériau à matrice cimentaire fibrée à ultra hautes performances (Bétons de Poudres Réactives), PhD thesis, Ecole Normale Supérieure de Cachan, 1996.

[BER 91] Y. B. BERTHAUD, E. RINGOT, D. FOKWAD, "A test for delaying localization in tension. Experimental investigation", *Cement Concrete Res.*, vol. 21, pp. 928-940, 1991.

[BER 96] R. B. BERTRANDY, C. PIKETTY, "Les granulats pour bétons", in J. BARON and J. P. OLLIVIER, *Les bétons: bases et données pour leur formulation*, Eyrolles, Paris, pp. 65-86, 1996.

[BIS 01] J. BISSCHOP, J. G. M. VAN MIER, "*Effect of aggregate size and paste volume on drying shrinkage microcracking in cement-based composites*", in F. ULM, Z.P. BAŽANT, F.H. WITTMANN, *Creep, Shrinkage & Durability Mechanics of Concrete and Other Quasi-Brittle Materials, Proc. of CONCREEP6@MIT*, Elsevier, Amsterdam, pp. 75-80, 2001.

[BOU 80] C. BOULAY, A. COLSON, "Un extensomètre à béton éliminant l'influence des déformations transversales sur la mesure des déformations longitudinales", *Mater. Struct.*, vol. 14, no. 79, pp. 35-38, 1980.

[BOU 89] C. BOULAY, "La boîte à sable pour bien écraser les bétons à hautes performances", ref. 3442, *Bulletin des Laboratoires des Ponts et Chaussées*, vol. 164, pp. 87-88, 1989.

[BOU 99] C. BOULAY, F. LE MAOU, S. RENWEZ, "Quelques pièges à éviter lors de la détermination de la résistance et du module en compression sur cylindres en béton", ref. 4262, *Bulletin des Laboratoires des Ponts et Chaussées*, vol. 220, pp. 63-74, 1999.

[BUR 97] N. BURLION, Compaction des bétons: éléments de modélisation et caractérisation expérimentale, PhD thesis, ENS Cachan, 1997.

[BUR 00] N. BURLION, F. BOURGEOIS, J. F. SHAO, "Coupling damage – drying shrinkage: experimental study and modelling", in V. BAROGHEL-BOUNY, P. C. AÏTCIN, *International RILEM Workshop on Shrinkage of Concrete, Shrinkage 2000*, RILEM Publications PRO 17, Paris, pp. 315-339, 2000.

[BUR 01a] N. BURLION, G. PIJAUDIER-CABOT, N. DAHAN, "Experimental analysis of compaction of concrete and mortar", *Int. J. Num. Anal. Methods Geomech.*, vol. 25, pp. 1467-1486, 2001.

[BUR 01b] N. BURLION, F. BOURGEOIS, J. F. SHAO, "Mechanical anisotropy induced by drying shrinkage: modelling and experiments", in R. DE BORST, J. MAZARS, G. PIJAUDIER-CABOT, J. VAN MIER, *Fracture Mechanics on Concrete and Concrete Structures, Proc. of FRAMCOS-4,* Balkema, The Hague, pp. 511-518, 2001.

[CAR 96] C. CARDE, Caractérisation et modélisation de l'altération des propriétés mécaniques due à la lixiviation des matériaux cimentaires, PhD thesis, INSA Toulouse, 1996.

[COL 85] F. COLLOMBET, Modélisation de l'endommagement anisotrope. Application au comportement du béton sous sollicitation multiaxiale, PhD thesis, Pierre and Marie Curie University (Paris VI) 1985.

[COU 95] O. COUSSY, *Mechanics of Porous Continua,* J. Wiley & Sons, Chichester, 1995.

[DAH 91] N. DAHAN, J. BOUTROUX, R. COUTY, A. BERTHAULT, "Compressibility of zinc single cristal under confining pressure", *High Pressure Res.*, vol. 7, pp. 225-228, 1991.

[DEL 92] F. DE LARRARD, R. LE ROY, "Relation entre formulation et quelques propriétés mécaniques des bétons à hautes performances", *Mater. Struct.*, vol. 25, pp. 464-475, 1992.

[DRE 98] G. D. DREUX, J. FESTA, *Nouveau guide du béton et de ses constituants*, Eyrolles, Paris, 1998.

[DUP 95] R. DUPAIN, R. LANCHON, J. C. SAINT ARROMAN, *Granulats, sols, ciments et bétons – caractérisation des matériaux de génie civil par les essais de laboratoire*, Editions Casteilla, Paris, 1995.

[GAC 72] H. GACHON, "Comportement du béton en sollicitation triaxiale", in *Proc. Int. Symposium RILEM*, Cannes, pp. 53-70, 1972.

[GET 96] R. GETTU, B. MOBASHER, S. CARMONA, "Testing of concrete under closed loop control", *Advn. Cem. Bas. Mater.*, vol. 3, pp. 54-71, 1996.

[GER 96] B. GERARD, Vieillissement des structures de confinement en béton: modélisation des couplages chimico-mécaniques, PhD thesis, ENS Cachan, 1996.

[GUO 87] Z. H. GUO, X. Q. ZANG, "Investigation of complete stress-deformation curves for concrete in tension", *ACI Mater. J.*, vol. 84, pp. 278-285, 1987.

[HEI 98] G. HEINFLING, Contribution à la modélisation numérique du comportement du béton et des structures en béton armé sous sollicitations thermo-mécaniques à haute température, PhD thesis, INSA of Lyon, 1998.

[HEU 01] F. H. HEUKAMP, F. J. ULM, J. T. GERMAINE, "Mechanical properties of calcium leached cement pastes: triaxial stress and influence of the pore pressure", *Cement and Concrete Research*, vol. 31, 767-774, 2001.

[HUG 70] B. P. HUGHES, J. E. ASH, "Anisotropy and failure criteria for concrete", *Mater. Struct.*, vol. 3, pp. 371-374, 1970.

[HUS 00] A. HUSSEIN, H. MARZOUK, "Behaviour of high-strength concrete under biaxial stresses", *ACI Mater. J.*, vol. 97, pp. 27-36, 2000

[JAM 84] P. JAMET, A. MILLARD, G. NAHAS, "Triaxial behaviour of a micro-concrete complete stress-strain for confining pressures ranging from 0 to 100 MPa", in *Proc. of Int. Conf. on Concrete under Multiaxial Conditions*, Paul Sabatier University, Toulouse, vol. 1, pp. 133-140, 1984

[JAN 97] D. C. JANSEN, S. P. SHAH, "Effect of length on compressive strain softening of concrete", *ASCE J. Engrg. Mech.*, vol. 123, pp. 25-35, 1997

[KOT 79] M. D. KOTSOVOS, "Effect of stress path on the behaviour of concrete under triaxial stress states", *ACI J.*, vol. 71, pp. 213-223, 1979

[KOT 83] M. D. KOTSOVOS, "Effect of testing techniques on the post-ultimate behavior of concrete in compression", *Mater. Struct.*, vol. 16, pp. 3-12, 1983

[KOT 86] M. D. KOTSOVOS, S. H. PERRY, "Behaviour of concrete subjected to passive confinement", *Mater. Struct.*, vol. 19, pp. 259-264, 1986

[KUP 69] H. B. KUPFER, H. K. HILSDORF, H. RUSCH, "Behaviour of concrete under biaxial stresses", *ACI J.*, vol. 66, 8, pp. 656-666, 1969

[KUP 73] H. B. KUPFER, "Das verhalten des betons unter mehrasiger Kurzzeitbelastung unter besouderer. Berük-Sichtigung der Zweiechsigen Beamspsuchung", *Deutscher ausschuss fur stahlbeton*, Ernst and Sohn, Berlin, Heft 229, pp. 1-105, 1973

[LAL 93] K. LAHLOU, P. C. AÏTCIN, Le confinement des bétons à hautes performances dans des tubes minces, derniers résultats, Internal Report, Civil Engineering Department, Applied Sciences Faculty, University of Sherbrooke, Quebec, Canada, 1993

[LAP 93] P. LAPLANTE, Propriétés mécaniques des bétons durcissants : analyse comparée des bétons classiques et à très hautes performances, PhD thesis, Ecole Nationale des Ponts et Chaussées, Paris, 1993

[LEB 01] C. LE BELLEGO, B. GERARD, G. PIJAUDIER-CABOT, "Mechanical analysis of concrete structures submitted to an aggressive water attack", in R. DE BORST, J. MAZARS, G. PIJAUDIER-CABOT and J. VAN MIER, *Fracture Mechanics on Concrete and Concrete Structures*, *Proc. of FRAMCOS-4*, Balkema, The Hague, pp. 239-246, 2001

[LIL 01] G. LILLIU, J. G. M. VAN MIER, "Experimental investigation of fracture processes in concrete cylinders subjected to torsion", in R. DE BORST, J. MAZARS, G. PIJAUDIER-CABOT and J. VAN MIER, *Fracture Mechanics on Concrete and Concrete Structures*, *Proc. of FRAMCOS-4*, Balkema, The Hague, pp. 395-402, 2001

[LUO 86] M. P. LUONG, "Un nouvel essai pour la mesure de la résistance à la traction", *Rev. Française de Geotechnique*, vol. 34, pp. 69-74, 1986

[MAL 92] Y. MALIER, *Les bétons à hautes performances*, ENPC editions, Paris, 1992

[MAZ 84] J. MAZARS, Application de la mécanique de l'endommagement au comportement non linéaire et à la rupture du béton de structure, PhD thesis, University of Paris 6, 1984

[MAZ 89] J. MAZARS, B. BERTHAUD, "Une technique expérimentale appliquée au béton pour créer un endommagement diffus et mettre en évidence son caractère unilatéral", *C.R.A.S.*, vol. 308, series II, pp. 579-584, 1989.

[MIE 84] J. G. M. VAN MIER, Strain softening of concrete under multiaxial loading conditions, PhD thesis, Eindhoven University of Technology, 1984.

[MIE 91] J. G. M. VAN MIER, R. A. VONK, "Fracture of concrete under multiaxial stress – recent developments", *Mater. Struct.*, vol. 24, pp. 61-65, 1991.

[MIE 93] J. G. M. VAN MIER, A. VERVUURT, "Micromechanical analysis and experimental verification of boundary rotation effects in uniaxial tension on concrete", in *IUTAM Symposium on Fracture of Brittle Disordered Materials: Concrete, Rocks, Ceramics*, Brisbane, pp. 20-24, 1993.

[MIE 96] J. G. M. VAN MIER, E. SCHLANGEN, A. VERVUURT, "Tensile cracking in concrete and sandstone: part2 – effect of boundary rotations", *Mater. Struct.*, vol. 29, no. 186, pp. 87-96, 1996.

[MIE 97a] J. G. M. VAN MIER *et al.*, "RILEM TC 148-SSC: strain – softening of concrete in uniaxial compression – final report", *Mater. Struct.*, vol. 30, p. 195-209, 1997.

[MIE 97b] J. G. M. VAN MIER, *Fracture Processes of Concrete*, CRC Press, Boca Raton, FL, 1997.

[NAH 86] G. NAHAS, Calculs à la ruine des structures en béton, PhD thesis, Pierre and Marie Curie University (Paris VI), 1986.

[NEC 00] W. NECHNECH, Contribution à la modélisation numérique du comportement du béton et des structures en béton armé soumises à des sollicitations thermiques et mécaniques couplées : une approche thermo-élasto-plastique endommageable, PhD thesis, INSA of Lyon, 2000.

[NEV 00] A. M. NEVILLE, *Propriétés des Bétons*, Eyrolles, Paris, 2000.

[PIJ 85] G. PIJAUDIER-CABOT, Caractérisation et modélisation du comportement du béton par un essai multiaxial automatique, PhD thesis, Pierre and Marie Curie University (Paris VI), 1985.

[PIJ 87] G. PIJAUDIER-CABOT, Non local fracture characteristics of strain-softening materials, PhD thesis, Northwestern University, 1987.

[POP 86] S. POPOVICS, "Effect of curing method and moisture condition on compressive strength of concrete", *ACI J.*, vol. 83, pp. 650-657, 1986.

[RAM 90] S. RAMTANI, Contribution à la modélisation du comportement multiaxial du béton endommagé avec description du caractère unilatéral, PhD thesis, Pierre and Marie Curie University (Paris VI), 1990.

[REI 94] H. W. REINHARDT, "Relation between the microstructure and structural performance of concrete", in A. AGUADO, R. GETTU, S. SHAH, *Proc. of the Int. RILEM Workshop on Technology Transfer on the New Trend in Concrete, ConTech '94*, Barcelona, pp. 19-32, 1994.

[RIL 72] RILEM TC14-CPC., "Concrete test methods", *Mater. Struct.*, vol. 5, pp. 395-411, 1972

[RIL 73] RILEM TC14-CPC., "Concrete test methods (following)", *Mater. Struct.*, vol. 6, p. 377-396, 1973

[ROS 70] I. ROSENTHAL, J. GLUCKLICH, "Strength of plain concrete under biaxial stress", *ACI J.*, pp. 903-914, 1970

[ROS 97] P. ROSSI, F. J. ULM, "Size effects in the biaxial tensile-compressive behaviour of concrete: physical mechanisms and modelling", *Mater. Struct.*, vol. 30, pp. 210-216, 1997

[SAO 88] C. SAOURIDIS, Identification et numérisation objectives des comportements adoucissants: une approche multi-échelle de l'endommagement du béton, PhD thesis, Pierre and Marie Curie University (Paris VI), 1988

[SCH 84] G. SCHICKERT, J. DANSSMANN, "Behaviour of concrete stressed by hygh hydrostatic compression", *Proc. of Int. Conf. on Concrete under Multiaxial Conditions*, Paul Sabatier University Editions, Toulouse, vol. 2, pp. 69-84, 1984

[SHA 87a] S. P. SHAH, R. SANKAR, "Internal cracking and strain softening response of concrete under uniaxial compression", *ACI Mater. J.*, vol. 84, pp. 200-212, 1987

[SHA 87b] J. F. SHAO, Etude du comportement d'une craie blanche très poreuse et modélisation, PhD thesis, University of Sciences and Technologies of Lille, 1987

[SKO 95] F. S. SKOCZYLAS, J. P. HENRI, "A study of the intrinsic permeability of granite to gas", *Int. J. Rock Mech. Min. Sci. & Geomech. Abstr.*, vol. 32, pp. 171-179, 1995

[TER 80] M. TERRIEN, "Emission acoustique et comportement mécanique post-critique d'un béton sollicité en traction", ref. 2398, *Bulletin de Liaison des Laboratoires des Ponts et Chaussées*, vol. 105, pp. 65-72, 1980

[TOR 87] J. M. TORRENTI, *Comportement multiaxial du béton: aspects expérimentaux et modélisation*, PhD thesis, Ecole Nationale des Ponts et Chaussées, Paris, 1987

[TOR 91] J. M. TORRENTI, B. DJEBRI, A. BASCOUL, J. L. GRANJU, "Comparative study of two biaxial presses for concrete", *Mater. Struct.*, vol. 24, pp. 52-60, 1991

[TOR 93a] J. M. TORRENTI, E. H. BENAIJA, C. BOULAY, "Influence of boundary conditions on strain softening in concrete compression test", *ASCE J. Engrg. Mech.*, vol. 119, pp. 2369-2384, 1993

[TOR 93b] J. M. TORRENTI, B. DJEBRI, A. BASCOUL, "Compression biaxiale du béton: influence du chemin de sollicitation sur les déformations", *Mater. Struct.*, vol. 26, pp. 181-184, 1993

[TOR 96] J. M. TORRENTI, Comportement mécanique du béton – Bilan de six années de recherche, *OA 23 Report*, LCPC, Paris, 1996

[TOR 99a] J. M. TORRENTI, O. DIDRY, J. P. OLLIVIER, PLAS F., *La dégradation des bétons, couplage fissuration – dégradation chimique*, Hermes, Paris, 1999

[TOR 99b] J. M. TORRENTI, P. DANTEC, C. BOULAY, J. F. SEMBLAT, “Projet de processus d’essai pour la détermination du module de déformation longitudinale du béton”, ref. 4263, *Bulletin des Laboratoires des Ponts et Chaussées*, vol. 220, pp. 79-81, 1999

[TOU 95] F. TOUTLEMONDE, Résistance aux chocs des structures en béton, PhD thesis, Ecole Nationale des Ponts et Chaussées, Paris, 1995

[TOU 99] F. TOUTLEMONDE, “Quelques réflexions sur la détermination du module d’élasticité du béton, en vue de l’élaboration d’un mode opératoire LPC”, ref. 4243, *Bulletin des Laboratoires des Ponts et Chaussées*, vol. 220, pp. 75-78, 1999

[ULM 99] F. J. ULM, M. PRAT, J. A. CALGARO, I. CAROLI, “Creep and shrinkage of concrete”, *Revue Française de Genie Civil*, vol. 3, pp. 3-4, 1999

[ULM 01] F. J. ULM, Z. P. BAŽANT, F. H. WITTMAN, *Creep, Shrinkage and Durability Mechanics of Concrete and Other Quasi-Brittle Materials*, Elsevier, Amsterdam, 2001

[VIL 65] G. W. D. VILE, “The strength of concrete under short term-static biaxial-stress”, *Proc. of the Int. Conf. on the Structure of Concrete*, London, pp. 275-288, 1965

[VIS 94] J. H. M. VISSER, J. G. M. VAN MIER, “Deformation controlled hydraulic fracture experiments on concrete”, in E. BOURDAROT, J. MAZARS and V. SAOUMA, *Dam Fracture and Damage*, Balkema, Rotterdam, pp. 133-141, 1994

[VLI 96] M. R. A. VAN VLIET, J. G. M. VAN MIER, “Experimental investigation of concrete fracture under uniaxial compression”, *Int. J. Mech. Cohesive-Frictional Mater.*, vol. 1, pp. 115-127, 1996

[VLI 00] M. R. A. VAN VLIET., J. G. M. VAN MIER, “Experimental investigation of size effect in concrete and sandstone under uniaxial tension”, *Engrg. Fract. Mech.*, vol. 65, pp. 165-188, 2000

[VON 89] R. VONK, H. RUTTEN, J. G. M. VAN MIER, H. FIJNEMAN, “Influence of boundary conditions on softening of concrete loaded in compression”, in SHAH S.P., SWARTZ S. and BAN H., *Fracture of Concrete and Rock: Recent Development*, Elsevier Applied Science, Cardiff, 1989

[WIN 84] H. WINKLER, “Fundamental investigations on the influence of test equipment on multiaxial test results of concrete”, in *Proc. of Int. Conf. on Concrete under Multiaxial Conditions*, Paul Sabatier University Editions, Toulouse, vol. 1, pp. 9-19, 1984

[WON 90] T. F. WONG, “Mechanical compaction and the brittle-ductile transition in porous sandstones”, in R. J. KNIPE, E. H. RUTTER *Deformation Mechanics, Rheology and Tectonics*, Geological Society Special Publication, London, vol. 54, pp. 111-122, 1990

[YUR 03] I. Y. YURTDAS, Couplage comportement mécanique et dessiccation des matériaux à matrice cimentaire : étude expérimentale sur mortiers, PhD thesis, University of Sciences and Technologies of Lille, 2003

[YUR 04] I. Y. YURTDAS, N. BURLION, F. SKOCZYLAS, “Experimental characterisation of the drying effect on uniaxial mechanical behaviour of mortar”, *Mater. Struct.*, Vol. 37, pp. 170-176, 2004

Chapter 2

Modeling the Macroscopic Behavior of Concrete

2.1. Introduction

A model is a theoretic framework that enables us to describe a process or relationships existing between various representative elements of a system. In this case, we are dealing with the stresses, deformations, and variables describing the non-linear behavior of material. The scales of modeling are multiple. A microscopic approach is applied if the aim is to model the physico-chemical mechanisms that govern the links between all the constitutive elements of concrete. If the focus is the global behavior of a civil engineering structure, the scale of modeling must be adapted to the dimensions of the said structure. The modeling tools that enable us to do this efficiently are based on a macroscopic or phenomenological approach. The latter provides the framework for the developments below. Two important theories are widely used in the field of civil engineering; the theory of plasticity and damage theory. Elasticity is present in both of these approaches.

The construction of a model is the result of the necessity to efficiently describe one or more physical mechanisms on the basis of simplifying hypotheses. The definition of the field of application of the model results from declining a certain number of hypotheses. This is why the validation stage, in which the results of the model are compared to experimental results, is important.

Chapter written by Jean-Marie REYNOUARD, Jean-François GEORGIN, Khalil HAIDAR and Gilles PIJAUDIER-CABOT.

The evolution in the behavior of quasi-brittle materials, especially concrete, submitted to quasi-static external mechanical loads, compression or tension, is initially characterized by a quasi-elastic phase, in which the material goes back to its initial state if it is not submitted to a load. Then, as the load increases, microcracks (the size of which depend on the material components) appear in a regular way and will coalesce to form a macrocrack (generally visible to the naked eye). This macrocrack will grow and spread until the material fractures. The appearance and development of these cracks play an essential role in the failure or collapse of concrete structures. Several approaches enable us to describe this behavior, of which the main ones are: the discrete and the continuous approach.

The *discrete approach* directly presents the appearance of the macrocrack as a geometrical discontinuity. The cracks are thus modeled by a separation between the edges of elements originally connected. This approach requires, as far as modeling is concerned, a continuous change in the topology of the structure. It can be used once the crack has appeared and predicts its evolution up to the complete failure or collapse of the structure.

Various calculation techniques are used, associated with a representation by finite elements: automatic remeshing techniques, discrete approaches based on the direct use of interface elements [CAM 96; HIL 76; ING 85].

Although this approach is close to reality, modeling becomes costly when the mesh and stiffness matrix need reconstructing. It would be unreasonable to use this approach to study the response of a large structure with a considerable number of cracks.

The *continuous approach* to behavior considers the cracked material as a continuous and homogenous environment [CHA 88; LEM 86]. The description of the cracking then consists of relationships between the stresses and deformations introducing, in particular, a softening post-peak behavior.

This kind of approach is widely used today because, as well as being highly efficient, it is relatively simple to use and does not require the *a priori* knowledge of the orientation of the cracks. The models associated with the continuous approach fall into two categories:

– the first category includes the models based on a total formulation, for which an injective relationship exists between the stresses and the total deformations. For example, the "fixed crack" models based on the elasticity theory, certain "rotating crack" models [COP 80; ROT 88], crack models based on a formulation originating in plasticity theory and the models of isotropic [BAZ 83; MAZ 89] or anisotropic elastic damage [FIC 96];

– the second category includes models based on an incremental formulation that uses a linear relationship between the stress rate and the deformation rate by the intermediary of a tangent module depending on the load history [FEE 95].

2.2. The discrete approach

This type of approach, based on fracture mechanics, began to be used at the end of the sixties to describe the collapse of concrete by Ngo and Scordelis [NGO 67]. Before describing this approach, we will give a historical presentation and elementary description of the mechanics of fracture and will end with a description of Hillerborg's fictitious crack model [HIL 76].

2.2.1. *Fracture mechanics – elements for the analysis of cracked areas*

Fracture mechanics was introduced by Griffith [GRI 21]. He showed that the fracture of a brittle elastic material could be characterized by a variable later called the critical energy release rate. The value of this variable, which is independent of the geometry of the structure, is an intrinsic characteristic of the material. The first theoretical developments of analysis of the fields of displacement, deformation, and stress around a crack were undertaken by Westergaard [WES 39]. This approach, called global, showed that the essential phenomena are situated near the tip of the crack, and that it is possible to study the macroscopic cracked area with the help of intrinsic variables. This is due to the strong concentration of stress at crack tips that linear elastic analysis translates into singularities in the field of stress.

The study of these singularities lead Irwin [IRW 57] to define stress intensity factors corresponding to special cinematics of the movement of cracks. This approach was subsequently generalized and constructed from the thermodynamics of irreversible processes during the seventies. The aim of fracture mechanics is to characterize the cracking behavior of structures with the help of quantifiable parameters in the engineering sense, especially stress, the size of the crack and the resistance to cracking of the material.

Fracture mechanics presupposes the existence of an initial crack in the structure studied and a system of external loads. The initial crack is either caused by damage under the effect of this solicitation, or a defect dating from the manufacture of the material or its implementation. Three successive zones can be briefly distinguished in a crack (Figure 2.1):

– the zone of microcracks (F): it can be found at the crack tip and in the wake of the crack left when it spreads. The stresses in this zone are very high, causing considerable damage to the material. The size of this zone is usually very small in the ductile fracture of metals, but for quasi-brittle materials, it is often very large;

– the close field zone (N): in which the fields of displacement, deformation and stress are continuous and have a formulation independent of the distant geometry of the structure. It is shown that in this zone (see for example [LEM 85]) the components of the field of stress are infinite near the edge of the crack ($r \rightarrow 0$). More precisely, the singularity is in ($r^{-1/2}$) in a linear elastic body;

– the far field zone (L): external, including the distant fields linked both to the singular zone and to the conditions at the limit concerning loading and displacement. In this zone, the fields of displacement, deformation, and stress vary insubstantially.

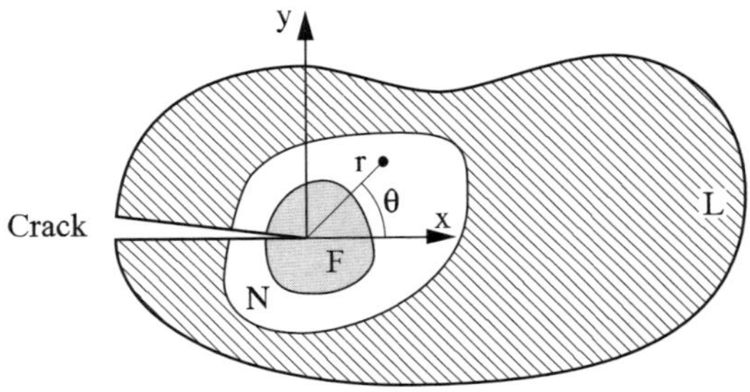

Figure 2.1. *Fracture process zone*

Various methods of analysis are used to study the fields of displacement, deformation and stress near a crack and especially in the zone (N) defined above. These methods are divided into two types of approach:

– direct approaches: based on the use of Airy functions; within these approaches, we note the Westergaard resolution [WES 39], the expansion of Williams [WIL 57] and more generally, the techniques used by Mushkelishvili;

– energy approaches: based on the energy analysis of the continuous environment containing a crack. This is a global balance report including the energy release rate due to a virtual growth of the crack. In these approaches, problems where the body is neither elastic, nor linear and homogenous can be treated.

2.2.2. *Discrete approach to cracking*

In this type of approach, the crack is introduced as a geometrical entity.

In the finite elements method, the crack evolves if the nodal force at the crack tip goes beyond the material stress limit. This node then divides into two nodes and the crack tip propagates to the following node. Thus the process is repeated as shown in Figure 2.2.

In this approach, the crack is forced to run along the edges of the finite elements. This requires prior knowledge of the direction in which the crack will develop. The use of automatic remeshing techniques [ING 85] enables this problem to be reduced, or even eliminated. However, numerical difficulties persist. These are mainly due to the continuous change in the topology of the mesh, which makes such an approach tedious and expensive.

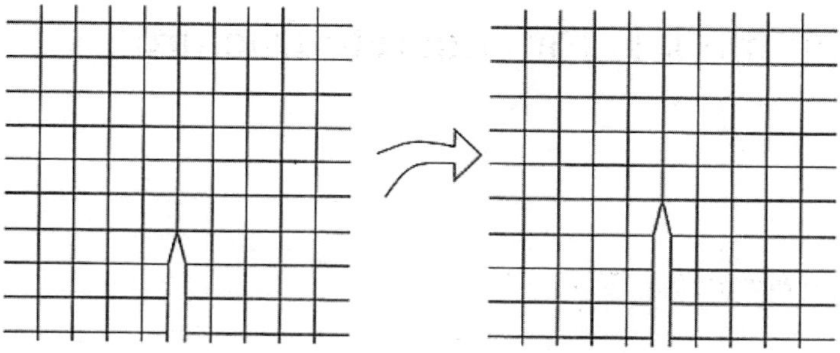

Figure 2.2. *Discrete modeling of cracking*

These problems have been overcome by mesh-free methods such as the Element Free Garlekin method [BEL 94]. Despite the efficiency of this method in several applications, difficulties related to its robust tridimensional implantation or to its computation cost, which is higher than that of classic finite element methods, have limited its use.

An original method, which satisfies the discrete spreading of cracks through elements, has been proposed by Belytschko and his colleagues [BEL 99; MOE 99].

This method exploits the properties of the partition of unity of finite element form functions [BAB 97]. The field of displacement, u, can be interpolated in the form:

$$u = \sum_{i=1}^{n} \varphi_i \left(\overline{a}_i + \sum_{j=1}^{m} \psi_j \widetilde{a}_{ij} \right) \qquad [2.1]$$

where φ_i is the form functions, n the total number of nodal points, $\overline{a}_i$ the degrees of nodal freedom, ψ_j the terms of growth of the base, $\widetilde{a}_{ij}$ the degrees of additional freedom at the node I, which represent the amplitude of j^e terms of the increase ψ_j. A crack is considered as a geometrical discontinuity of the field of displacement. The field of displacement, u, which contains only one discontinuity Γ_d (Figure 2.3), can be written as the sum of two continuous fields of displacement $\bar{u}$ and $\tilde{u}$ separated by the *Heaviside* function H_{Γ_d}:

$$u = \bar{u} + H_{\Gamma_d}\tilde{u} \quad [2.2]$$

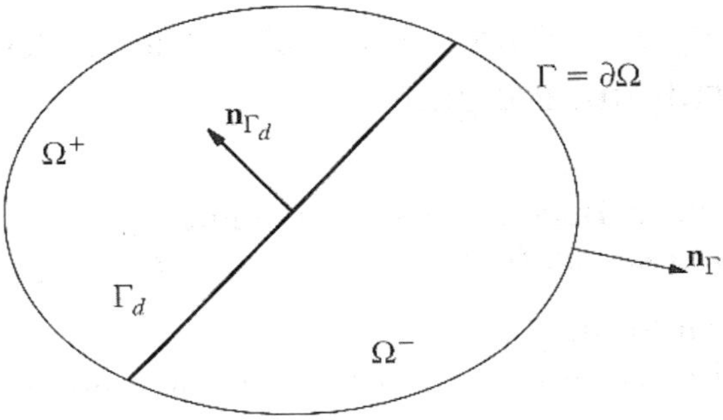

Figure 2.3. *A solid composed of continuous fields of displacement on either side of the discontinuity Γ_d*

This decomposition of the displacement field has a structure similar to the interpolation given in equation [2.1]. This is directly noticeable when the equation [2.2] is rewritten in the form:

$$u = N\left(\bar{a} + H_{\Gamma_d}\tilde{a}\right) = N\bar{a} + H_{\Gamma_d}N\tilde{a} = \bar{u} + H_{\Gamma_d}\tilde{u} \quad [2.3]$$

wherein N contains the functions of standard forms, $\bar{a}$ and $\tilde{a}$ contain the conventional and additional degrees of nodal freedom, respectively. Consequently, the properties of the partition of unity of shape functions can be used directly to incorporate discontinuities, and thus to model discrete cracking, so as to preserve the discontinuous character of the cracks. In the initial contribution of Belytschko *et al.* [BEL 99] and Moës *et al.* [MOE 99], the discrete crack was introduced using linear fracture mechanics. This required special form functions to simulate singularity at the crack tip. For this reason, the terms of increase of the base were introduced in the form:

$$\psi = \left(\sqrt{r}\,cos(\theta/2), \sqrt{r}\,sin(\theta/2), \sqrt{r}\,sin(\theta/2)sin(\theta), \sqrt{r}\,cos(\theta/2)sin(\theta)\right) \quad [2.4]$$

wherein r is the distance from the crack tip, and θ is measured from the current position of propagation of the crack. Far from the crack tip, the *Heaviside* functions were employed as explained above.

2.2.3. *Hillerborg's fictitious crack model*

The use of linear fracture mechanics is limited when it comes to describing fracture behavior in concrete structures. This is mainly due to the presence of a microcrack zone at the crack tip which is large, relative to the size of the structure for this type of material. Hillerborg *et al.* [1976] were among the first to propose a non-linear model for concrete, based on fracture mechanics. They based their reasoning on work carried out by Dugdale [DUG 60] and Barenblatt [BAR 62], concerning the plastic zone at the crack tip. They introduced the notion of the fictitious crack with a behavior between applied stress and the opening of the crack measured by means of a tensile strength test.

It is a non-linear model in which the crack is modeled in two parts as shown in Figure 2.4. The first corresponds to the physical crack (real crack) for which the stress is null; the second corresponds to the fictitious crack for which the stress is not null and depends on the crack opening.

Crack propagation takes place when the stress at the crack tip reaches maximum tensile stress, f_t. In this case, stress distribution varies all along the microcrack zone as shown in Figure 2.4.

Cohesion pressure σ acts on the crack faces. It is used to model the various hardening mechanisms in the microcrack zone. It is the essential parameter in the fictitious crack model. Hillerborg [HIL 76] proposed to deduct this stress distribution by means of a uniaxial tensile strength test. This cohesive stress tends to close the crack and is supposed to be a decreasing function either of the length of the microcrack zone noted, t [REI 84], or of the crack opening noted, w [HOR 91]. Reinhardt [REI 84] reached the following equation based directly on the Dugdale/Barenblatt model:

$$\frac{\sigma(t)}{f_t} = 1 - \gamma\left(\frac{(a+t)-x}{t}\right)^n \qquad [2.5]$$

where γ is a factor that varies between 0 and 1. The exponent n plays an important role in determining the form of the distribution of the cohesive stress all along the fictitious crack tip. It can vary from a rectangular shape (constant pressure following the Dugdal/Barenblatt model when $n \to \infty$) to a perfectly brittle behavior (following linear elastic fracture mechanics for $n = 0$).

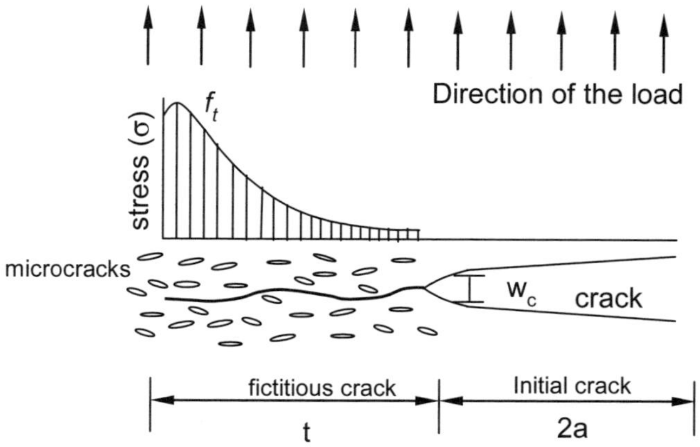

Figure 2.4. *Hillerborg's model*

Other empirical relationships have been proposed, which relate the cohesive stress distribution to the crack opening w. Amongst these, let us note Hordijk [HOR 91]:

$$\frac{\sigma(w)}{f_t} = \left[1 + \left(c_1 \frac{w}{w_c}\right)^3\right].e^{-c_2 \frac{w}{w_c}} - \frac{w}{w_c}\left(1 + c_1^3\right).e^{-c_2} \qquad [2.6]$$

where w_c corresponds to the maximum crack opening corresponding to cohesive stress, which is null (for example, at the real crack tip). c_1 and c_2 are two parameters that depend on the shape of the stress distribution softening curve. For $w<w_c$, the crack corresponds to a microcracked zone with remaining ligaments for stress transfer.

Two main parameters are directly related to the fictitious crack model. The area below the cohesive stress distribution diagram (Figure 2.3), which corresponds to the fracture energy per unit of crack face, is absorbed so as to overcome the various hardening mechanisms in the microcrack zone and thus open the crack from w=0 to w=w_c. It is given by equation [2.7]:

$$G_f = \int_0^{w_c} \sigma(w)dw \qquad [2.7]$$

The second parameter is called the "brittleness index" or characteristic length, l_{ch}, given by equation [2.8]:

$$l_{ch} = \frac{EG_f}{f_t^2} \quad [2.8]$$

2.3. Continuous approach

2.3.1. *Damage mechanics*

The second large family of theoretical approaches enabling us to model the non-linear fracture behavior of concrete is the continuum damage mechanics approach [CHA 88; KAC 58; LEM 90; RAB 69].

This type of approach macroscopically translates the influence of deterioration due to the appearance, growth, and coalescence of microflaws in the material, by introducing an internal variable representative of this state of deterioration called the "damage variable". The crack is then schematized by a completely damaged zone of material (Figure 2.5).

The redistribution of the stresses, which result from the concentration of the deformations and the evolution of the damage, takes place in a partially damaged region, called the *Fracture Process Zone*. Damage evolution in this microcracked zone determines the direction of propagation in the rest of the structure of the crack.

This approach is used in the study of structures of a size above that of the heterogeneities of the material, that of a representative elementary volume (REV) depending on the material. The constitutive relations are established with reference to this volume and attributed during calculation at each point of the structures concerned.

2.3.1.1. *Thermodynamic bases*

The models formulated in the framework of continuum damage mechanics are based on the formalism of the thermodynamics of irreversible processes [LEM 90], that is, the use of state variables, observable or internal, that define the state of the closed system at any given moment.

The present study will be limited to a quasi-brittle material of elastic behavior, with a supposed isotropic initial state.

The damage is represented by a scalar, *d*, which varies between 0 (healthy material) and 1 (fractured material). This variable is considered as a state variable. We suppose the existence of the thermodynamic potential in the form Ψ (ε_e, *T*, *d*, V_i), wherein ε_e is the tensor of elastic deformations, *T* the temperature and V_i the other internal variables of the problem (Table 2.1). σ is the stress tensor, s is the entropy of the system, Y is the damage associated variable and the Va are the variables associated with the internal variables Vi.

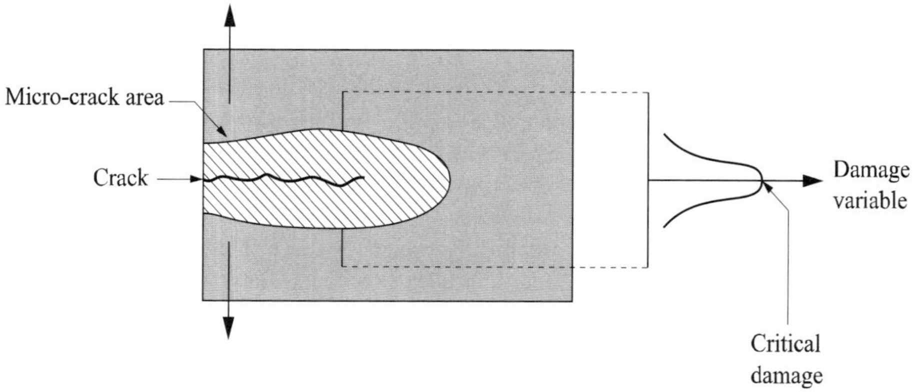

Figure 2.5. *Crack defined as a completely damaged zone*

Leaving aside thermal effects, and assuming the elasticity effects associated with damage are uncoupled from the variables V_i, the potential can be split in a potential of elastic state ψ_e, and a potential associated with the variables V_i. The elastic state potential ψ_e will be studied below. The chosen form is:

$$\Psi(\varepsilon_e, d) = \frac{1}{2}\varepsilon_e : E(d) : \varepsilon_e \qquad [2.9]$$

with $E(d) = (1 - d)E_0$, wherein $E(d)$ is the stiffness matrix at a given stage of damage *d* and E_0 is the initial stiffness matrix of the material.

Considering reversible evolutions, i.e. without dissipation, and using Clausius-Duhem's inequality, we deduct the expression of the stress tensor:

$$\sigma = \rho \frac{\partial \Psi_e}{\partial \varepsilon}. \qquad [2.10]$$

In the same way, the variable Y associated with the damage is written:

$$Y = \rho \frac{\partial \Psi_e}{\partial d} . \qquad [2.11]$$

It can be noted that this quantity corresponds precisely to the energy dissipated per unit of damage rate. This is why it is called elastic energy release rate.

In this particular case, Clausius-Duhem's inequality is reduced to:

$$-Y\dot{d} \geq 0 . \qquad [2.12]$$

This defines the laws of state of the model. Noting that $-Y$ is a positive definite quadratic form, this inequality requires that damage should grow or remain constant.

State variables		Associated variables
Observable	Internal	
ε		σ
T		S
	D	Y
	V_i	V_a

Table 2.1. *State variables and their associated variables*

Threshold function

The behavior law will evolve when a criterion is reached. It is generally written in the form:

$$f(\sigma, Y, V_a, s) = 0 \qquad [2.13]$$

This function, also called the threshold function, defines a load surface. Within this surface there is no irreversible evolution. On the surface, these evolutions are possible. Concerning damage, the threshold function is written:

$$f(Y) = Y - Z - Y_0 \qquad [2.14]$$

Z is the work hardening variable, by analogy with plasticity. $Z=Y_0$ initially and $Z= Sup(Y,Y_0)$ subsequently. Y_0 is the initial damage threshold. Similar to plasticity, in the framework of standard generalized materials with an associated law $\varphi = \lambda.f$, wherein φ is a convex potential, the following evolution laws are introduced:

$$\dot{d} = \dot{\lambda}\frac{\partial f}{\partial Y} \qquad [2.15]$$

$$\dot{z} = -\dot{\lambda}\frac{\partial f}{\partial Z} \qquad [2.16]$$

z and Z are linked by the work hardening module $H_e = -E_0/\gamma$, wherein γ is the work hardening parameter characteristic of the material:

$$\dot{z} = H_e \dot{Z} \qquad [2.17]$$

λ and f verify the Kuhn-Tucker conditions:

$$\dot{\lambda} \geq 0, f \leq 0 \quad and \quad \dot{\lambda} f = 0 \qquad [2.18]$$

2.3.1.2. *Example 1: Mazars's damage model*

Mazars proposed a model based on a local approach to damage mechanics in order to describe the fracture behavior of concrete [MAZ 84].

Stress is determined by the following relationship:

$$\sigma_{ij} = (1-d)E_0\varepsilon_{kl} \qquad [2.19]$$

where σ_{ij} and ε_{kl} are the stress and deformation tensor components, respectively, E_0 is the initial elasticity module and d the damage variable (i, j, k, l $\in$ [1,3]). We remain in the same conditions as those defined in the preceding section, that is the material is supposed to have an isotropic damageable elastic behavior, d is a scalar, the residual deformations are not described. We also suppose that the cracks are produced by extension of the material.

A new variable is introduced, $\tilde{\varepsilon}$, called equivalent deformation, which represents the intensity of local deformations. It is defined by:

$$\tilde{\varepsilon} = \sqrt{\sum \langle \varepsilon_i \rangle_+^2} \text{ with } \langle \varepsilon_i \rangle_+ = 0 \text{ if } \varepsilon_i < 0 \text{ and } \langle \varepsilon_i \rangle_+ = \varepsilon_i \text{ if } \varepsilon_i \geq 0 \quad [2.20]$$

ε_i represents the principal deformation in the direction i. The evolution of the scalar damage is defined from a threshold function:

$$f(\tilde{\varepsilon}, d) = \tilde{\varepsilon} - \kappa(d) \quad [2.21]$$

to $\kappa(d=0) = \kappa_0$ where $\kappa_0 = \dfrac{f_t}{E_0}$ is the initial damage threshold. Once this threshold is reached, κ is equal to the maximal value reached by $\tilde{\varepsilon}$ during the load history of the material.

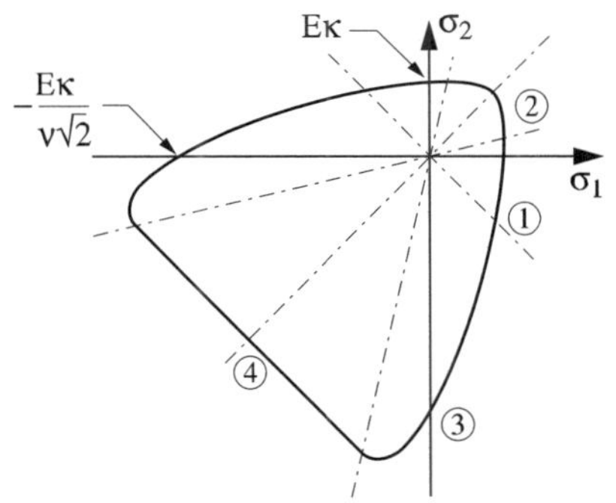

Figure 2.6. *Mazars' model: outline of the threshold surface in the plane σ_3=0 [MAZ 84]*

Figure 2.6 presents the contour of the threshold surface $\tilde{\varepsilon} - \kappa = 0$ described in the stress plane $\sigma_1 - \sigma_2$.

Two distinct evolution laws are proposed for the development of damage under tensile and compressive loads. Global damage is given by a linear relationship between these two:

$$d = \alpha_t d_t + \alpha_c d_c \quad [2.22]$$

where d_t and d_c are the contributions to d of the damage mechanisms in direct tension (d_t) and the mechanisms induced under the solicitation of compression (d_c). α_t is defined as a function of the stress state. In pure tension α_t=1, α_c=0 and in pure compression α_t=0, α_c=1. The evolution of the threshold function is defined as follows:

$$\text{if } f(\tilde{\varepsilon}) = 0 \text{ and } \quad \dot{f}(\tilde{\varepsilon}) = 0$$

$$\text{then } \begin{cases} d = h(\kappa) \\ \kappa = \tilde{\varepsilon} \end{cases} \quad \text{with } d \geq 0$$

$$\text{otherwise } \begin{cases} \dot{d} = 0 \\ \dot{\kappa} = 0 \end{cases} \qquad [2.23]$$

Damage evolution laws, written $h(\kappa)$ in equation [2.23], are given by the following formulae:

$$d_t = 1 - \frac{\kappa_0(1 - A_T)}{\kappa} - \frac{A_T}{exp(B_T(\kappa - \kappa_0))}$$

$$d_c = 1 - \frac{\kappa_0(1 - A_C)}{\kappa} - \frac{A_C}{exp(B_C(\kappa - \kappa_0))} \qquad [2.24]$$

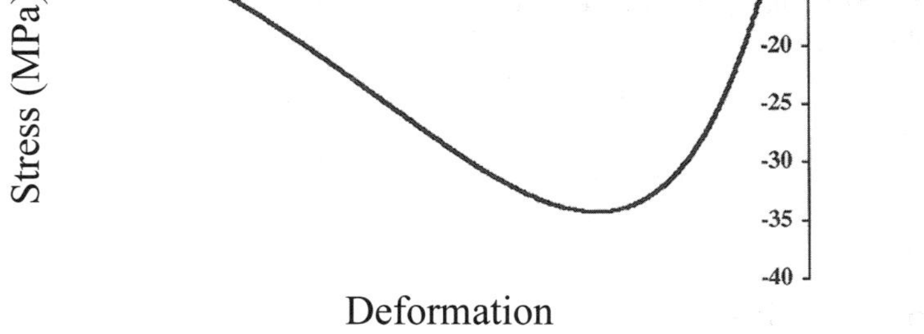

Figure 2.7. *Mazars' model, uniaxial behavior in tension and compression*

There are eight parameters to identify. The Young's modulus and the Poisson's ratio are measured using a uniaxial compression test. A direct tension test or a three-point flexion test enables the damage parameters under tension to be determined (κ_0, A_T, B_T). The A_C and B_C parameters are identified from the response of the material under uniaxial compression.

Finally β must be identified from the response of the material to shearing.

As this type of test is difficult to carry out, it is recommended that the value β=1 is used. Table 2.2 presents the standard values of the parameters of the damage model for ordinary concrete.

$E_0 \approx 30{,}000\text{-}40{,}000$ MPa
$\nu_0 \approx 0.2$
$\kappa_0 \approx 1\times10^{-4}$
$0.7\leq A_T\leq 1.2$
$10^4\leq B_T\leq 5\times10^4$
$1\leq A_C\leq 1.5$
$10^3\leq B_C\leq 2\times10^3$
$1\leq\beta\leq 1.05$

Table 2.2. *Standard values of Mazars's damage model parameters*

Figure 2.7 shows the uniaxial response of this model, in compression and tension, with the following parameters: E_0=30,000 MPa, ν_0=0.2, κ_0=0.0001, A_T=1, B_T=15,000, A_C=1.2, B_C = 1,500, β=1.

This model has been improved to obtain a finer description of the material under certain solicitations, for example, Laborderie [LAB 91] and Pontirolli [PON 95] introduce the phenomena of crack closure and Ragueneau [RAG 99] introduces internal dissipation in the material. Some of these models will be developed in Chapter 4.

2.3.1.3. *Taking the localization of deformations into account*

It has been shown that the models developed within the framework of continuum damage mechanics using a local approach are not adapted to materials with a softening behavior [BAZ 86; BOR 91; SIM 89]. The reasons are essentially physical. In fact, the development of the preceding models is mainly aimed at their use in a finite element type code. The structure studied is discretized according to a mesh in which the number of elements depends on the chosen fineness. Calculations revealed that the localization of the deformations takes place in one finite element only. Thus, the finer the mesh, the smaller the localized zone, which tends towards zero for an extremely fine mesh and the total energy dissipated upon fracture, tends to vanish, which is physically unacceptable [BAZ 86; NEE 88]. In order to avoid these numerical problems, various approaches have been developed. These techniques eliminate mesh dependence, while increasing calculation time.

2.3.1.3.1. Non-local approach to damage

In this approach, the actual value of a variable at a point x is considered as the weighted average for this same variable calculated near point x. This theory was introduced by Pijaudier-Cabot and Bazant [BAZ 88; PIJ 87]. They applied it to the damage model presented in section 2.3.1.2. The non-local effect is considered by relating the damage variable to the equivalent non-local deformation $\bar{\varepsilon}$. This deformation is written as a weighted average of the local equivalent deformations $\tilde{\varepsilon}$ [2.20] in a volume V nearby the point where the behavior law is calculated. The volume V is assumed to be fixed The equivalent nonlocal deformation $\bar{\varepsilon}$ is calculated according to the equation:

$$\bar{\varepsilon}(x) = \frac{1}{V_r(x)} \int_{\Omega} \psi(x-s)\tilde{\varepsilon}(s)ds$$

with:

$$V_r(x) = \int \psi(x-s)ds \quad [2.25]$$

where Ω is the volume of the structure, V_r is the representative volume at point x, and $\psi(x-s)$ is the weight function. The weight function is arbitrarily defined by a Gaussian equation:

$$\psi(x-s) = exp\left(\frac{4\|x-s\|^2}{l_c^2}\right) \quad [2.26]$$

where l_c is a new parameter of the damage model called internal length. In the case of concrete, l_c is generally considered equal to three times the size of the largest aggregate [BAZ 89]. $\bar{\varepsilon}$ is the variable that controls damage growth according to the following conditions. The evolution of the threshold function is defined according to:

$$F(\bar{\varepsilon}) = \bar{\varepsilon} - \kappa \quad [2.27]$$

and the damage evolution is still:

$$F(\bar{\varepsilon}) = 0 \text{ and } \dot{F}(\bar{\varepsilon}) = 0$$

$$\text{then } \begin{cases} d = h(\kappa) \\ \kappa = \bar{\varepsilon} \end{cases} \text{ with the condition } d \geq 0 \qquad [2.28]$$

$$\text{otherwise } \begin{cases} \dot{d} = 0 \\ \dot{\kappa} = 0 \end{cases} .$$

An important point in the use of the nonlocal damage model is the determination of the internal length, l_c. This parameter cannot be directly measured. Two methods enable it to be identified:

– from the data resulting from scale effect tests [LEB 03];

– from two uniaxial tensile tests on samples reinforced by fibers [BAZ 89].

The numerical study of bifurcations and loss of stability in this approach was carried out by Pijaudier-Cabot and Huerta [PIJ 91], and Pijaudier-Cabot and Bodé [PIJ 92]. The criterion of deformation localization was developed by Pijaudier-Cabot and Benallal [PIJ 93]. Certain improvements have been made to this model to take into account the evolution in internal length according to the state of damage at each point of the material, based on micromechanical arguments [HAI 02; PIJ 04]. This parameter was so far considered as a constant intrinsic parameter of the model.

2.3.1.3.2. Gradient approach

This type of approach consists of introducing the second gradient of the variable in the expression that defines the evolution of damage [BOR 92]. This method was applied to damage by Peerlings [PEE 95]. It was applied to the damage model presented in section 2.3.1.2. These models use a Taylor series development truncated to the order of 2:

$$\tilde{\varepsilon}(x+s) = \tilde{\varepsilon}(x) + \frac{\partial \tilde{\varepsilon}(x)}{\partial x} s + \frac{\partial^2 \tilde{\varepsilon}(x)}{2\partial x^2} \frac{s^2}{2!} + ... \qquad [2.29]$$

Local deformation is then replaced by an average deformation, ε, on an area near x (equation [2.25]) and the expression of ε becomes:

$$\bar{\varepsilon}(x) = \tilde{\varepsilon}(x) + c^2 \nabla^2 \tilde{\varepsilon}(x) \qquad [2.30]$$

c is a parameter that depends on the type of weight function. The units of c are square meters, and it can be considered as the square of the internal length.

The implementation of such a model is difficult because it requires continuity in the interpolation of the displacement field. This difficulty can be solved by the following consideration:

$$\nabla^2 \bar{\varepsilon}(x) = \nabla^2 \tilde{\varepsilon}(x) + \underbrace{\nabla^2 (c^2 \nabla^2 \tilde{\varepsilon}(x))}_{\text{Negligible}} \qquad [2.31]$$

By considering the second term of the right-hand expression as negligible, we deduce:

$$\nabla^2 \bar{\varepsilon}(x) \approx \nabla^2 \tilde{\varepsilon}(x) \qquad [2.32]$$

and therefore:

$$\bar{\varepsilon}(x) - c^2 \nabla^2 \bar{\varepsilon}(x) = \tilde{\varepsilon}(x). \qquad [2.33]$$

Equation [2.33] is strictly equivalent to an integral formulation with a special choice of weight function.

The implemetation of the gradient model actually becomes similar to that of a (local) thermomechanical model in which the nonlocal deformation replaces temperature in a given node [PIJ 00]. This type of model has shown good results in predicting fracture in concrete structures [PEE 95]. Different versions of gradient damage models exist: Frémond *et al.* [FRE 95] based their studies on the principle of virtual work, adding the work of internal forces, which introduces higher-order terms in equation [2.29].

2.3.1.4. *Example 2: the diffuse cracking model*

Diffuse cracking models include two sorts of models: fixed crack models and rotating crack models [ROT 88]. In the first, the normal direction of the cracking is fixed after it starts.

Conversely, in the second model, the direction of the crack can turn during cracking, following the axes of the main deformation reference point. The relation between stress and strain is vectorial.

2.3.1.4.1. Fixed crack model

Before cracking, the material is considered as isotropic, linear elastic. For example, in plane stress, the relation between stress and strain is:

$$\begin{bmatrix} \sigma_{xx} \\ \sigma_{yy} \\ \sigma_{zz} \end{bmatrix} = \frac{E_0}{1-\nu^2} \begin{bmatrix} 1 & \nu & 0 \\ \nu & 1 & 0 \\ 0 & 0 & (1-\nu) \end{bmatrix} \begin{bmatrix} \varepsilon_{xx} \\ \varepsilon_{yy} \\ \varepsilon_{zz} \end{bmatrix} \quad [2.34]$$

As soon as the highest principal stress is greater than the tensile fracture stress, a crack starts perpendicular to the direction of this stress. The isotropic behavior law is then replaced by an orthotropic behavior law in the frame of reference of the initial crack. This relation is expressed in (n,t) directions (Figure 2.8; where n represents the normal direction and t the direction tangential to the crack), which is the orthotropic frame of reference:

$$\sigma_{nt} = s_{E_{nt}} \varepsilon_{nt} \quad [2.35]$$

where $s_{E_{nt}}$ is the secant stiffness matrix. This matrix has been written in several ways by different authors. The most common way of writing it for a plane stress or deformation is [BAZ 83]:

$$s_{E_{nt}} = \begin{bmatrix} \dfrac{\mu E}{1-\nu^2 \mu} & \nu \dfrac{\mu E}{1-\nu^2 \mu} & 0 \\ \nu \dfrac{\mu E}{1-\nu^2 \mu} & \dfrac{E}{1-\nu^2 \mu} & 0 \\ 0 & 0 & \beta G \end{bmatrix} \quad [2.36]$$

where β represents the shear reduction modulus ($0 \leq \beta \leq 1$) and μ is the factor of normal reduction.

From equation [2.35], the constitutive relation in the global system (of coordinates (x,y)) can be determined by using the following relationships:

$$\begin{cases} \sigma_{xy} = T(\theta_0)^t \, \sigma_{nt} \\ \varepsilon_{xy} = T(\theta_0)^t \, \varepsilon_{nt} \end{cases} \quad [2.37]$$

where θ_0 is the angle existing between the local orthotropic frame of reference (n,t) and the global one (x,y). This angle is constant. σ_{xy} and σ_{nt} are the stress tensor in the global frame of reference and the stress in the orthotropic frame of reference (Figure 2.8), respectively. $T(\theta_0)$ is the passage matrix; it is given by:

$$T(\theta_0)=\begin{bmatrix} cos^2\theta_0 & sin^2\theta_0 & -sin\theta_0 \, cos\theta_0 \\ sin^2\theta_0 & cos^2\theta_0 & sin\theta_0 \, cos\theta_0 \\ 2\,sin\theta_0 \, cos\theta_0 & -2\,sin\theta_0 \, cos\theta_0 & cos^2\theta - sin^2\theta_0 \end{bmatrix} \quad [2.38]$$

by using equations [2.35] and [2.36], the relation between the stress and the strain can be written in the global frame of reference:

$$\sigma_{xy} = T(\theta_0)^t s_{E_{nt}} T(\theta_0)\varepsilon_{xy} \text{ with } \theta_0\text{=constant} \quad [2.39]$$

The tangent matrix is obtained by differentiating the equation above:

$$\dot{\sigma}_{xy} = \left\{T(\theta_0)^t s_{E_{nt}} T(\theta_0)\right\}\dot{\varepsilon}_{xy} + \left\{T(\theta_0)^t \frac{\partial s_{E_{nt}}}{\partial t} T(\theta_0)\right\}\varepsilon_{xy} \quad [2.40]$$

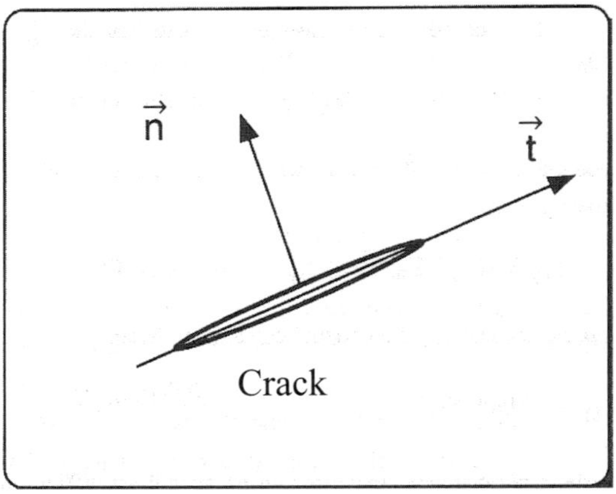

Figure 2.8. *Orthotropic frame of reference*

This approach has two major disadvantages: first, it is difficult to easily combine the cracking phenomenon with another non-linear and incremental phenomenon (plasticity) because the formulation is written in a total stress format. Second, the crack plane direction is fixed whatever the load. This brings about a stress locking phenomenon [ROT 92a]. To solve this second problem, de Borst and Nauta [BOR

85], and Rots [ROT 88] have proposed a model in which the total strain is decomposed into a concrete part ε^{c0} and a cracking part ε^{cr} (Figure 2.9).

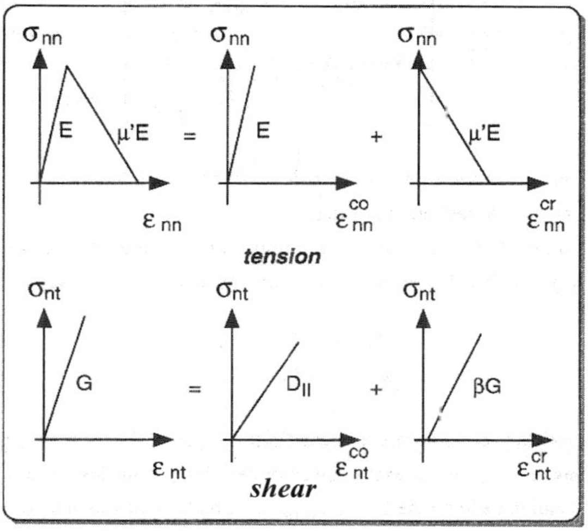

Figure 2.9. *Decomposition of normal deformations and shear into a concrete part and a cracking part*

In an incremental formulation, we can write:

$$\dot{\varepsilon} = \dot{\varepsilon}^{co} + \dot{\varepsilon}^{cr} \qquad [2.41]$$

$\dot{\varepsilon}^{cr}$ can be decomposed into several parts representing cracking:

$$\dot{\varepsilon}^{cr} = \dot{\varepsilon}^{cr1} + \dot{\varepsilon}^{cr2} + \ldots \qquad [2.42]$$

This is the multiple fixed cracks concept for which $\dot{\varepsilon}^{cr1}$ represents the deformation rate caused by the first crack, $\dot{\varepsilon}^{cr2}$ represents the deformation rate caused by the second crack, etc.

A great advantage of this method is the possibility of treating the behavior of the undamaged concrete separately from that of the crack. The main disadvantage of this model is the introduction of new parameters which must be selected by the user (such as the critical angle from which a new crack originates). The various choices of these parameters can considerably influence the result of the calculation as the tangent stiffness of the crack system depends strongly on these parameters.

In what follows we are going to see how this model was generalized to a rotating crack concept.

2.3.1.4.2. Rotating crack model

The rotating crack model was proposed by Cope [COP 80]. This model can be considered as the limit case for the multiple fixed cracks concept for a critical angle of zero [ROT 88]. The constitutive relation has the same form as that described in equation [2.39]. However, for the rotating crack model, the angle between the local orthotropic frame of reference and the global frame of reference is no longer constant. The direction associated with the greatest main stress and the normal direction of the crack, remain aligned during the whole cracking process. The behavior equation is written:

$$\sigma_{xy} = T(\theta_0)^t s_{E_{nt}} T(\theta_0)\varepsilon_{xy} \text{ with } \theta_0 \neq \text{constant} \quad [2.43]$$

We can note that the local reference point (n,t) is now the same as the main stress tensor reference point. Furthermore, with regard to equation [2.43], it appears that the main directions of the stress and strain tensors coincide during the entire cracking process. The tangent matrix of the rotating crack model is obtained by differentiating equation [2.43]:

$$\dot{\sigma}_{xy} = \left\{ T(\theta_0)^t s_{E_{nt}} T(\theta_0) + \frac{\partial T(\theta)^t}{\partial \theta} \right\} \dot{\varepsilon}_{xy} + \left\{ T(\theta_0)^t \frac{\partial s_{E_{nt}}}{\partial t} T(\theta_0) \right\} \varepsilon_{xy} \quad [2.44]$$

We can observe that, compared with equation [2.40], there is an extra term which corresponds to the rotation of the main directions. When this equation is developed in the local frame of reference, we obtain:

$$\begin{bmatrix} \dot{\sigma}_{nn} \\ \dot{\sigma}_{tt} \\ \dot{\sigma}_{nt} \end{bmatrix} = \begin{bmatrix} \frac{\mu E}{1-\nu^2 \mu} & \nu \frac{\mu E}{1-\nu^2 \mu} & 0 \\ \nu \frac{\mu E}{1-\nu^2 \mu} & \frac{E}{1-\nu^2 \mu} & 0 \\ 0 & 0 & \frac{(\sigma_{nn} - \sigma_{tt})}{2(\varepsilon_{nn} - \varepsilon_{tt})} \end{bmatrix} \begin{bmatrix} \dot{\varepsilon}_{nn} \\ \dot{\varepsilon}_{tt} \\ \dot{\varepsilon}_{nt} \end{bmatrix} \quad [2.45]$$

This shows that the variation in the shear term of the stiffness tensor $(\sigma_{nn}-\sigma_{tt})/2(\varepsilon_{nn}-\varepsilon_{tt})$ appears automatically upon rotation of the principal axes of the stress and strain tensors and when their coaxiality is required. In order to maintain this coaxiality, it is required that the stiffness of the incremental total shear of the crack system and the concrete βG be equal to the term of shear stiffness of equation [2.45]:

$$\beta G = \frac{(\sigma_{nn} - \sigma_{tt})}{2(\varepsilon_{nn} - \varepsilon_{tt})} \quad [2.46]$$

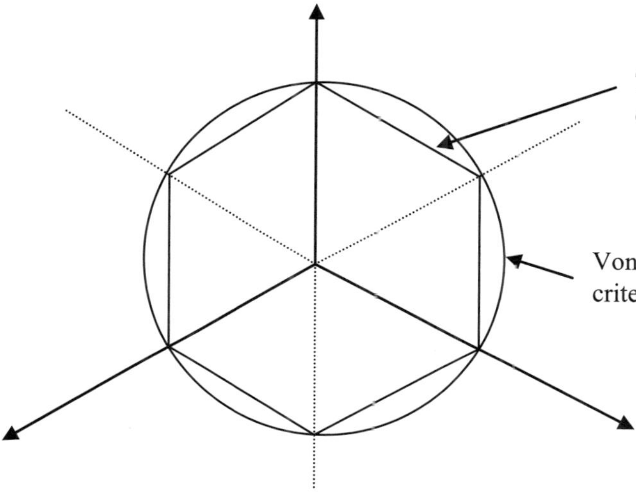

Figure 2.10. *Intersection of the criteria with the deviatoric plane*

2.3.2. *Plasticity theory*

Originally, plasticity theory was used to describe the behavior of metals presenting permanent deformations when they are submitted to mechanical stress. The microscopic mechanism responsible for the plasticity of metals resides in the respective slip of molecular planes leading to irreversibility. Concerning metals, plasticity is an irreversible process that is engendered by shear. Tresca in 1864 stated the criterion of maximum shear at which plastic deformation occurs. In terms of main stress, the following criterion was obtained:

$$|\sigma_1 - \sigma_2| = 2\tau_{\max} \quad \text{and} \quad |\sigma_1 - \sigma_3| = 2\tau_{\max} \quad \text{and} \quad |\sigma_2 - \sigma_3| = 2\tau_{\max} \text{ with } \sigma_1 \geq \sigma_2 \geq \sigma_3 \quad [2.47]$$

This criterion is composed of six planes parallel to the hydrostatic axis, and therefore, presents some angular points. Von Mises proposed an approximation to

Tresca's criterion (Figures 2.10 and 2.11) which is a hexahedral prism, by a criterion of cylindrical form passing through the six summits of Tresca's criterion.

The mathematical expression of the cylinder with radius R circumscribed to the hexahedral prism is:

$$(\sigma_1 - \sigma_2)^2 + (\sigma_2 - \sigma_3)^2 + (\sigma_3 - \sigma_1)^2 = 2R^2 \quad [2.48]$$

$$J2 = \tfrac{1}{6}\left[(\sigma_1 - \sigma_2)^2 + (\sigma_2 - \sigma_3)^2 + (\sigma_3 - \sigma_1)^2\right] \quad [2.49]$$

the following expression is obtained:

$$\sqrt{3\,J_2} - R = 0 \quad [2.50]$$

Computing expression [2.48] defining the multiaxial behavior (σ_1, σ_2, σ_3) in the case of uniaxial behavior (σ_{eq} , 0, 0) leads to the value of R:

$$R = \sigma_{eq} \quad [2.51]$$

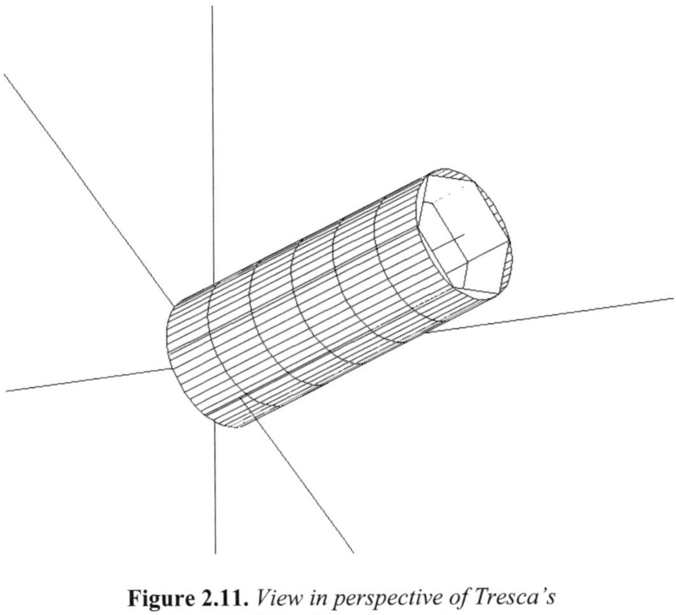

Figure 2.11. *View in perspective of Tresca's and Von Mises's criteria*

J_2, the second invariant of the deviatoric stress tensor. The Von Mises criterion is then defined as follows:

$$\sqrt{3\,J_2} - \sigma_{eq} = 0 \qquad [2.52]$$

This describes quite well the ultimate behavior of metals, which is independent of hydrostatic pressure, unlike concrete. This is why, in the case of concrete, a term that is a function of the first invariant of the stress tensor, I_1, is added in mathematical expression [2.52] leading to the Drucker-Prager criterion, the expression of which is:

$$\sqrt{3J_2} + \alpha I_1 - R = 0$$

with:

$$\alpha = \frac{2\sin(\phi)}{3-\sin(\phi)} \text{ and } R = \frac{6c\,\cos(\phi)}{3-\sin(\phi)} \qquad [2.53]$$

φ and c represent the angle of friction and the cohesion of the material, respectively. Cohesion is the maximum tangential stress at zero normal stress. This formulation is generally used to describe the behavior of soils.

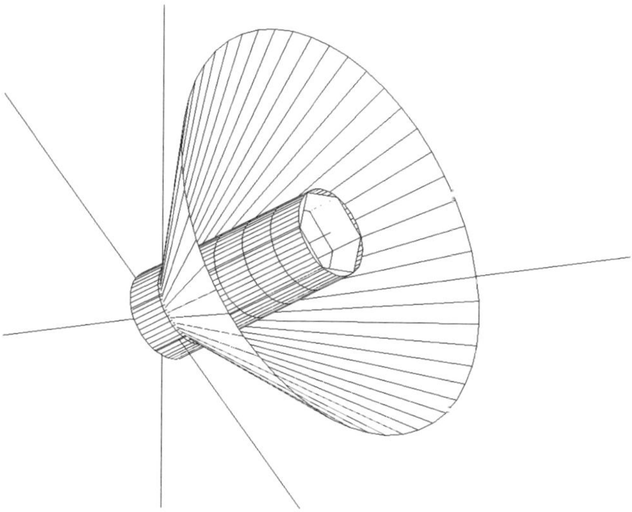

Figure 2.12. *Perspective view of the criteria of Tresca, Von Mises and Drucker-Prager*

In the case of concrete, the criterion parameters are determined according to the mechanical resistance characteristics, which are uniaxial resistance, f_c, and bi-axial resistance, βf_c, in the case of plane stress, or:

$$\alpha = \frac{1-\beta}{2\beta-1} \qquad [2.54]$$

$$R = \frac{\beta}{2\beta-1} f_c \qquad [2.55]$$

Other more complex criteria can be found in the literature, such as those of Willam-Warnke, Ottosen, Chen and Chen, etc. Indeed, in order to describe more precisely the ultimate behavior of concrete in states of tridimensional stress, a third term can be added in expression [2.53], which is a function of the third invariant, J_3, of the deviatoric stress tensor.

All failure criteria define, in the principal stresses in space, a domain beyond which the material is considered as fractured. In order to describe irreversible processes, the mathematical formulation of these criteria has been extended to those of the load surface, f. This surface, which consequently no longer represents a failure limit but a threshold of plastic deformation, defining the domain of elastic behavior, evolves according to a work hardening law up to a maximum threshold characterizing the failure of the material.

In the case of Von Mises's criterion, the evolution direction of the load surface is orthogonal to the hydrostatic axis. Work hardening takes place without plastic variation of volume. In the case of Drucker-Prager's criterion, the direction of the work hardening is orthogonal to the longitudinal profile of the cone (Figure 2.12). The plastic volumetric strains occur.

2.3.2.1. *Theoretical aspects*

Plasticity theory enables us to describe the irreversible character of plastic deformations which will be called $\{\varepsilon_p\}$. This notation is used in order to recall throughout the mathematical developments that $\{.\}$ is a tensor including six components.

The partition of the total deformation is decomposed into a reversible elastic part and an irreversible plastic part. The energy that results from the elastic deformations is stored then released during any discharge, while the plastic deformations lead to a dissipation of energy in the form of heat and are, therefore, irreversible.

$$\{\dot{\varepsilon}\} = \{\dot{\varepsilon}^e\} + \{\dot{\varepsilon}^p\} \quad [2.56]$$

The evolution of plastic deformations is defined by a flow rule, which can be said to be associated (standard materials) or non-associated (non-standard materials).

2.3.2.1.1. Standard materials

This corresponds to most metals. The flow rule verifies the principle of maximum plastic work. Plastic flow is normal to plastic potential, which is identical to the yield surface:

$$\{\dot{\varepsilon}^p\} = \dot{\lambda}\left\{\frac{\partial f}{\partial \sigma}\right\} = \dot{\lambda}\{n\} \quad [2.57]$$

2.3.2.1.2. Non-standard materials

This is often the case in geomaterials. Soils for example have a different slip mechanism, they are strongly dilatant. Plastic flow is normal to a plastic potential, g, which is different from the plasticity surface, f, to represent dilatancy:

$$\{\dot{\varepsilon}^p\} = \dot{\lambda}\left\{\frac{\partial g}{\partial \sigma}\right\} = \dot{\lambda}\{m\} \quad [2.58]$$

A work hardening law that characterizes the evolution of plastic strains in concrete under stress is yet to be defined. According to thermodynamics, the relation between the work hardening force, R, and its associated plastic variable, κ, must be expressed. Two hypotheses are currently admitted to evaluate this plastic variable: the plastic work hypothesis and the cumulated plastic deformation:

– plastic work:

$$R\,\dot{\kappa} = \{\sigma\}^T\{\dot{\varepsilon}^p\} \quad [2.59]$$

– cumulated plastic deformation:

$$\dot{\kappa} = \sqrt{\frac{2}{3}\{\dot{\varepsilon}^p\}^T Q\{\dot{\varepsilon}^p\}} \quad [2.60]$$

These equations enable the unknowns in the models to be completely solved: given an initial state and an increment of deformations, the increments of elastic deformation and plastic deformation, the stress increments and the new values for internal variables (accumulated plastic deformation) can then be calculated.

In the rest of this chapter, we are going to describe the implementation, accuracy, and robustness of this procedure. These qualities are essential in order to generalize the use of these models in finite element computations, particularly for industrial applications on large structures requiring the solution of a very high number of unknowns.

2.3.2.2. *Numerical approach and solution*

In the case of plasticity, two states characterize the behavior of the material. It can either be elastic and the stress is inside the load surface, f, or plastic and the state of stress is represented in the stress space by a point on the load surface, f. *Criterion* is the term usually used to denote the surface, which indicates the state of the material for which the fracture has begun. Whereas the term *loading surface* is used for the threshold reached during work hardening. In isotropic work hardening, the loading surface is obtained by a homothetic transformation of the criterion. As mentioned above, the evolution of plastic deformation results from a plastic potential, g. In associated plasticity, the load surface is assumed as the plastic potential, so f=g. Therefore the following notations are required:

- g $\rightarrow$ the plastic potential;
- f $\rightarrow$ the criterion or the load surface;
- κ $\rightarrow$ the work hardening variable;
- λ $\rightarrow$ the plastic multiplier.

The load surface depends both on the state of stress, and on the work hardening variable. Thus the following definitions are obtained:

– elastic behavior is defined by:

$$f(\sigma,\kappa)<0 \tag{2.61}$$

$$\Delta\lambda = 0 \tag{2.62}$$

– plastic behavior is defined by:

$$f(\sigma,\kappa)=0 \tag{2.63}$$

$$\Delta\lambda > 0. \tag{2.64}$$

The increment of total deformation is decomposed into a reversible plastic part and an irreversible plastic part:

$$\{\Delta\varepsilon\}=\{\Delta\varepsilon^{e}\}+\{\Delta\varepsilon^{p}\}. \tag{2.65}$$

The initial state of the material is known and characterized by an ensemble of variables indexed n. The application of a deformation increment, $\Delta\varepsilon$, leads to an *a priori* unknown final state, the values of which are indexed n + 1.

$$\begin{array}{ll} \text{Initial state} & \text{Final state} \\ \left\{\begin{array}{l} \{\sigma_n\} \\ \{\varepsilon_n\} \\ \kappa_n \end{array}\right. \rightarrow & \left\{\begin{array}{l} \{\sigma_{n+1}\} = \{\sigma_n\} + [D^e]\{\Delta\varepsilon^e\} \\ \{\varepsilon_{n+1}\} = \{\varepsilon_n\} + \{\Delta\varepsilon\} \\ \kappa_{n+1} = \kappa_n + \Delta\kappa \end{array}\right. \end{array} \quad [2.66]$$

To estimate the final state of the material, it is first assumed that it behaves elastically. Therefore, the following conditions must be checked:

$$f(\sigma_{n+1}, \kappa_n) < 0 \quad [2.67]$$

$$\{\Delta\varepsilon^p\} = 0 \quad [2.68]$$

If the conditions are not met, this implies that plasticity occurred. Finding the final state consists of determining the stress state, σ_{n+1}, and the work hardening variable, κ_{n+1}, which satisfy mathematical expression [2.69] of the load surface:

$$f(\sigma_{n+1}, \kappa_{n+1}) = 0 \quad [2.69]$$

The work hardening variable is related to the plastic multiplier by the work hardening law defined by the work hardening function called *h*:

$$\Delta\kappa = \Delta\lambda\, h(\sigma, \kappa) \quad [2.70]$$

The form of the work hardening function results implicitly from the choice made on the hypothesis of plastic work [2.59] or the cumulated deformation [2.60].

The parameters entering in this function are obtained from experimental data. Concerning steel, a tensile test is the most usual. For concrete, uniaxial (or multiaxial) compression tests are generally used.

Determining the final state is equivalent to determining the plastic multiplier, λ_{n+1}, or its increment, $\Delta\lambda = \lambda_{n+1} - \lambda_n$:

$$\{\sigma_{n+1}\} = \{\sigma^e\} - \Delta\lambda\, [D^e]\left\{\frac{\partial g}{\partial\sigma}\right\} \quad [2.71]$$

$$\left\{\varepsilon^p{}_{n+1}\right\} = \left\{\varepsilon^p{}_n\right\} + \Delta\lambda \left\{\frac{\partial g}{\partial \sigma}\right\} \quad [2.72]$$

$$\kappa_{n+1} = \kappa_n + \Delta\lambda\ h(\sigma,\kappa) \quad [2.73]$$

2.3.2.3. *Determination of the plastic multiplier*

Equations [2.71] and [2.73] lead equation [2.69] to be written in the following form [PRA 89]:

$$f(\Delta\lambda) = 0 \quad [2.74]$$

The Newton-Raphson method is used to solve the non-linear equation [2.74] with one unknown:

$$\Delta\lambda^{(i+1)} = \Delta\lambda^{(i)} - \left(\frac{\partial f}{\partial \lambda}\right)^{-1} f\left(\Delta\lambda^{(i)}\right) \quad [2.75]$$

Three strategies are possible for the evaluation of the plastic multiplier. Let us consider Taylor's series development to a first order of the load surface $f(\sigma,\kappa) = 0$ at a given point noted * in the principal stress space. Therefore, we have:

$$f_{n+1} \approx f_* + \left\{\frac{\partial f}{\partial \sigma}\right\}_*^T \left(\{\sigma\}_{n+1} - \{\sigma\}_*\right) + \left.\frac{\partial f}{\partial \kappa}\right|_* \left(\kappa_{n+1} - \kappa_*\right) \quad [2.76]$$

The choice of the point where the criterion is linearized enables the three abovementioned approaches to be differentiated. We can consider the initial point indexed, n, the point corresponding to the trial state (elastic predictor) indexed, e, or the current point indexed, n + 1.

2.3.2.3.1. Linearization of the criterion around the initial point

We then have:

$$\{\sigma\}_* = \{\sigma\}_n \quad [2.75]$$

$$\kappa_* = \kappa_n \quad [2.76]$$

In this case, Taylor's series development to a first order becomes:

$$f_{n+1} \approx f_n + \left\{\frac{\partial f}{\partial \sigma}\right\}_n^T \left(\{\sigma\}_{n+1} - \{\sigma\}_n\right) + \left.\frac{\partial f}{\partial \kappa}\right|_n \left(\kappa_{n+1} - \kappa_n\right) \qquad [2.77]$$

To simplify we note: $f_t = f\left(\{\sigma\}_t , \kappa_t\right)$

As the initial and final states are plastically admissible, that is $f_n=0$ and $f_{n+1}=0$, expression [2.77] can be formulated:

$$\Delta f \approx \left\{\frac{\partial f}{\partial \sigma}\right\}_n^T \{\Delta\sigma\} + \left.\frac{\partial f}{\partial \kappa}\right|_n \Delta\kappa = 0 \qquad [2.78]$$

Equation [2.71] is rewritten:

$$\{\Delta\sigma\} = \left[D^e\right]\left(\{\Delta\varepsilon\} - \Delta\lambda\left\{\frac{\partial g}{\partial \sigma}\right\}\right) \qquad [2.79]$$

and taking into account equation [2.70], the increment of the plastic multiplier is then expressed by:

$$\Delta\lambda = \frac{\left\{\frac{\partial f}{\partial \sigma}\right\}_n^T \left[D^e\right]\Delta\varepsilon}{-\left.\frac{\partial f}{\partial \kappa}\right|_n h(\sigma,\kappa) + \left\{\frac{\partial f}{\partial \sigma}\right\}_n^T \left[D^e\right]\left\{\frac{\partial g}{\partial \sigma}\right\}_n} \qquad [2.80]$$

The expression of the plastic flow thus obtained can be substituted in equation [2.71], which is then written:

$$\{\Delta\sigma\} = \left[\left[D^e\right] - \frac{\left\{\frac{\partial f}{\partial \sigma}\right\}_n^T \left[D^e\right]\left\{\frac{\partial g}{\partial \sigma}\right\}_n}{-\frac{\partial f}{\partial \kappa} h(\sigma,\kappa) + \left\{\frac{\partial f}{\partial \sigma}\right\}_n^T \left[D^e\right]\left\{\frac{\partial g}{\partial \sigma}\right\}_n}\right]\{\Delta\varepsilon\} \qquad [2.81]$$

This expression defines the tangent matrix:

$$\{\Delta\sigma\} = \left[D^{ep}\right]\{\Delta\varepsilon\} \qquad [2.82]$$

This is an explicit approach because the terms that constitute the tangent matrix are calculated at the previous step n.

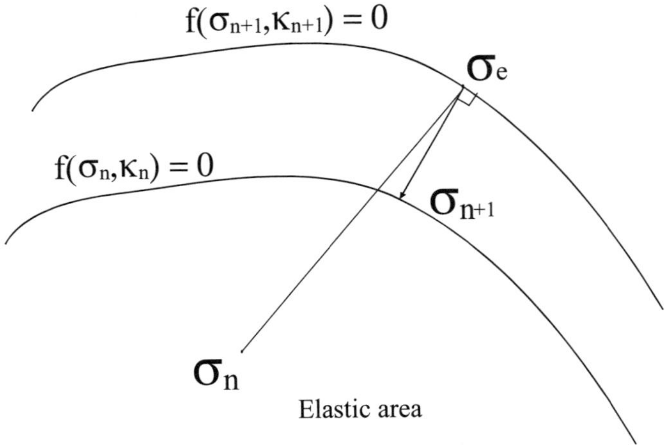

Figure 2.13. *Return mapping scheme by estimation of the plastic multiplier at the elastic point*

This technique has the disadvantage of requiring very small increments to obtain acceptable accuracy. We will return later to the notion of accuracy of the iterative algorithm of resolution of the non-linear equation [2.74].

In addition, this approach does not allow the treatment of materials that present softening behavior. Therefore, it is of limited use for concrete.

2.3.2.3.2. Linearization of the criterion around the elastic point

We have:

$$\{\sigma\}_* = \{\sigma\}_e \tag{2.83}$$

$$\kappa_* = \kappa_n \tag{2.84}$$

The development in Taylor's series [2.76] of the first order is written:

$$f_{n+1} \approx f_e + \left\{\frac{\partial f}{\partial \sigma}\right\}_e^T \left(\{\sigma\}_{n+1} - \{\sigma\}_e\right) + \left.\frac{\partial f}{\partial \kappa}\right|_e \left(\kappa_{n+1} - \kappa_n\right) \tag{2.85}$$

with:

$$f_t = f(\{\sigma\}_t, \kappa_t)$$

Using the fact that $f_{n+1}=0$ and equations [2.71] and [2.73], the expression of the plastic multiplier is obtained:

$$\Delta\lambda = \frac{f_e}{-\left.\frac{\partial f}{\partial \kappa}\right|_e h(\sigma,\kappa) + \left\{\frac{\partial f}{\partial \sigma}\right\}_e^T [D^e] \left\{\frac{\partial g}{\partial \sigma}\right\}_e} \qquad [2.86]$$

This technique, although explicit, enables softening behavior to be described. When the derivatives of the surfaces f and g to be introduced in the expression of the multiplier are linear in stress, the return on the load surface is direct. In the opposite case, the plastic return is no longer radial and becomes dependent on the size of the time step.

2.3.2.3.3. Linearization of the criterion around the current point

The iterative procedure presented by Ortiz *et al.* [ORT 86] then Simo *et al.* [SIM 88] consists of linearizing the criterion in Taylor's series around the current point. It is now necessary to introduce the indexes (i) that correspond to the iteration number of the iterative process. The return mapping relations are expressed by:

$$\{\sigma^{(i+1)}\}_{n+1} = \{\sigma^{(i)}\}_{n+1} - \{\Delta\lambda^{(i)}\}[D^e]\left\{\frac{\partial g}{\partial \sigma}^{(i)}\right\}_{n+1} \qquad [2.87]$$

$$\kappa_{n+1}^{(i+1)} = \kappa_{n+1}^{(i)} + \Delta\lambda^{(i)} h(\sigma,\kappa) \qquad [2.88]$$

The linearization of criterion in Taylor's series to the first order is written:

$$f_{n+1}^{(i+1)} \approx f_{n+1}^{(i)} + \left\{\frac{\partial f}{\partial \sigma}^{(i)}\right\}_{n+1}^T \left(\{\sigma^{(i+1)}\}_{n+1} - \{\sigma^{(i)}\}_{n+1}\right) + \left.\frac{\partial f}{\partial \kappa}\right|_{n+1}^{(i)} \left(\kappa_{n+1}^{(i+1)} - \kappa_{n+1}^{(i)}\right) \qquad [2.89]$$

At the convergence $f_{n+1}^{(i+1)} = 0$, equations [2.87], [2.88], and [2.89] lead to the following expression of the plastic multiplier:

$$\Delta\lambda^{(i)} = \frac{f_{n+1}^{(i)}}{-\left.\frac{\partial f}{\partial \kappa}\right|_{n+1}^{(i)} h(\sigma,\kappa) + \left\{\frac{\partial f}{\partial \sigma}^{(i)}\right\}_{n+1}^T [D^e] \left\{\frac{\partial g}{\partial \sigma}^{(i)}\right\}_{n+1}} \qquad [2.90]$$

Figure 2.14 gives a geometrical interpretation of the iterative process of this method.

2.3.2.3.4. Generalization of the resolution algorithm

We are going to consider the methods of integration of incremental laws in a more general way. Two large families of algorithms are routinely used. They coincide with the three methods presented above in particular cases. They can be placed in the following framework:

$$\{\sigma\}_{n+1} = \{\sigma\}_n - \Delta\lambda \left[D^e\right] m_\alpha \qquad [2.91]$$

$$\kappa_{n+1} = \kappa_n + \Delta\lambda\, h_\alpha \qquad [2.92]$$

with $m_\alpha = \left\{\dfrac{\partial f}{\partial \sigma}\right\}_\alpha$

These two families of algorithms differ in the calculation of the terms m_α and h_α These methods consist of establishing an estimation of these terms based on a proportion between their initial value of degree α=0 and the final value of degree α=1.

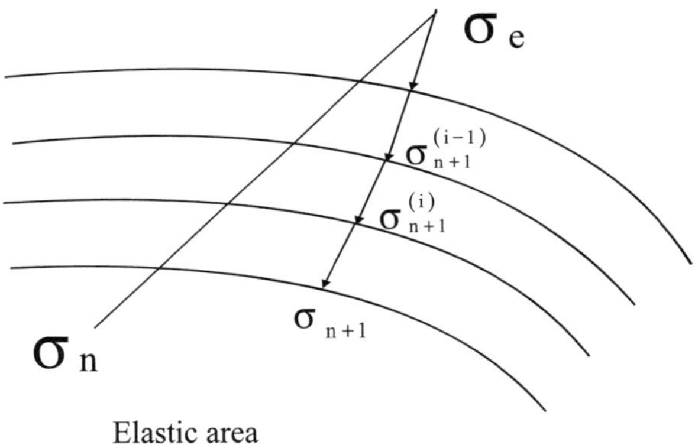

Figure 2.14. *Return mapping by estimation of the plastic multiplier at the current point*

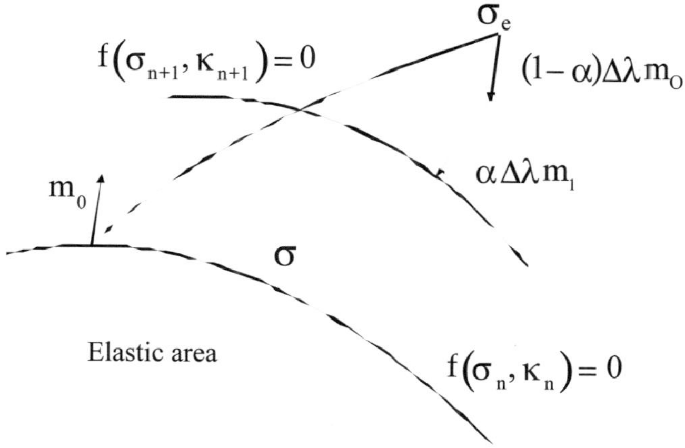

Figure 2.15. *Plastic return by the generalized trapezoidal rule*

It is a compromise between a resolution method with an explicit scheme and a method with an implicit scheme to optimize the estimation of the plastic multiplier responsible for the plastic convergence.

These two families of algorithms are defined below and a geometrical interpretation is presented in Figures 2.15 and 2.16, respectively.

Generalized trapezoidal rule

$$m_\alpha = (1-\alpha)m_0 + \alpha m_1 \quad [2.93]$$

$$h_\alpha = (1-\alpha)h_0 + \alpha h_1 \quad [2.94]$$

Generalized midpoint rule

$$m_\alpha = m[(1-\alpha)\sigma_0 + \alpha\sigma_1, (1-\alpha)\kappa_0 + \alpha\kappa_1] \quad [2.95]$$

$$h_\alpha = h[(1-\alpha)\sigma_0 + \alpha\sigma_1, (1-\alpha)\kappa_0 + \alpha\kappa_1] \quad [2.96]$$

The generalized midpoint technique coincides with classical algorithms in some particular cases of α:

$\alpha = 0$: explicit method or *Forward Euler*;

$-\alpha = \frac{1}{2}$: *Mean normal*;

$-\alpha = 1$: implicit method or *Backward Euler*.

The diversity of the methods of integration of constitutive laws makes the choice of one or the other of them difficult. However, two criteria enable this choice to be made: the accuracy of the method and the stability of the iterative process. Several studies have been carried out on this subject particularly by Ortiz *et al.* [ORT 85] and Simo *et al.* [SIM 86; SIM 88; SIM 91].

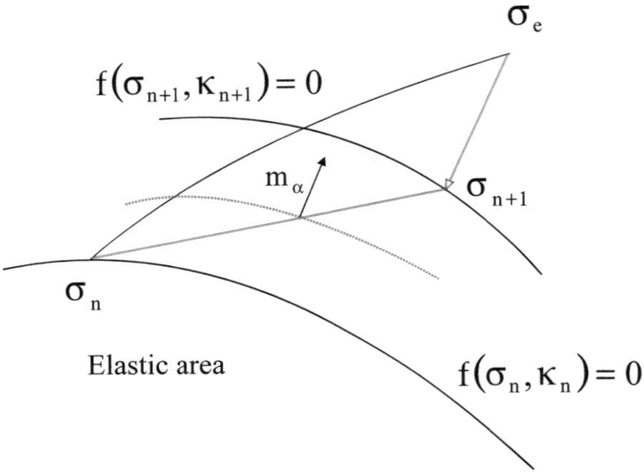

Figure 2.16. *Return mapping by the generalized midpoint rule*

The accuracy of the algorithm

It can be shown by mathematical analysis that the choice of the numerical value α=0.5 for the parameter of the algorithm leads to a second-order accuracy.

However, the numerical examples show that in the presence of strong deformation gradients, higher values for α can improve the accuracy of the algorithm.

The stability of the numerical process

Stability analysis consists of finding in what conditions finite disturbance of the initial stress state is attenuated in the final stress state. A disturbance of the stress state can be translated mathematically by two states that are very close to each other, that is to say, $\sigma^{(1)}$ and $\sigma^{(2)}$.

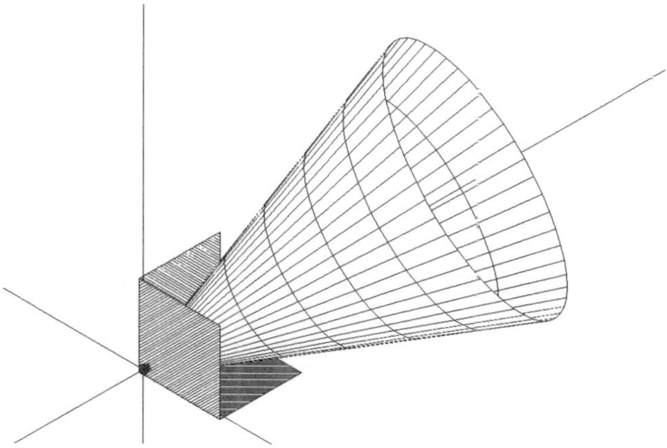

Figure 2.17. *Perspective of the plasticity criteria of the Drucker-Prager and Rankine type*

A mathematical distance, $d(x,y)$, is then defined between two points x and y. The stability of the algorithm implies that the distance between the two final states be smaller than that between the two initial states:

$$d\left(\sigma_{n+1}^{(2)}, \sigma_{n+1}^{(1)}\right) \leq d\left(\sigma_{n}^{(2)}, \sigma_{n}^{(1)}\right) \quad [2.97]$$

Studies in the case of each family have shown rather important differences.

Generalized trapezoidal rule

$$\left\| d\sigma_{n+1} \right\| \leq \left| \frac{1-\alpha}{\alpha} \right| k \left\| d\sigma_{n} \right\| \quad [2.98]$$

The coefficient, k, expresses the distortion of the criterion. It is dependent on the second derivatives of the criterion relative to the stress. Its value is unity for constant curves of the Von Mises type, or infinite for corners or angular points of criteria, such as in the case of a Drucker-Prager and Rankine coupling. The algorithm is unconditionally stable when the following equation is proved:

$$\left| \frac{1-\alpha}{\alpha} \right| k \leq 1 \quad [2.99]$$

Thus, the minimum value $\alpha_{\min}$ of α varies from 0.5 for k=1 to 1 for k=∞.

Generalized midpoint rule

$$\|d\sigma_{n+1}\| \leq \left|\frac{1-\alpha}{\alpha}\right| \|d\sigma_n\| \qquad [2.100]$$

In this case, the minimum value, α_{min}, of α is explicitly 0.5.

To conclude, it can be said that the explicit method (*Forward Euler*) should be avoided because it is of mediocre accuracy and is conditionally stable. Furthermore, the plastic return solution may not exist if the deformation increments are too important. In general, it has been shown [RUN 88] that all methods are unconditionally stable if $\alpha \geq \alpha_{cr}$. In the case of the generalized midpoint rule, on a: $\alpha_{cr}=0.5$. Whereas in the case of the generalized trapezoidal rule, $\alpha_{\mathbf{min}}$ depends strongly on the load criterion. In particular, the presence of corners (angular points) in the criterion requires the value $\alpha=1$ to obtain unconditional numerical stability. Furthermore, it is shown in [RUN 88] using numerical examples that in the presence of very severe deformations, higher values of α in the projection method lead to more accurate results. Therefore, it is preferable to consider the value $\alpha=1$ and the generalized midpoint rule.

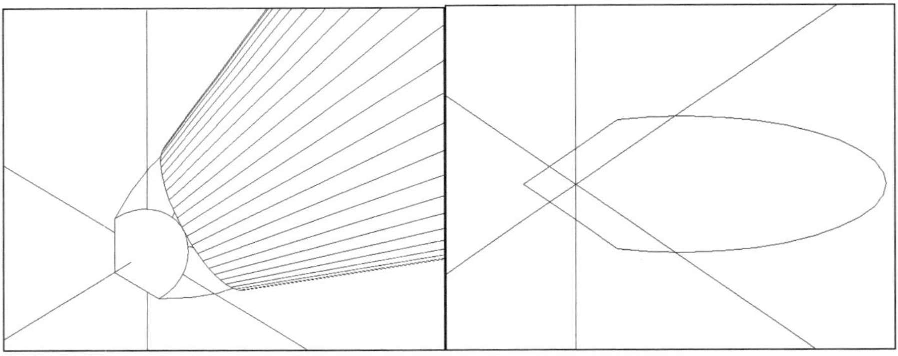

a) b)

Figure 2.18. *Multisurface plasticity criterion used for concrete (a) perspective and b) plane views)*

2.3.2.4. *A two-dimensional mechanical behavior model for concrete*

The behavior of concrete is characterized by a non-symmetric response between tension and compression. In order to describe this kind of behavior, Feenstra *et al.* [FEE 95] defined the load surface based on the coupling of two surfaces of plasticity. A surface of the Drucker-Prager type and a surface of the Rankine type were chosen to describe the behavior of compression and tension, respectively. In Figure 2.17 a geometrical representation of these two surfaces can be observed.

The trace of the load surface (Drucker-Prager + Rankine) in the principal stress plane ($\sigma_3 = 0$) is represented in Figure 2.18.

2.3.2.4.1. Resolution for multisurface criteria

The particularity of multisurface criteria is in the existence of corners, which lead to normal non-unicity on the work hardening surface. The ambiguity regarding the plastic flow can be lifted by Koiter's [KOI 53] and Maier's [MAI 69] proposition considering the individual contribution of each loading surface. Therefore, the plastic deformation increment is written:

$$\{\Delta\varepsilon^p\} = \Delta\lambda_1\left\{\frac{\partial g_1}{\partial\sigma}\right\} + \Delta\lambda_2\left\{\frac{\partial g_2}{\partial\sigma}\right\} \qquad [2.101]$$

Figure 2.19 defines a geometrical representation of the resultant of the induced plastic flow. Not knowing which surface is active at the beginning of a deformation increment is a major difficulty in the use of such criteria. Indeed, one cannot rapidly determine *a priori* whether a loading surface should be taken into account or not in the plastic resolution.

Even if the two loading surfaces are violated by the elastic predictor as illustrated in Figure 2.19, at the end of the increment the final point may find itself inside a surface. Consequently, the material does not undergo work hardening in compression or tension according to whether the Drucker-Prager or Rankine criterion is applied.

The definition of the active surfaces is two-phase. First, the number of active surfaces is supposed to be equal to the number of surfaces violated by the plastic predictor. Then the plastic return is initiated by an iterative procedure in which the appearance of a negative value of the plastic multiplier $\Delta\lambda_1$ or $\Delta\lambda_2$ leads to the deactivation of the corresponding surface f_1 or f_2.

2.3.2.4.2. Multicriteria resolution

In the case of one activated criterion only, the non-linear relation is to be resolved into $\Delta\lambda$:

$$f(\Delta\lambda) = 0 \qquad [2.102]$$

In order to do this, a Newton-Raphson type iterative process can be used:

$$\Delta\lambda^{(i+1)} = \Delta\lambda^{(i)} - J_{(i)}^{-1}.f(\Delta\lambda^{(i)}) \qquad [2.103]$$

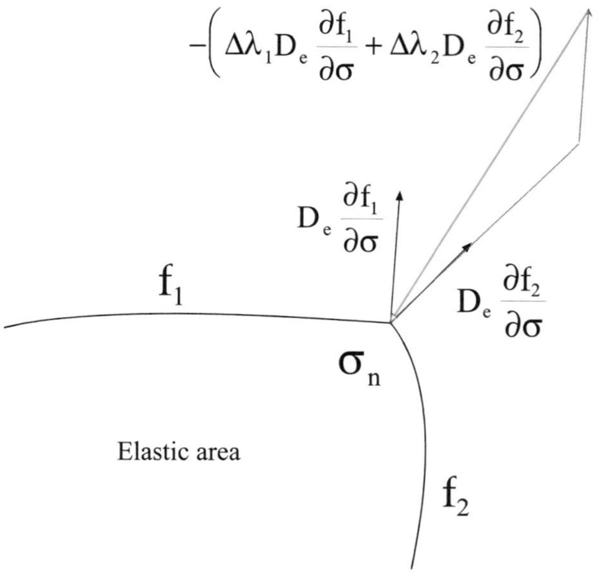

Figure 2.19. *Plastic flow of the two loading surfaces*

The Jacobian matrix in equation [2.103] is written classically:

$$J^{(i)} = \frac{f\left(\Delta\lambda^{(i-1)}\right) - f\left(\Delta\lambda^{(i)}\right)}{\Delta\lambda^{(i-1)} - \Delta\lambda^{(i)}} \qquad [2.104]$$

The resolution method presented corresponds to the so-called secant method. Indeed, the plastic multiplier is calculated at step (i+1) with the help of the secant derivative calculated relative to the preceding steps (i) and (i–1). Figure 2.20 gives a geometrical representation of the resolution process associated with this method.

It should be noted that the methods presented in section 2.3.2.3.4 to estimate the plastic multiplier at the current point are actually equivalent to the tangent method. Indeed, the plastic multiplier is obtained by linearization of the criterion at the current point. Figure 2.21 gives a geometrical representation of the iterative process of this method.

In the case of two surfaces being activated, the non-linear system to be resolved with two unknowns is the following:

$$\begin{aligned} f_1(\Delta\lambda_1,\Delta\lambda_2) &= 0 \\ f_2(\Delta\lambda_1,\Delta\lambda_2) &= 0 \end{aligned} \qquad [2.105]$$

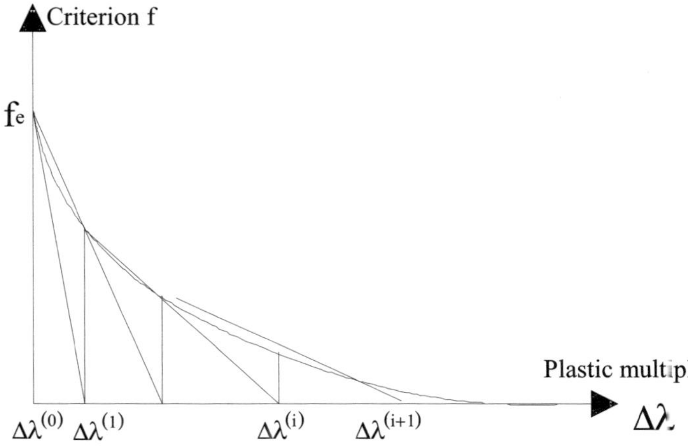

Figure 2.20. *Geométrical representation of the secant method*

The iterative Newton-Raphson type resolution process is again used. The vectorial equation is written:

$$\begin{Bmatrix} \Delta\lambda_1 \\ \Delta\lambda_2 \end{Bmatrix}^{(i+1)} = \begin{Bmatrix} \Delta\lambda_1 \\ \Delta\lambda_2 \end{Bmatrix}^{(i)} - [J]^{-1}_{(i)} \cdot \begin{Bmatrix} f_1(\Delta\lambda_1,\Delta\lambda_2) \\ f_2(\Delta\lambda_1,\Delta\lambda_2) \end{Bmatrix}^{(i)} \qquad [2.106]$$

The new estimation of the Jacobian matrix (2×2) is made by means of a Broyden method:

$$[J]_i = [J]_{i-1} + \frac{(\{s\}_{i-1} - [J]_{i-1}\{y\}_{i-1})\{s\}^T_{i-1}[J]_{i-1}}{((s)_{i-1},[H]_{i-1}\{y\}_{i-1})} (\{s\}_{i-1},\{s\}_{i-1}) \qquad [2.107]$$

Where (a, b) represents the scalar product $a^T\, b$:

$$\begin{aligned} \{s\}_{i-1} &= \{\Delta\lambda\}^i - \{\Delta\lambda\}^{i-1} \\ \{y\}_{i-1} &= \{f(\Delta\lambda^i)\} - \{f(\Delta\lambda^{i-1})\} \end{aligned} \qquad [2.108]$$

with:

$$\{\Delta\lambda\} = \begin{Bmatrix} \Delta\lambda_1 \\ \Delta\lambda_2 \end{Bmatrix} \text{ and } \{f(\Delta\lambda)\} = \begin{Bmatrix} f_1(\Delta\lambda_1, \Delta\lambda_2) \\ f_2(\Delta\lambda_1, \Delta\lambda_2) \end{Bmatrix}$$

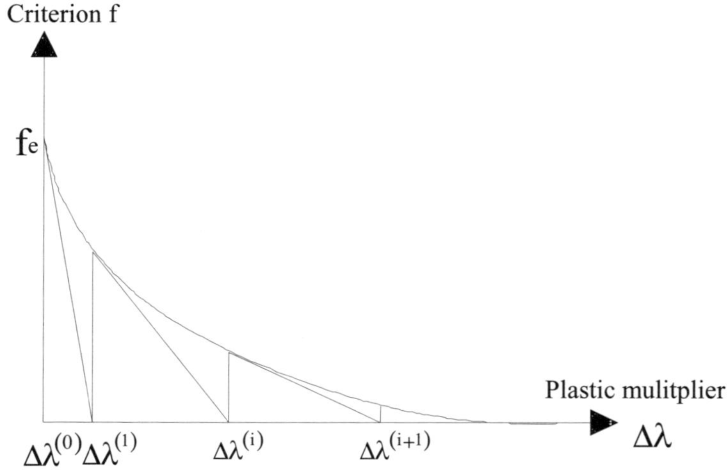

Figure 2.21. *Geometrical representation of the tangent method*

These resolution methods nevertheless require knowledge of the initial values of the zero index of the Jacobian matrix with single or double criteria. Therefore, these values are estimated by a linearization of the criteria around the trial stat in Taylor's series. According to equation [2.85], we have the following relation:

$$f_{n+1} \approx f_e + \left\{\frac{\partial f}{\partial \sigma}\right\}_e^T (\{\sigma_{n+1}\} - \{\sigma_e\}) + \left.\frac{\partial f}{\partial \kappa}\right|_e (\kappa_{n+1} - \kappa_n) \qquad [2.109]$$

which is equivalent to:

$$f_{n+1} - f_e \approx -\Delta\lambda(\left\{\frac{\partial f}{\partial \sigma}\right\}_e^T [D^e] \left\{\frac{\partial f}{\partial \sigma}\right\}_e + \left.\frac{\partial f}{\partial \kappa}\right|_e h) \qquad [2.110]$$

hence an estimation of the initial Jacobian matrix is obtained:

$$[J^{(0)}] = \left[\frac{\partial f}{\partial \lambda}\right]^{(0)} \approx \left[\frac{df}{d\lambda}\right]^{(0)} = -\left(\left\{\frac{\partial f}{\partial \sigma}\right\}_e^T [D^e] \left\{\frac{\partial f}{\partial \sigma}\right\}_e + \left.\frac{\partial f}{\partial \kappa}\right|_e h\right). \qquad [2.111]$$

In the case of two criteria being activated, the Jacobian matrix is no longer defined by a scalar but by a 2×2 matrix, which is expressed mathematically by:

$$[J]^{(i)} = \begin{bmatrix} \dfrac{\partial f_1}{\partial \lambda_1} & \dfrac{\partial f_1}{\partial \lambda_2} \\ \dfrac{\partial f_2}{\partial \lambda_1} & \dfrac{\partial f_2}{\partial \lambda_2} \end{bmatrix} \qquad [2.112]$$

In the same way as for one criterion, the initial Jacobian is estimated by developing the two criteria to the first order in a Taylor's series. For the criterion indexed 1:

$$f_{1n+1} \approx f_{1e} + \left\{\frac{\partial f_1}{\partial \sigma}\right\}_e^T \left(\{\sigma\}_{n+1} - \{\sigma\}_e\right) + \left.\frac{\partial f_1}{\partial \kappa_1}\right|_e \left(\kappa_{1\,n+1} - \kappa_{1n}\right) + \left.\frac{\partial f_1}{\partial \kappa_2}\right|_e \left(\kappa_{2\,n+1} - \kappa_{2n}\right) \qquad [2.113]$$

As before, we obtain:

$$f_{1_{n+1}} \approx f_{1_e} + \left\{\frac{\partial f_1}{\partial \sigma}\right\}_e^T \left(-\Delta\lambda_1 \left[D^e\right]\left\{\frac{\partial f_1}{\partial \sigma}\right\}_e - \Delta\lambda_2 \left[D^e\right]\left\{\frac{\partial f_2}{\partial \sigma}\right\}_e\right) + \left.\frac{\partial f_1}{\partial \kappa_1}\right|_e \Delta\lambda_1 + \left.\frac{\partial f_1}{\partial \kappa_2}\right|_e \Delta\lambda_2$$

$$f_{1n+1} \approx f_{1e} + \left(-\left\{\frac{\partial f_1}{\partial \sigma}\right\}_e^T \left[D^e\right]\left\{\frac{\partial f_1}{\partial \sigma}\right\}_e + \left.\frac{\partial f_1}{\partial \kappa_1}\right|_e\right)\Delta\lambda_1 + \left(-\left\{\frac{\partial f_1}{\partial \sigma}\right\}_e^T \left[D^e\right]\left\{\frac{\partial f_2}{\partial \sigma}\right\}_e + \left.\frac{\partial f_1}{\partial \kappa_2}\right|_e\right)\Delta\lambda_2$$

and for the criterion indexed 2, we have:

$$f_{2_{n+1}} \approx f_{2_e} + \left(-\left\{\frac{\partial f_2}{\partial \sigma}\right\}_e^T \left[D^e\right]\left\{\frac{\partial f_1}{\partial \sigma}\right\}_e + \left.\frac{\partial f_1}{\partial \kappa_1}\right|_e\right)\Delta\lambda_1 + \left(-\left\{\frac{\partial f_2}{\partial \sigma}\right\}_e^T \left[D^e\right]\left\{\frac{\partial f_2}{\partial \sigma}\right\}_e + \left.\frac{\partial f_1}{\partial \kappa_2}\right|_e\right)\Delta\lambda_2$$

Finally, an estimation of the Jacobian matrix at the initial point defining the elastic jump is either:

$$J^{(0)} = \begin{bmatrix} -\left\{\dfrac{\partial f_1}{\partial \sigma}\right\}_e^T \left[D^e\right]\left\{\dfrac{\partial f_1}{\partial \sigma}\right\}_e + \left.\dfrac{\partial f_1}{\partial \kappa_1}\right|_e & -\left\{\dfrac{\partial f_1}{\partial \sigma}\right\}_e^T \left[D^e\right]\left\{\dfrac{\partial f_2}{\partial \sigma}\right\}_e \\ -\left\{\dfrac{\partial f_2}{\partial \sigma}\right\}_e^T \left[D^e\right]\left\{\dfrac{\partial f_1}{\partial \sigma}\right\}_e & -\left\{\dfrac{\partial f_2}{\partial \sigma}\right\}_e^T \left[D^e\right]\left\{\dfrac{\partial f_2}{\partial \sigma}\right\}_e + \left.\dfrac{\partial f_2}{\partial \kappa_2}\right|_e \end{bmatrix} \qquad [2.114]$$

or

$$\left[J^{(0)}\right]=\begin{bmatrix} -m_1\left[D^e\right]m_1+H_1 & -m_1\left[D^e\right]m_2 \\ -m_2\left[D^e\right]m_1 & -m_2\left[D^e\right]m_2+H_2 \end{bmatrix} \quad [2.115]$$

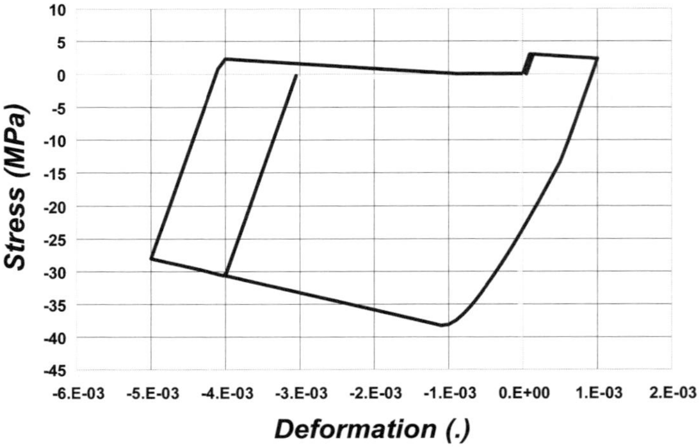

Figure 2.22. *Multisurface plasticity model prediction in tension and compression with loading and unloading*

2.3.2.4.3. 1D behavior of the model

Figure 2.22 shows the uniaxial behavior of this model, in compression and tension, with the following parameters: E_0=32,000 MPa, ν_0=0.2, f_c=38.3 MPa, f_t=3 MPa, and β=1.16.

2.4. Conclusion

We have presented in the preceding sections two fundamentally different types of models: damage models, whereby the non-linear behavior of the material is due to the degradation of the elastic behavior, and plasticity models, whereby the non-linearity is due to the growth in plastic deformations. These two types of behavior model are not opposed; on the contrary, they complement each other for a better description of the behavior of concrete. Indeed, experimental data (for example those given in Chapter 1) clearly show that both a degradation of material stiffness and the appearance of permanent deformations are observed. According to the type of solicitation applied, one or another of the effects will be important. Under tension, damage is the main factor. Under compression, plastic deformations are more severe and the loss of stiffness more moderate.

(1) Updating of the geometry: $\{\varepsilon_{n+1}\}=\{\varepsilon_n\}+\{\Delta\varepsilon\}$	
(2) Elastic predictor: $\{\sigma^e\}=\{\sigma_n\}+[D^e]\{\Delta\varepsilon\}$ $\{\varepsilon_{n+1}^{p(0)}\}=\{\varepsilon_n^p\}$ $\kappa_{n+1}^{(0)}=\kappa_n$	
(3) Criterion activated or not: $f(\sigma_e,\kappa_{n+1})\leq 0\,?$	If so, n=n+1 go to (1) If not, go to (4)
(4) Initialization i = 0 $J^{(0)}=-\left(\left\{\frac{\partial f}{\partial\sigma}\right\}_e^T[D^e]\left\{\frac{\partial f}{\partial\sigma}\right\}_e+\left.\frac{\partial f}{\partial\kappa}\right\|_e h\right)$ $\Delta\lambda^{(i)}=-J_{(0)}^{-1}.f(\sigma_e,\kappa_{n+1})$	
(5) Updating of the state of the material: $\{\sigma_{n+1}^{(i)}\}=\{\sigma(\Delta\lambda^{(i)})\}$ $\kappa_{n+1}^{(i)}=\kappa_{n+1}^{(i-1)}+\Delta\lambda^{(1)}$	
(6) Plastic convergence? $\left\|f(\sigma_{n+1}^{(i)},\kappa_{n+1}^{(i)})\right\|\leq tol$	If so, n=n+1, go to (1) If not, go to (7)
(7) New estimation of the variables: $J^{(i)}=\frac{f(\Delta\lambda^{(i-1)})-f(\Delta\lambda^{(i)})}{\Delta\lambda^{(i-1)}-\Delta\lambda^{(i)}}$ $\Delta\lambda^{(i+1)}=\Delta\lambda^{(i)}-J_{(i)}^{-1}.f(\Delta\lambda^{(i)})$	Go to (5)

Table 2.3. *Resolution algorithm in the case of only one activated criterion*

(1) Updating of the geometry: $\{\varepsilon_{n+1}\}=\{\varepsilon_n\}+\{\Delta\varepsilon\}$	
(2) Elastic predictor: $\{\sigma^e\}=\{\sigma_n\}+[D^e]\{\Delta\varepsilon\}$ $\kappa_{1n+1}^{(0)}=\kappa_{1n}$ $\{\varepsilon_{n+1}^{p(0)}\}=\{\varepsilon_n^p\}$ $\kappa_{2n+1}^{(0)}=\kappa_{2n}$	
(3) Criterion activated or not: $f_1(\sigma_e,\kappa_{1n+1})\leq 0$? (a) $f_2(\sigma_e,\kappa_{2n+1})\leq 0$? (b)	If yes in (a) and (b), n = n + 1, go to (1) If no in (a) and yes in (b), or yes in (a) and no in (b), go to Table 2.1 in (4) If no in (a) and no in (b), go to (4)
(3) Initialization I = 0 $[J^{(0)}]=\begin{bmatrix}-m_1[D^e]m_1+H_1 & -m_1[D^e]m_2\\ -m_2[D^e]m_1 & -m_2[D^e]m_2+H_2\end{bmatrix}$ $\begin{Bmatrix}\Delta\lambda_1\\ \Delta\lambda_2\end{Bmatrix}^{(i)}=-J_{(0)}^{-1}\cdot\begin{Bmatrix}f_1(\sigma_e,\kappa_{1n})\\ f_2(\sigma_e,\kappa_{2n})\end{Bmatrix}$	If $\Delta\lambda_1\leq 0$ and $\Delta\lambda_2>0$, go to Table 2.1 in (4) If $\Delta\lambda_1>0$ and $\Delta\lambda_2\leq 0$, go to Table 2.1 in (4) If $\Delta\lambda_1\leq 0$ and $\Delta\lambda_2\leq 0$, n = n + 1 go to (1)
(4) Update the material state $\{\sigma_{n+1}^{(i)}\}=\{\sigma(\Delta\lambda_1^{(i)},\Delta\lambda_2^{(i)})\}$ $\kappa_{1n+1}^{(i)}=\kappa_{1n+1}^{(i-1)}+\Delta\lambda_1^{(1)}$ $\kappa_{2n+1}^{(i)}=\kappa_{2n+1}^{(i-1)}+\Delta\lambda_2^{(1)}$	
(5) Plastic convergence? $\lvert f_1(\sigma_{n+1}^{(i)},\kappa_{1n+1}^{(i)})\rvert\leq tol$?(c) $\lvert f_2(\sigma_{n+1}^{(i)},\kappa_{2n+1}^{(i)})\rvert\leq tol$?(d)	If yes in (c) and yes in (d) n = n + 1, go to (1) If no in (c) and yes in (d), or yes in (c) and no in (d), or no in (c) and no in (d), go to (7)
(6) New estimation of the variables $[J]_i=[J]_{i-1}+\dfrac{(\{s\}_{i-1}-[J]_{i-1}\{y\}_{i-1})\{s\}_{i-1}^T[J]_{i-1}}{((s)_{i-1},[H]_{i-1}\{y\}_{i-1})}(\{s\}_{i-1},\{s\}_{i-1})$ $\begin{Bmatrix}\Delta\lambda_1\\ \Delta\lambda_2\end{Bmatrix}^{(i+1)}=\begin{Bmatrix}\Delta\lambda_1\\ \Delta\lambda_2\end{Bmatrix}^{(i)}-J_{(i)}^{-1}\cdot\begin{Bmatrix}f_1(\Delta\lambda_1,\Delta\lambda_2)\\ f_2(\Delta\lambda_1,\Delta\lambda_2)\end{Bmatrix}^{(i)}$	go to (5)

Table 2.4. *Resolution algorithm in the case of two activated criteria*

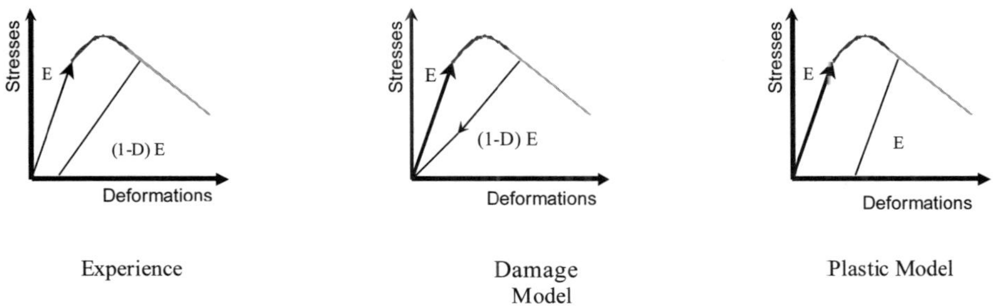

Figure 2.23. *Unloading gradient: experimental response, damage and plasticity models*

Therefore, it is a good idea to combine these two types of behavior law to obtain a model that is more representative of the experimental response of the material, as shown in Figure 2.23.

This combination can be carried out in several ways. The first possibility is to index damage growth on the amount of plastic deformation observed in the material (for example on the accumulated plastic deformation). This method is directly inspired by ductile damage models used for metals [LEM 85]. It is very easy to use because it is sufficient to add a relation defining damage growth, which is often explicitly defined, to the plasticity laws.

Unfortunately, the physical justifications for such a law do not exist. Regarding steel, it is generally accepted that cavity growth (damage) results from the accumulation of plastic deformations, this explanation is much more difficult to accept for concrete. In this kind of material, plastic deformation is due rather to internal slipping between between the lips of microcracks generated by the manufacturing process or by mechanical loads. One would consequently tend to think that it is the plastic deformation that is made possible by the existence of damage rather than vice-versa. That leads, for example, to the behavior laws presented in Chapter 4.

It should be noted, however, that these explanations remain particularly simplistic. Currently there is no consensus regarding the best way to couple damage and plasticity to represent the behavior of concrete. Therefore, upon closing this chapter, we shall present a method with the advantage of numerically decoupling damage and plasticity. The effects related to plasticity (irreversible deformations) and to damage (softening) are jointly described by the model. From the total deformation tensor, ε, an effective stress, σ', is calculated with the plasticity relations. Then, knowing the elastic-plastic decomposition of the deformations, the damage and the "real" stress σ are obtained.

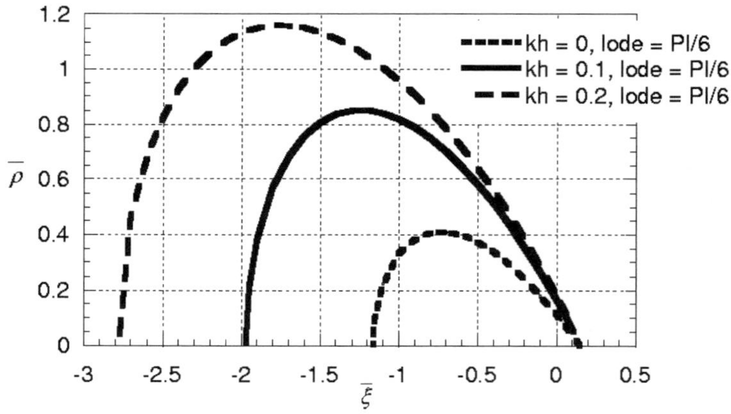

Figure 2.24. *Evolution of the threshold surface under simple compression [JAS 06]*

The main advantage of this approach is its adaptability to a large number of constitutive relations. We present here, as an example, the model proposed recently by Jason [JAS 04; JAS 06]. It also gives a glimpse of current possibilities concerning modeling of the quasi-static instantaneous behavior of concrete.

Concerning the "plasticity" part of the model, the threshold surface was chosen to reach several objectives. The volumetric behavior must be correctly simulated with the appearance of a change from contraction towards dilation observed under compression. This condition prevents Von Mises type laws being used as they are only a function of the second invariant and their volumetric response is elastic. Particular attention is also paid to confinement simulation. For high levels of hydrostatic pressure, plastic effects appear experimentally. This supposes a closed function according to the first invariant (the existence of a plastic threshold in confinement) and eliminates Drucker-Prager type equations.

The plasticity model is defined by the following equations:

$$\begin{aligned} d\varepsilon_{ij} &= d\varepsilon_{ij}^{e} + d\varepsilon_{ij}^{p} \\ \sigma_{ij}^{t} &= C_{ijkl}^{0}\varepsilon_{kl}^{e} \\ d\varepsilon_{ij}^{p} &= d\lambda . m_{ij}\left(\sigma^{t}, k_{h}\right) \\ dk_{h} &= d\lambda . h\left(\sigma^{t}, k_{h}\right) \end{aligned} \qquad [2.116]$$

where the effective stress is given by:

$$\sigma_{ij}^t = \frac{\sigma_{ij}}{(1-d)} \qquad [2.117]$$

The effective stress can be interpreted as the stress supported by the material, once the defects have been removed (or between the defects induced by damage). ε, ε^e, ε^p are the total, elastic, and plastic strains, respectively. Contrary to the preceding sections, we have chosen to use an indexed notation. This enables the detail of the equations to be presented, without the notations being too cumbersome and difficult to understand. λ is the plastic multiplier and k_h the work hardening variable ($0 \leq k_h \leq 1$). m and h are defined by:

$$m_{ij} = \frac{\partial f}{\partial \sigma_{ij}^t}, \quad h = \frac{\sqrt{\frac{2}{3}\frac{\partial f}{\partial \sigma_{ij}^t}.\frac{\partial f}{\partial \sigma_{ij}^t}}}{\zeta(\sigma^t)} \qquad [2.118]$$

f is the load surface and ζ is a function of the effective stress:

$$f = \bar{\rho}^2(J_2^t) - \frac{k(k_h, I_1^t)\bar{\rho}_c^2(I_1^t)}{r^2(\theta, I_1^t)} \qquad [2.119]$$

I^t_1 and J^t_2 are the invariants of the effective stress tensor. This enables damage to be eliminated from plasticity equations. θ is the Lode angle. The evolution of the plastic multiplier is given by the relations:

$$f(\sigma^t, k_h) \leq 0, \quad d\lambda \geq 0, \quad f(\sigma^t, k_h).d\lambda = 0 \qquad [2.120]$$

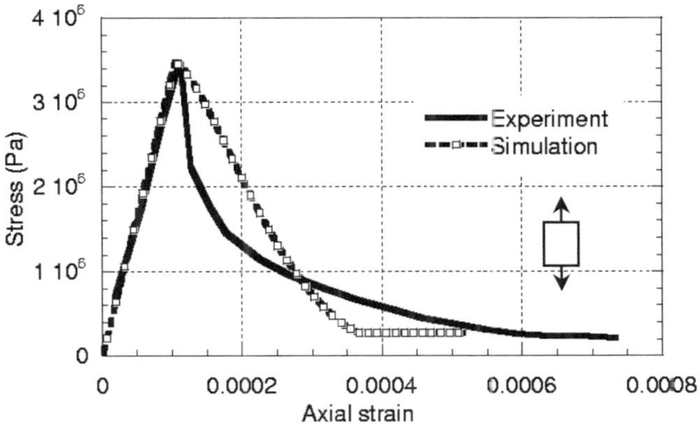

Figure 2.25. *Response of the elasto-endoplastic model under uniaxial tension [JAS 06]*

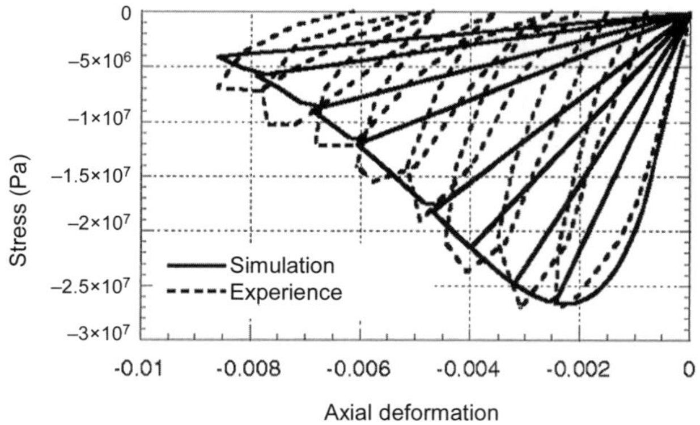

Figure 2.26. *Response of the damage model under uniaxial compression [JAS 06]*

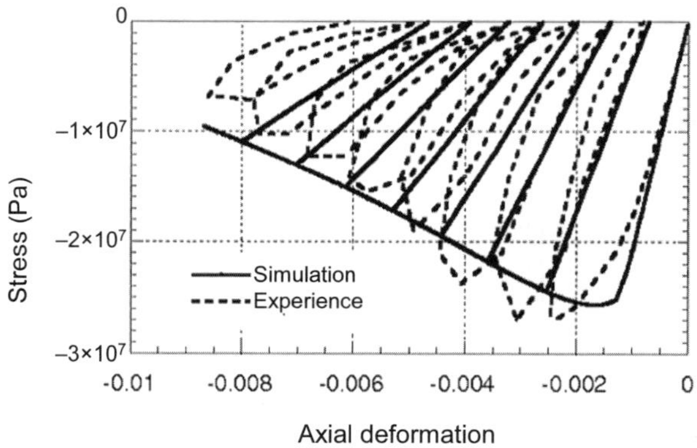

Figure 2.27. *Response of the elasto-endoplastic model coupled to damage under uniaxial compression [JAS 06]*

Figure 2.24 shows the evolution of the threshold surface relative to the work hardening parameter under simple compression. The damage model is that initially developed in [MAZ 84]. The equations were given at the beginning of this chapter (equations [2.19]-[2.24]).

Figure 2.25 gives the response of the elastic-endoplastic model under uniaxial tension. In this situation, damage is dominant and plasticity has little influence on the model response.

However, the influence of plasticity is much greater under simple compression. Figure 2.26 represents the response of the damage model. It can be observed that during unloading, the behavior law predicts a return to zero deformation. This is not consisten with experimental data [SIN 64].

Figure 2.27 gives the response of the model coupling plasticity and damage. It can be observed that it better respects the experimental response of the material. Plastic deformations appear in softening phase in particular, and damage is evaluated more realistically.

Figure 2.28 illustrates the difference between these two models in terms of volumetric response.

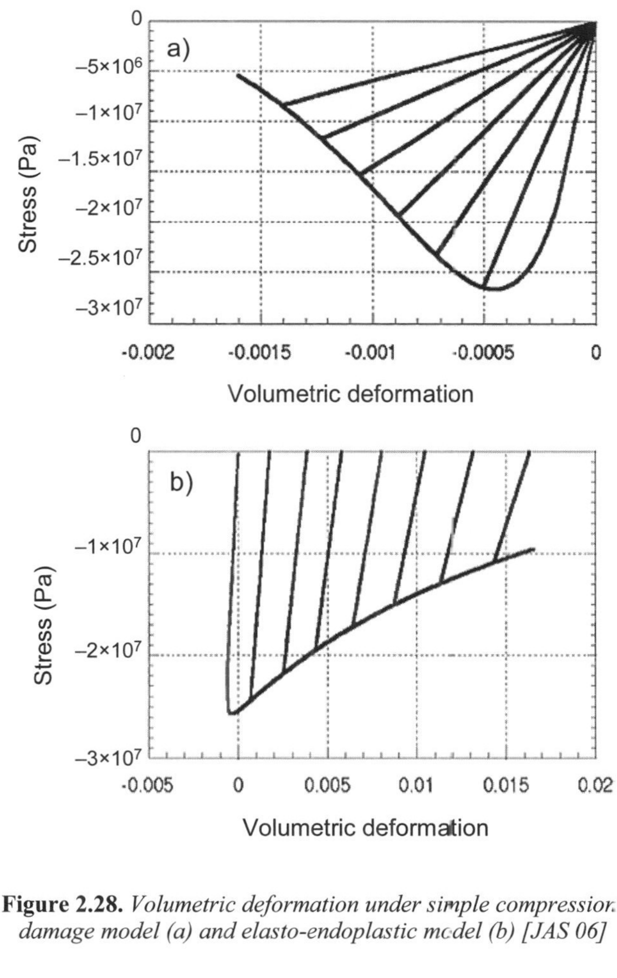

Figure 2.28. *Volumetric deformation under simple compression: damage model (a) and elasto-endoplastic model (b) [JAS 06]*

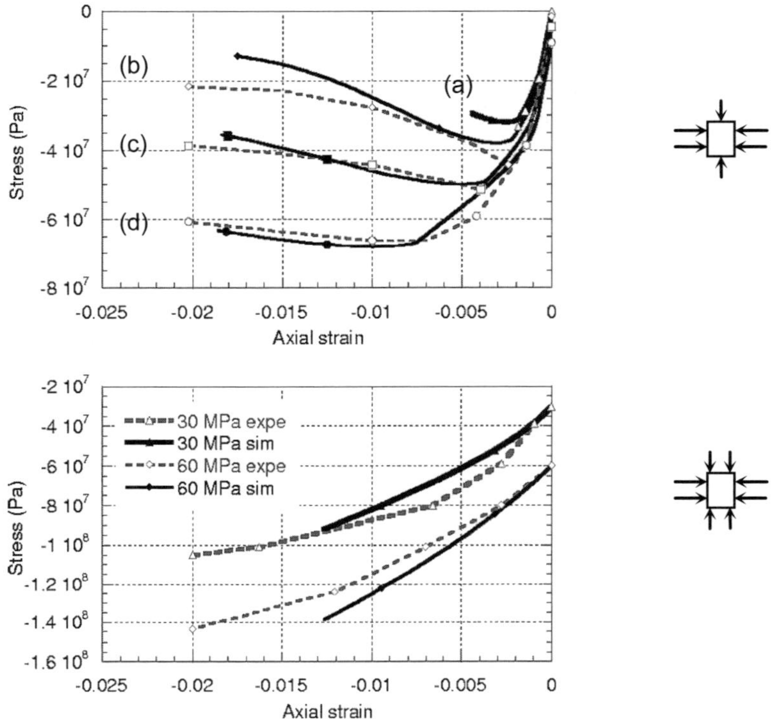

Figure 2.29. *Results of the elasto-endoplastic model for several triaxial tests at different levels of confinement [JAS 06]. The dotted line curves indicate the experimental results*

With the simple damage model, the volumetric deformations remain negative. Conversely, introducing plasticity generates a behavior change from contraction towards dilatancy observed experimentally (see Chapter 1).

The elasto-endoplastic model can also describe the behavior of concrete under confined compression. Figure 2.29 shows the response of this model during compressive tests on cylindrical samples submitted to lateral pressure. The combination of plasticity and damage enables the ductile-brittle transition to be represented. In cases of weak confinement, damage induces softening behavior. This disappears in cases of high confinement. Only plastic deformations subsist and the material becomes ductile, with positive work hardening induced by the plastic behavior.

Compared to "simple" models of plasticity or damage, this model presents an extra level of complexity, which enables material behavior to be better represented.

This complexity has a cost: the number of parameters of the model increases because both those of the plasticity model and those of the damage model must be determined. The amount of experimental data required to carry out its calibration will, therefore, also increase.

The physical interpretation of each parameter also becomes more delicate, as the coupling of plasticity and damage effects influences their definition.

Let us underline in conclusion that if it is always possible to improve constitutive laws, such as those presented in this chapter, it should only be done if each improvement is absolutely necessary. The anisotropy induced by damage (the microcracks are often oriented to the load applied) can be taken into account. For instance, here again, the complexity of the formulation and implementation of the behavior law in a calculation code will increase. The same can be said of the taking into account of crack closure or hysteresis observed during loading-unloading cycles. These models will be briefly presented in Chapter 4. In many current applications, simple models enable adequate simulations to be carried out. In any case they can give a first approximation which should not be neglected.

2.5. Bibliography

[BAB 97] BABUSKA I., MELENK J.-M., "The partition of unity method", *Int. J. Num. Meth. Eng.*, vol. 40, pp. 727-758, 1997.

[BAR 62] BARENBLATT G.-I., "The mathematical theory of equilibrium cracks in brittle fracture", *Adv. Appl. Mech.*, vol. 7, pp. 55-129, 1962.

[BAZ 83] BAŽANT Z.-P., "Fracture in concrete and reinforced concrete", *Symposium on Mechanics of Geomaterials, Rocks, Concretes, soils*, Northwestern University, pp. 281-316, 1983.

[BAZ 86] BAŽANT Z.-P., "Mechanics of distributed cracking", *Appl. Mech. Rev.*, ASME, vol. 39, pp. 675-705, 1986.

[BAZ 87] BAŽANT Z.-P., PFEIFFER P.-A., "Determination of fracture energy from size effect and brittleness number", *ACI Mater. J.*, vol. 84, pp. 463-480, 1987.

[BAZ 88] BAŽANT Z.-P., PIJAUDIER-CABOT G., "Non local continuum damage, localisation instability and convergence", *J. Appl. Mech.* vol. 55, pp. 287-293, 1988.

[BAZ 89] BAZANT Z.-P., PIJAUDIER-CABOT G., "Measurement of characteristic length of nonlocal continuum", *J. Eng. Mech.*, vol. 115, pp. 755 -767, 1989.

[BEL 94] BELYTSCKO T., LU Y.-Y., GU L., "Element-free Garlekin methods", *Int. J. Num. Meth. Eng.*, vol. 37, pp. 229-256, 1994.

[BEL 99] BELYTSCKO T., BLACK T., "Elastic crack growth in finite element with minimal remeshing", *Int. J. Num. Meth. Eng.*, vol. 45, pp. 601-620, 1999.

[BOR 85] DE BORST R., NAUTA P., "Non-orthogonal cracks in a smeared finite element model", *Eng. Comput.*, vol. 2, pp. 35-46, 1985.

[BOR 92] DE BORST R., MÜHLHAUS H.-B., "Gradient dependent plasticity: formulation and algorithmic aspects", *Int. J. Num. Meth. Eng.*, vol. 35, pp. 521-540, 1992.

[CAM 96] CAMACHO G.-T., ORTIZ M., "Computational modeling of impact damage in brittle materials", *Int. J. Solids Structures*, vol. 33, pp. 247-258, 1996.

[CHA 88] CHABOCHE J.-L., "Continuum damage mechanics. Part I . General concepts; Part II. Damage growth, crack initiation, and crack growth", *J. Appl. Mech.*, vol. 55, pp. 59-64 and pp. 65-72, 1988.

[COP 80] COPE R.-J., RAO P.-V., CLARK L.A., NORRIS P., "Modelling of reinforced concrete behaviour for finite element analysis of bridge slabs", in C. TAYLOR *et al.*, *Numerical Methods for Non-Linear Problems*, vol. 1, Pineridge Press, Swansea, pp. 457- 470, 1980.

[DUG 60] DUGDALE D.-S., "Yielding of steel sheets containing slits", *J. Mech. Phys. Solids*, vol. 8, pp. 100-108, 1960.

[FEE 95] FEENSTRA P.-H., DE BORST R., "A plasticity model and algorithm for mode-I cracking in concrete", *Int. J. Meth. Eng.*, vol. 38, pp. 2509-2529, 1995.

[FIC 96] FICHANT S., Endommagement et anisotropie induite du béton des structures.Modélisations approchées, PhD thesis, Ecole Normale Supérieure, Cachan, 1996.

[FRE 95] FREMOND M., NEDJAR B., "Damage in concrete: the unilateral phenomenon", *Nucl. Eng. Design*, vol. 156, pp. 323-335, 1995.

[GRI 21] GRIFFITH A.-A., "The phenomena of rupture and flow in solids", *Phil. Trans*, vol. 221A, pp. 179-180, 1921.

[HAI 02] HAIDAR K., Modélisation de l'endommagement des structures en béton. Approches numériques et effet de la microstructure sur les propriétés de rupture, PhD thesis, Ecole Centrale de Nantes and University of Nantes, France, 2002.

[HIL 76] HILLERBORG A., MODEER M., PETERSSON P.-E., "Analysis of crack formation and crack growth in concrete by means of fracture mechanics and finite elements", *Cement Concr. Res.*, vol. 6, pp. 773-781, 1976.

[HOR 91] HORDIJK D., Local approach to fracture of concrete, PhD thesis, Delft University, 1991.

[ING 85] INGRAFFEA A.-R., SAOUMA V., "Numerical modeling of discrete crack propagation in reinforced and plain concrete", in G.C. SIH and A. DI TOMMASO, *Fracture Mechanics of Concrete: Structural Application and Numerical Calculation*, Martinus Nijhoff Publishers, Dordrecht, pp. 171-225, 1985.

[IRW 57] IRWIN G.-R., "Analysis of stress and strains near the end of crack traversing a plate", *J. Appl. Mech.*, vol. 24, pp. 351-364, 1957.

[JAS 04] JASON L., Relation endommagement-perméabilité pour les bétons. Application au calcul de structure, PhD thesis, École Centrale de Nantes and University of Nantes, 2004.

[JAS 06] JASON L., HUERTA A., PIJAUDIER-CABOT G., GHAVAMIAN S., "An elastic plastic damage formulation for concrete: application to elementary and structural tests", *Comp. Meth. Appl. Mech. Eng.*, vol. 195, pp. 7077-7092, 2006.

[KAC 58] KACHANOV, L.-M., "On the time to failure under creep conditions", *Izv. Akad. Nauk. SSSR*, Otd. Tekhn. Nauk., vol. 8, pp. 26-31, 1958 [in Russian].

[KOI 53] KOITER, W.-T., "Stress-strain relations, uniqueness and variational theorems for elastic-plastic materials with a singular yield surface", *Q. Appl. Math*, vol. 3, 1953.

[LAB 91] LABORDERIE C., Phénomènes unilatéraux dans un matériau endommageable. Modélisation et application à l'analyse des structures en béton, PhD thesis, Ecole Normale Supérieure de Cachan, 1991.

[LEB 03] LE BELLEGO C., DUBE J.-F., PIJAUDIER-CABOT G., GERARD B., "Calibration of non local damage model from size effect tests", *Eur. J. Mech. A/solids*, vol. 22, pp. 33-46, 2003.

[LEM 85] LEMAITRE J., CHABOCHE J.-L., *Mécanique des Milieux Solides*, Dunod, Paris, 1985.

[LEM 86] LEMAITRE J., "Local approach to fracture", *Eng. Fract. Mech.*, vol. 25, pp. 523-537, 1986.

[LEM 90] LEMAITRE J., CHABOCHE J.-L., *Mechanics of Solid Materials*, Cambridge University Press, Cambridge, 1990.

[MAI 69] MAIER G., "Linear flow-laws of elastoplasticity: a unified general approach", *Lincei-Rend. Sci. Fis. Mater. Nat.*, vol. 47, pp. 266-276, 1969.

[MAZ 84] MAZARS J., Application de la mécanique de l.endommagement au comportement non linéaire et à la rupture du béton de structure, PhD thesis, University Paris VI, France, 1984.

[MAZ 89] MAZARS J., PIJAUDIER-CABOT G., "Continuum damage theory application to concrete", *ASCE J.Eng. Mech.*, vol. 115, pp. 345-365, 1989.

[MOE 99] MOES N., DOLBOW J., BELYTSCKO T., "A finite element method for crack growth without remeshing", *Int. J. Num. Meth. Eng.*, vol. 46, pp. 131-150, 1999.

[NEE 88] NEEDLEMAN A., "Material rate dependence and mesh sensitivity in localization problems", *Comput. Meth. Appl. Mech. Eng.*, vol. 67, pp. 69-85, 1988.

[NGO 67] NGO D., SCORDELIS A.C., "Finite element analysis of reinforced concrete beams", *J. Am. Concr. Inst.*, vol. 64, pp. 152-163, 1967.

[ORT 85] ORTIZ, M., POPOV E.-P., "Accuracy and stability of integration algorithms forelastoplastic constitutive relations", *Int. J. Num. Meth. Eng.*, vol. 21, pp. 1561-1576, 1985.

[ORT 86] ORTIZ, M., SIMO J.-C., "An analysis of a new class of integration algorithms for elastoplastic constitutive relations", *Int. J. Num. Meth. Eng.*, vol. 23, pp. 353-366, 1986.

[PEE 95] PEERLINGS R., DE BORST R., BREKELMANS W., DE VREE J., "Computational modelling of gradient-enhanced damage for fracture and fatigue problems", *Proceedings of Computational Plasticity IV*, pp. 975-986, 1995.

[PIJ 87] PIJAUDIER-CABOT G., BAZANT Z.-P., "Nonlocal damage theory", *J. Eng. Mech., ASCE*, vol. 113, pp. 1512-1533, 1987.

[PIJ 91] PIJAUDIER-CABOT G., HUERTA A., "Finite element analysis of bifurcation in nonlocal strain softening solids", *Comp. Meth. Appl. Mech. Eng.*, vol. 90, pp. 905-919, 1991.

[PIJ 92] PIJAUDIER-CABOT G., BODE L., "Localisation in a nonlocal continuum", *Mech. Res. Com.*, vol. 19, pp. 145-153, 1992.

[PIJ 93] PIJAUDIER-CABOT G., BENALLAL A., "Strain localization and bifurcation in anonlocal continuum", *Int. J. Solids Struct.*, vol. 30, pp. 1761-1775, 1993.

[PIJ 00] PIJAUDIER-CABOT G., DE BORST R., Enriched damage models for continuum failure analysis, in G.-A. MAUGIN *et al.*, *Continuum Thermomechanics, The Art of Modelling*, Kluwer Publisher, 2000.

[PIJ 04] PIJAUDIER-CABOT G., HAIDAR K., DUBE J.-F., "Nonlocal damage model with evolving internal length", *Int. J. Num. Anal. Meth. Geomechan.*, vol. 28, pp. 633- 652, 2004.

[PON 95] PONTIROLLI C., Comportement au souffle des structures en béton armé, PhD thesis, École Normale Supérieure de Cachan, France, 1995.

[PRA 89] PRAMONO, E., WILLAM K., "Implicit integration of composite yield surfaces with corners", *Eng. Comput.*, vol. 6, pp. 186-197, 1989.

[RAB 69] RABOTNOV Y.-N., *Creep Problems in Structural Members*, North-Holland, Amsterdam, 1969.

[RAG 99] RAGUENEAU F., Fonctionnement dynamique des structures en béton. Influence des comportements hystériques locaux, PhD thesis, École Normale Supérieure de Cachan, France, 1999.

[REI 84] REINHARDT H.-W., "Fracture mechanics of an elastic softening material like concrete", *Heron*, no. 29, 1984.

[ROT 88] ROTS J.-G., Computational modeling of concrete fracture, PhD thesis, Delft University, Netherlands, 1988.

[ROT 92] ROTS J.-G., "Removal of finite elements in smeared crack analysis", *Proceedings of COMPLAS III*, pp. 669-680, 1992.

[RUN 88] RUNESSON K., STURE S., WILLAM K., "Integration in computational plasticity", *Comp. Struct.*, vol. 30, pp. 119-130, 1988.

[SIM 86] SIMO J.-C., TAYLOR R.-L., "A return mapping algorithm for plane stress elastoplasticity", *Int. J. Num. Meth. Eng.*, vol. 22, pp. 649-670, 1986.

[SIM 88] SIMO J.-C., KENNEDY J.-G., GOVINDJEE S., "Non-smooth multisurface plasticity and viscoplasticity. Loading/unloading conditions and numerical algorithms", *Int. J. Num. Meth. Eng.*, vol. 26, pp. 2161-2185, 1988.

[SIM 89] SIMO J.-C., Strain softening and dissipation: a unification of approaches, in J. MAZARS and Z.-P. BAZANT, *Cracking and Damage*, Elsevier Science, Oxford, pp. 440-461, 1989.

[SIM 91] SIMO J.-C., GOVINDJEE S., "Non-linear B-stability and symmetry preserving return mapping algorithms for plasticity and viscoplasticity", *Int. J. Num. Meth. Eng.*, vol. 31, pp. 151-176, 1991.

[SIN 64] SINHA B.-P., GERSTLE K.-H., TULIN L.-G., "Stress strain relations for concrete under cyclic loading", *ACI Journal*, pp. 195-211, 1964.

[WES 39] WESTERGAARD H.-M., "Bearing pressure and cracks", *J. Appl. Mech.*, vol. 6, pp. A49-A53, 1939.

[WIL 57] WILLIAMS M.-L., "On the stress distribution at the base of a stationary crack", *J. Appl. Mech.*, vol. 24, pp. 109-114, 1957.

Chapter 3

Failure and Size Effect of Structural Concrete

3.1. Introduction

Traditional laboratory tests are aimed at the characterization of the mechanical response of materials, assuming of course that the constitutive relations deduced from the tests are not dependent on the type of test performed. A second assumption is that the response of the material should be independent of the size of the structure, the considered structure being a laboratory specimen or a real-size structure. This is not always true, at least some deviations to this basic principle have been observed quite a number of times and we are going to discuss this phenomenon called *structural size effect* in this chapter.

Structural size effect is defined here as the dependence of the mechanical response on the size of the structure. This is a well known phenomenon that has been observed for more than 500 years [VIN 1500s; GAL 38; MAR 86; YOU 07; GRI 21]. L'Hermite [HER 73] demonstrated that geometrically similar concrete beams subjected to three-point bending exhibited a structural size effect. From experimental results, the material strength has been computed using standard elasticity and the results show that the larger the beam, the smaller the material strength (Figure 3.1).

Analyses of different types of structures show that the critical stress intensity factor, or fracture toughness, K_{IC} also changes with the size and geometry of a

---

Chapter written by Gilles PIJAUDIER-CABOT and Khalil HAIDAR.

concrete structure. Usually, K_{IC} is considered as constant, it is a material property at least according to Linear Elastic Fracture Mechanics (LEFM). For the design of concrete structures, such an assumption may not be conservative: the fracture toughness decreases as the size of the structure increases and extrapolation from laboratory experiments to real-size structures may become delicate in the absence of rational theories explaining and describing structural size effect.

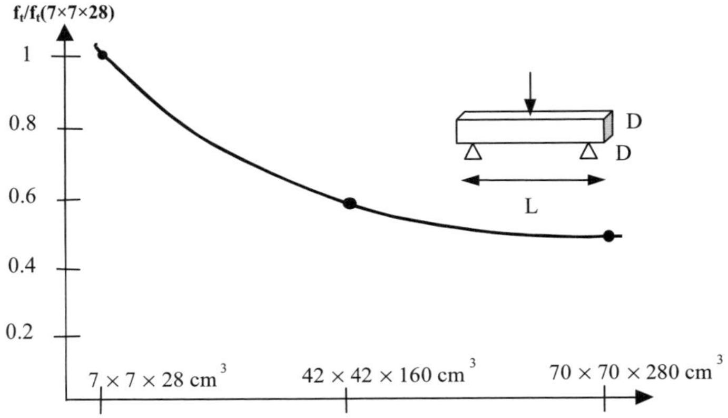

Figure 3.1. *Size effect observed on three-point bending beams. f_t is the tensile strength computed on the bottom fiber according to elasticity at peak load [HER 73]*

Based on the results of Mariotte [MAR 86], Weibull [WEI 39] proposed a probabilistic theory constructed on the weakest link principle. The strength of a structure is the smallest strength of the elements from which the structure is assembled. As the probability of finding a weak element of any given strength decreases as the volume of the structure decreases, the apparent strength of a structure increases as its size decreases. This probabilistic model relies on a specific distribution of the local strength, more precisely on a specific description of the tail of this distribution. Most of the studies on size effect conducted up to the 1980s have been based on Weibull distribution (see e.g. [ZAI 74; MIH 77, 81]).

In the weakest link theory, structures are considered to fail at crack initiation and the fracture process zone is considered to be very small – negligible compared to the size of the crack and to the size of the structure. Originally, the weakest link theory was a one-dimensional theory. Redistribution of stress during failure cannot occur. Extensions to two or three-dimensional cases require either that redistribution of stress does not occur (as stated above), or some refinement that accounts for a specific distribution scheme.

The above assumptions hold for materials with a rather small grain size, such as ceramics or alloys. In the case of concrete or rocks, the size of the fracture process zone is not negligible (several grain sizes). Bazant [BAZ 84] proposed a deterministic size effect law that is based on the redistribution of elastic energy near the tip of a crack in the process zone.

Inspired by several works dealing with the fractality of cracks [MAN 84; BRO 87; LON 91; SAO 94], Carpenteri *et al.* [CAR 95a-c] devised a size effect law based on the geometry of the crack and fracture surface.

This chapter proposes an overview of these three theories:

– the probabilistic theory due to Weibull [WEI 39];

– the deterministic theory due to Bazant [BAZ 84, 99a, 99b];

– the fractal size effect theory due to Carpenteri *et al.* [CAR 95a-c].

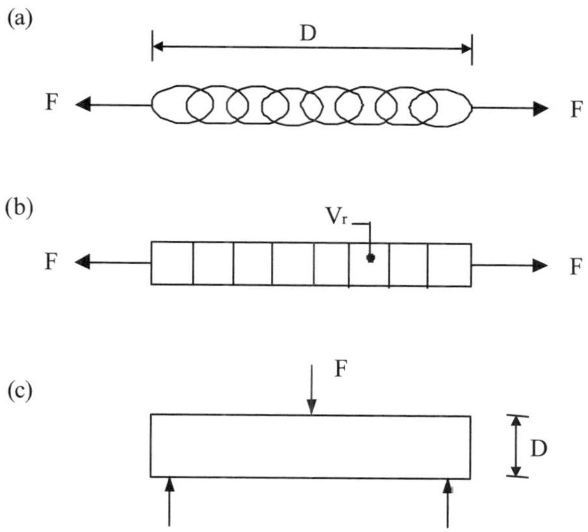

Figure 3.2. *Different structures on which the weakest link theory is applied: a) one-dimensional discrete chain, b) continuous one-dimensional bar, c) three-point bend*

We will outline the limits of each theory, but it is not the purpose of this chapter to advocate one over the others. Some may be combined, the probabilistic and deterministic theories, for instance. Very often, current experimental data, at least on concrete structures, do not provide a definite means to prefer one theory compared with another. In the final section of this chapter, we will focus on deterministic size effect because this is used in non-local failure models.

There are other sources of size effect, which will not be considered in this chapter: boundary effects, Poisson's effect induced by the transition of crack propagation condition from plane stress to plane strain within the thickness of the structure, and time dependent effects due to transport of mass within the material shall not be discussed. For a more comprehensive treatment on structural size effect, the reader may refer to the books by Bazant and co-workers [BAZ 98, 04].

3.2. Probabilistic structural size effect

The foregoing probabilistic structural size effect law is based on the idea of random structural strength combined with Weibull distribution of probability of failure (see Weibull [WEI 39, 49, 51, 56]). It is a weakest link theory as illustrated with the following examples dealing with discrete and continuous structural descriptions. We will start with one-dimensional examples and then we will discuss extensions to the multidimensional case.

3.2.1. *Basics of the weakest link theory*

Let us consider a one-dimensional chain made of N elements (Figure 3.2a). Each element of the chain has a different strength. This distribution of strength is characterized by the cumulative distribution of probability $P_1(\sigma)$ which represents the probability that in an element the stress be higher than the admissible strength. The probability of survival of this element is $(1 - P_1(\sigma))$. The survival probability of the entire chain loaded by the stress σ, is:

$$1 - P_f = (1 - P_1)\ (1 - P_1)\ldots\ (1 - P_1) = (1 - P_1)^N \qquad [3.1]$$

P_f is the probability of failure of the chain loaded by the stress σ. Let us now take the logarithm of this expression:

$$ln(1 - P_f) = N \times ln(1 - P_1) \qquad [3.2]$$

As P_1 is usually quite small, we have $ln(1 - P_1) \approx -P_1$ and the probability of failure of the chain becomes:

$$P_f(\sigma) = 1 - exp[-N \times P_1(\sigma)] \qquad [3.3]$$

Equation [3.3] can be extended to a continuous bar as shown in Figure 3.2b. Let us define $N=V/V_r$ where V is the volume of the bar and V_r the material representative volume. We substitute now in equation [3.3]:

$$P_f(\sigma) = 1 - exp\left[-\frac{V}{V_r}P_1(\sigma)\right] \qquad [3.4]$$

where $P_f(\sigma)$ is the probability of failure of a representative volume for the load level defined by stress σ. Within this theory, V_r is the smallest part of the structure in which a stress can be defined statistically, at least this is the present meaning of the representative volume.

The theory may be extended to quasi-brittle materials, which possess a large fracture process zone. However, in these material models there is a characteristic – internal – length, which modifies the meaning of the representative volume. According to the Dugdale model [HIL 76], the characteristic length defines the size of the process zone ahead of the crack tip. It is this size that is representative and that defines the smallest volume in which a failure criterion may be defined locally. Equation [3.4] is modified and becomes:

$$P_f(\sigma) = 1 - exp[-c(\sigma)\cdot V] \qquad [3.5]$$

where $c(\sigma)$ is an empirical function, a spatial concentration function, defined by Weibull [WEI 39, 51]:

$$c(\sigma) = \frac{1}{V_0}\left\langle\frac{\sigma - \sigma_1}{\sigma_0}\right\rangle^m \qquad (\sigma \geq \sigma_l \cong 0) \qquad [3.6]$$

and $\langle x \rangle$ is the positive part of x. m, σ_0 and σ_l are material constants (m = Weibull modulus $m \in [5,50]$; σ_0 = scale parameter ; σ_l = failure threshold). V_0 is a reference volume. If the failure threshold is taken equal to zero, equation [3.6] becomes:

$$P_f(\sigma) = 1 - exp\left[-\frac{V}{V_0}\left\langle\frac{\sigma}{\sigma_0}\right\rangle^m\right] \qquad [3.7]$$

which defines the Weibull distribution of probability. Note that this Weibull distribution may also be recovered as a special case of extreme value statistics.

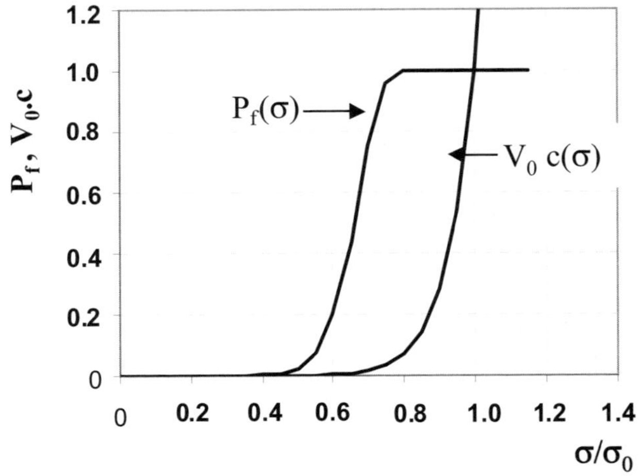

Figure 3.3. *Weibull distribution (m = 12, V/V0 = 100 and σ_l = 0)*

Figure 3.3 shows the form of the distribution function with $\sigma_l = 0$ and $m = 12$ (an appropriate set of parameters for concrete [ZEC 77]).

Given the probability of failure in Equation [3.7], the average failure stress $\overline{\sigma}_u$ on a structure of volume V is:

$$\overline{\sigma}_u = \int_{-\infty}^{-\infty} \sigma.dP_f(\sigma) = \frac{V.m}{V_0\sigma_0^m}\int_0^{\infty} \sigma^m \exp\left[-\frac{V.\sigma^m}{V_0\sigma_0^m}\right] d\sigma = \sigma_0\ \Gamma\left(1+\frac{1}{m}\right)\left(\frac{V_0}{V}\right)^{1/m} \qquad [3.8]$$

where $\Gamma(p)$ is the Gama function:

$$\Gamma(p) = \int_0^{\infty} u^{p-1}e^{-u}du \quad \text{with } u = (V/V_0)\times(\sigma/\sigma_0)^m \qquad [3.9]$$

For values of the Weibull modulus between 5 and 50, Planas proposed an approximation of the Gama function with an accuracy greater than or equal to 0.5%:

$$\Gamma\left(1+\frac{1}{m}\right) \approx 0.6366^{1/m} \qquad (5 \leq m \leq 50) \qquad [3.10]$$

The coefficient of variation ω of the failure stress distribution is given as a function of the variance s, $\omega = s/\overline{\sigma}_u$:

$$s^2 = \int_{-\infty}^{\infty} \left[\sigma^2 . dP_f(\sigma)\right] - \overline{\sigma}_u^2 = \Gamma\left(1+\frac{2}{m}\right)\left(\frac{V_0}{V}\right)^{\frac{2}{m}} \sigma_0^2 - \overline{\sigma}_u^2 \qquad [3.11]$$

Upon substitution of σ_0 as a function of $\overline{\sigma}_u$ from equation [3.8] and with equation [3.10], we have:

$$s^2 = \overline{\sigma}_u^2 \left[\frac{\Gamma(1+2/m)}{\Gamma^2(1+1/m)} - 1\right] \qquad [3.12]$$

and the coefficient of variation ω reads:

$$\omega = \left[\frac{\Gamma\left(1+2m^{-1}\right)}{\Gamma^2\left(1+m^{-1}\right)} - 1\right]^{1/2} . \qquad [3.13]$$

Note that the coefficient of variation depends on the Weibull modulus only, and not on the volume V of the structure, as opposed to the failure stress $\overline{\sigma}_u$.

3.2.2. *Generalized probabilistic size effect law*

Equation [3.3] may be further generalized to the case of a continuous structure, with a non-homogeneous state of stress, e.g. a bending beam (Figure 3.2c). For this, we assume that the structure is decomposed into sub-volumes $\Delta V_{(j)}$ in which the stress is homogeneous $\sigma_{(j)}$ ($j = 1, \ldots, N$). The stress $\sigma_{(j)}$ depends on the external loading defined by a "stress" σ_N: $\sigma_{(j)} = \sigma_{(j)}(\sigma_N)$. The probability of survival of the structure subjected to stress σ_N is the superposition of the survival probabilities of each part of the structure:

$$1 - P_f(\sigma_N) = \prod_{j=1}^{N} \exp\left[-c\left[\sigma_{(j)}(\sigma_N)\right] \Delta V_{(j)}\right] = \exp\left[-\sum_{j=1}^{N} c\left[\sigma_{(j)}(\sigma_N)\right] \Delta V_{(j)}\right] \qquad [3.14]$$

Let us now consider that the size of each sub-volume tends to zero, and that the number of sub-volumes becomes infinite. The stress $\sigma_{(j)}$ is now a continuous function $\sigma(x, \sigma_N)$. Equation [3.14] becomes:

$$P_f(\sigma_N) = 1 - exp\left[-\int_V c\left[\sigma(x, \sigma_N)\right] dV(x)\right] \qquad [3.15]$$

Note that the distribution function $c(x, \sigma_N)$ is a scalar valued quantity defined from a tensorial stress. Hence, it is a function of the principal stresses σ^I, σ^{II}, σ^{III}:

$$c(x, \sigma_N) = c(\sigma^I, \sigma^{II}, \sigma^{III}) \qquad [3.16]$$

Freudenthal [FRE 68] proposed the following expression:

$$c(x, \sigma_N) = \frac{1}{V_0 \sigma_0^m}\left(\langle \sigma^I \rangle^m + \langle \sigma^{II} \rangle^m + \langle \sigma^{III} \rangle^m\right) = \frac{1}{V_0 \sigma_0^m} \sum_{P=1}^{III} \langle \sigma^P \rangle^m \qquad [3.17]$$

If we introduce the definition of an effective stress $\tilde{\sigma}$ as

$$\tilde{\sigma} = \left[\sum_{P=1}^{III} \langle \sigma^P \rangle^m\right]^{\frac{1}{m}} \qquad [3.18]$$

The spatial concentration function becomes:

$$c(x, \sigma_N) = \frac{1}{V_0}\left(\frac{\tilde{\sigma}}{\sigma_0}\right)^m \qquad [3.19]$$

and the probability of failure is:

$$P_f(\sigma_N) = 1 - \exp\left[-\int_V \left[\frac{\tilde{\sigma}(x)}{\sigma_0}\right]^m \frac{dV(x)}{V_0}\right] \qquad [3.20]$$

In the case of a structure subjected to proportional loading, the distribution of stress may be defined, in a system x of coordinates (x_1, x_2, x_3), as

$$\sigma(x, \sigma_N) = \sigma_N . s\left(\frac{x}{D}\right) = \sigma_N . s\left(\frac{x_1}{D}, \frac{x_2}{D}, \frac{x_3}{D}\right) \qquad [3.21]$$

s is a tensorial valued function describing the distribution of stress for a unit nominal (applied) stress σ_N. D is a characteristic dimension of the structure. Using equation [3.18], the effective uniaxial stress is:

$$\tilde{\sigma} = \tilde{\sigma}\left[\sigma_N . s\left(\frac{x}{D}\right), m\right] = \sigma_N . \tilde{\sigma}\left[s\left(\frac{x}{D}, m\right)\right] = \sigma_N . \tilde{s}\left(\frac{x}{D}, m\right) \qquad [3.22]$$

Note that $\tilde{s}\left(\frac{x}{D},m\right)$ is a function of the geometry of the structure and of the Weibull modulus. It does not depend on the load level, i.e. on the nominal stress σ_N. The probability of failure is obtained from equation [3.20]:

$$P_f(\sigma_N)=1-exp\left\{-\frac{1}{V_0}\left(\frac{\sigma_N}{\sigma_0}\right)^m\int_V\left[\tilde{s}\left(\frac{x}{D},m\right)\right]^m dV(x)\right\} \quad [3.23]$$

The integral in this equation has the dimension of a volume. It is called the uniaxial equivalent volume, V_N:

$$V_N=\int_V\left[\tilde{s}\left(\frac{x}{D},m\right)\right]^m dV(x) \quad [3.24]$$

It is the volume of a one-dimensional structure equivalent to the considered structure. Equation (3.24) is re-written as:

$$P_f(\sigma_N)=1-exp\left[-\frac{V_N}{V_0}\left(\frac{\sigma_N}{\sigma_0}\right)^m\right] \quad [3.25]$$

which is similar to equation [3.7] if σ and V are replaced by σ_N and V_N. The average nominal failure stress $\overline{\sigma}_{Nu}$ becomes:

$$\overline{\sigma}_{Nu}=\left(\frac{V_0}{V_N}\right)^{\frac{1}{m}}\overline{\sigma}_0 \quad [3.26]$$

with a coefficient of variation defined in equation [3.13].

The Weibull size effect law derives directly from the above equation. Consider a reference dimension denoted as D_0 with a reference average failure stress $\overline{\sigma}^{0}_{Nu}$. For a structure of size D, the average failure stress is:

$$\overline{\sigma}_{Nu}=\overline{\sigma}^{0}_{Nu}\left(\frac{V_N(D_0)}{V_N(D)}\right)^{\frac{1}{m}} \quad [3.27]$$

In order to further simplify this expression, we may consider that the uniaxial equivalent volume defined in equation [3.24] is proportional to the volume of the structure *when geometrically similar structures are considered.* In this special case and for a three-dimensional structure, we have:

$$\bar{\sigma}_{Nu} = \bar{\sigma}^{0}_{Nu}\left(\frac{D_0}{D}\right)^{\frac{3}{m}} \qquad [3.28]$$

and in the case of a two-dimensional structure, we have a different size effect law:

$$\bar{\sigma}_{Nu} = \bar{\sigma}^{0}_{Nu}\left(\frac{D_0}{D}\right)^{\frac{2}{m}} \qquad [3.29]$$

In the two-dimensional case and when the Weibull modulus is m=20, the decrease of nominal stress at failure is 15% when the size of the structure D is multiplied by 5. If the size is divided by 5, the increase of failure strength is of 17%.

A salient characteristic of this size effect law is that it does not contain an internal length because it is a power law. In a physical theory where the scaling law is expressed as a power law, there is no characteristic length. The structure size is compared to a reference, but the comparison is performed through a ratio and the actual size of the reference does not appear in the scaling law.

3.3. Deterministic size effect

A second source of size effect, besides a random distribution of strength, is due to the redistribution of stresses ahead of the crack tip, in the fracture process zone. This size effect applies typically to quasi-brittle materials which possess a fracture process zone whose size may not be considered as negligible [BAZ 84].

Let us consider a rectangular plate that is initially subjected to a uniform stress. The stress distribution is a function of the nominal stress σ_N (Figure 3.4). In this plate, the crack and its process zone form a band of width h_f. Its propagation over a unit length requires the energy G_f. Conservation of energy is assumed, which means that the energy released by the structure is equal to the energy consumed for the propagation of the crack and its process zone. When the length of this band increases by Δa, the corresponding energy released in the structure is assumed to be that in the dense cross-hatched zone (Figure 3.4). Clearly, the energy released in a small specimen will be smaller than that in a large geometrically similar specimen for the same increase of the crack band length.

Quantitatively, the energy released in the zone depicted in Figure 3.4 is a function of the area of this surface ($h_f.\Delta a + 2.k.a_0.\Delta a$) multiplied by the thickness of the plate b and by the energy density $\sigma_N^2 / 2E$, where E is the Young's modulus of the material. Conservation of energy yields:

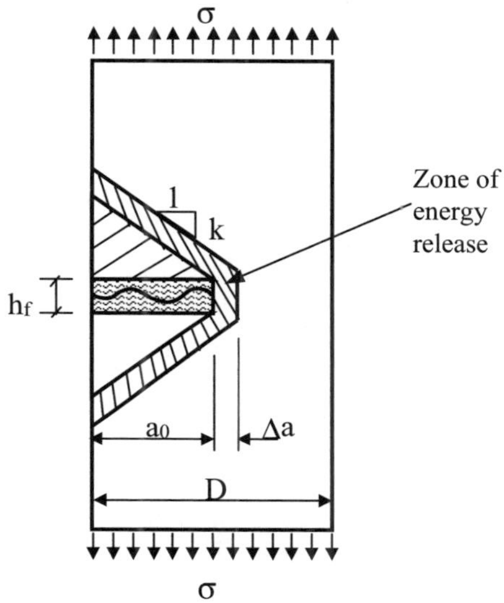

Figure 3.4. *Schematic redistribution of stresses and energy release during crack propagation (after [BAZ 04])*

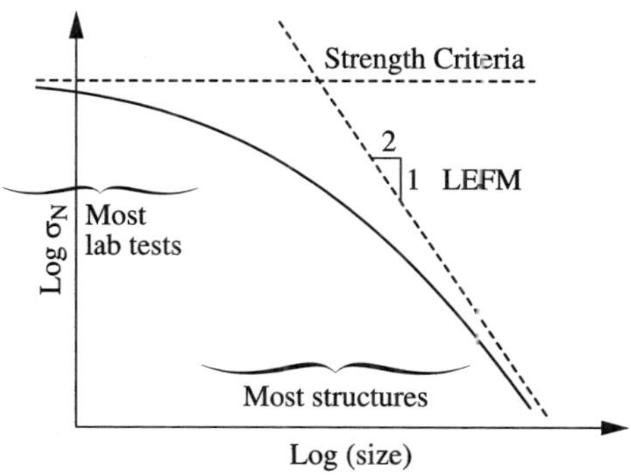

Figure 3.5. *Bazant's size effect law (after [BAZ 04])*

$$b\left(h_f \Delta a + 2ka_0 \Delta a\right)\frac{\sigma_N^2}{2E} = G_f b \Delta a \qquad [3.30]$$

Solving for the nominal stress we have [BAZ 83, 84]:

$$\sigma_N = \frac{Bf_t}{\sqrt{1 + D / D_0}} \qquad [3.31]$$

with $Bf_t = \sqrt{\dfrac{2G_f E}{h_f}}$ = constant and $D_0 = \dfrac{h_f D}{2ka_0}$ = constant.

D_0 depends on the geometry of the structure, but not on its actual size if geometrically similar structures are considered (D/a_0 = constant, a_0 being the initial crack length at the onset of failure). h_f is assumed to be constant (material property). f_t is the tensile strength of the material. As we will see further, the two constants B and D_0 may be fitted from experiments involving specimens of different sizes.

It has been verified that this size effect formula, first introduced by Bazant (see [BAZ 98, 04]), describes quite consistently size effect for numerous geometries (tension, three-point bending, compression, splitting failure, etc.) and different materials (rocks, concrete, ceramics, composites, etc.) [BAZ 99a,b].

Usually, equation [3.31] is plotted in a log-log coordinate system (Figure 3.5). On the horizontal axis, we have log (D/D_0) and on the vertical axis we have log (σ_N/Bf_t). For large values of D/D_0, this equation is such that (σ_N/Bf_t) is proportional to $(D/D_0)^{-1/2}$.

$$\sigma_N \approx \frac{Bf_t D_0^{1/2}}{D^{1/2}} \qquad [3.32]$$

This corresponds exactly to LEFM where $K_I \propto \sigma_N \sqrt{D}$. Substitution in equation [3.32] yields:

$$K_I = K_{IC} \propto Bf_t \sqrt{D_0} = \text{constant} \qquad [3.33]$$

When $D/D_0 \ll 1$, the nominal stress σ_N is approximately equal to Bf_t. It is size independent and, therefore, this is close to classical strength of the material. These two limit cases correspond to the straight lines in Figure 3.5. It can be seen that the size effect law is a match for these two asymptotic cases. In between, the variation of the nominal stress with the size is non-linear. When the size of the structure is divided by 5, the nominal stress may increase by as much as 30%, and when the size of the structure is multiplied by 5 it may decrease by as much as 55%. This size effect is stronger than that due to randomness of the material strength as described by Weibull and fitted on concrete specimens (with $m \approx 10$).

The determination of the two coefficients entering in the size effect law follows from a recommendation by Rilem [RIL 90]. It is performed from size effect tests on geometrically similar specimens. A minimum of three different sizes is required, but four different sizes is recommended as shown in Figure 3.6.

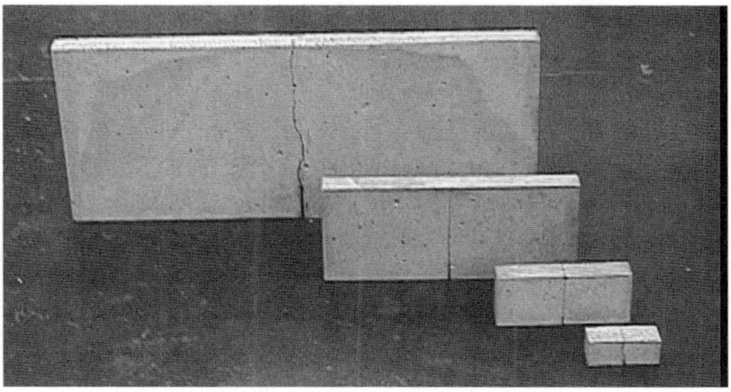

Figure 3.6. *Geometrically similar three-point bending beams notched in the center*

Equation [3.31] is rewritten in the simple form:

$$Y = AX + C \qquad [3.34]$$

where: $X = D$; $Y = (f_t/\sigma_N)^2$; A and C are constants to be fitted:

$$Bf_t = 1/\sqrt{C}\ ; D_o = C/A \qquad [3.35]$$

These two constants are easily obtained by linear regression. From these constants, the material strength and that fracture energy may be computed [BAZ 87]:

$$G_f = \frac{g(\alpha_0)}{EA} \text{ and } c_f = \frac{g(\alpha_0)\ C}{g'(\alpha_0)\ A} \qquad [3.36]$$

$\alpha_0 = a_0/D$, $g(\alpha)$ and $g'(\alpha) = dg(\alpha)/d\alpha$ are functions that depend on the geometry of the specimen only. For prismatic bending beams of span l and height D we have:

$$g(\alpha) = \left(\frac{l}{D}\right)^2 \Pi\alpha[1.5f(\alpha)]^2 \qquad [3.37]$$

$f(\alpha)$ can be found in the handbook of Tada *et al.* [TAD 85]: for $l/D = 2.5$

$$f_{2.5}(\alpha) = \frac{1.0 - 2.5\alpha + 4.49\alpha^2 - 3.98\alpha^3 + 1.33\alpha^4}{(1-\alpha)^{3/2}} \qquad [3.38]$$

for $l/D = 4$

$$f_4(\alpha) = \frac{1.99 - \alpha(1-\alpha)(2.15 - 3.93\alpha + 2.7\alpha^2)}{\Pi^{0.5}(1+2\alpha)(1-\alpha)^{3/2}} \qquad [3.39]$$

$f(\alpha)$ may be linearly interpolated for intermediate cases.

The above size effect formula applies for structures containing a long crack that fail at the inception of crack propagation. It has been generalized by Bazant [BAZ 95; 97], who proposed a "universal" size effect formula that applies to structures with and without notches or initial cracks:

$$\sigma_N = Bf_t\left[1 + \left(\frac{D}{D_0}\right)^r\right]^{-1/2r}\left[1 + s\frac{2l_f D_0}{(2\gamma l_f + D)(D + D_0)}\right]^{1/s} \qquad [3.40]$$

σ_N is the nominal stress at failure of a structure of size D, B and D_0 are the constants defined previously, f_t is the tensile strength, l_f is the thickness of a boundary layer in which damage develops prior to cracking in structures without notches and γ is a constant. Bazant suggested that $r = s = 1$, which is the most appropriate set of values in comparisons with experiments (Figure 3.7).

3.4. Fractality and size effect

The fractal character of crack surfaces has motivated several theoretical developments aimed at connecting fractality and size effect [CAR 94a, b; CAR 95a, b, c]. The key idea is that the fractal character of the macrocrack surface or the fractality of the microcrack distribution induces some size effect due to the non-integer dimension number of the exponent in the power law connecting the description of cracking to the size of the structure. Let us first recall some of the basis of fractals.

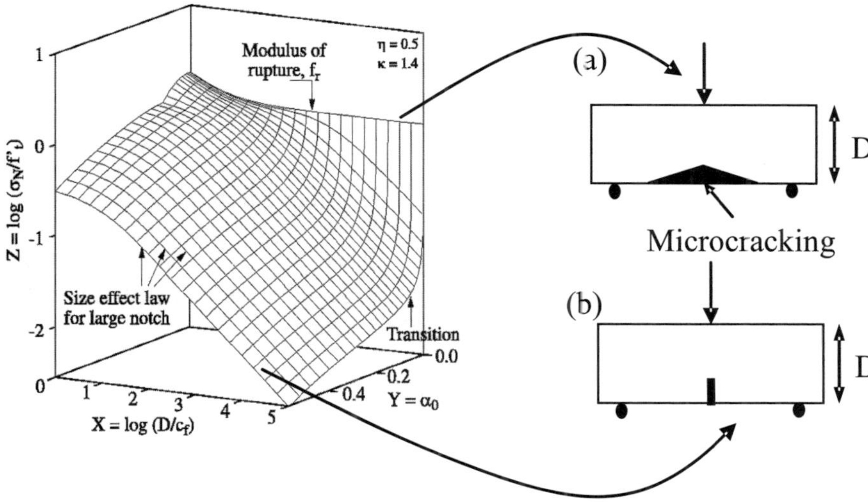

Figure 3.7. *Universal size effect law proposed by Bazant which applies a) to notched structures and b) to structures without notches (after [BAZ 04])*

3.4.1. *Basis of the fractal approach*

Most of the geometry-related quantities in the mechanics of materials are self similar (for metals see [MAN 84], for concrete see [SAO 90]). For the sake of simplicity, let us consider the length L of a line defined by the segment AB in Figure 3.8. This length is measured with a resolution denoted as u. Segments of length less that u are considered straight and above this measure, the tortuosity of the line is followed in order to compute its length. If the line is a true straight line, the measure of the length does not depend on the resolution:

$$L(u) = L(u_0) \quad [3.41]$$

where u_0 is a resolution of reference defined arbitrarily. Self-similarity means that the product $L(u)u^{\phi}$ is a characteristic quantity:

$$L(u) = L(u_0)\left[\frac{u_0}{u}\right]^{\phi} \quad [3.42]$$

ϕ is an exponent (to be measured) which depends on the self-similar property $L(u)$ considered.

This equation may be rewritten as:

$$\frac{L(u)}{u} = N(u) = n(u_0)\left[\frac{u_0}{u}\right]^{d_f} \qquad [3.43]$$

$d_f = 1 + \phi$ is the fractal dimension. Unless ϕ=0, which means that AB is really a straight line whatever the value of u, the length is size dependent: when D increases, the length does not increase similarly. The fractal dimension is usually not an integer. If ϕ=0 then the measure of the length is Euclidian. When ϕ>0, fractality is called invasive and when ϕ<0, fractality is called lacunar.

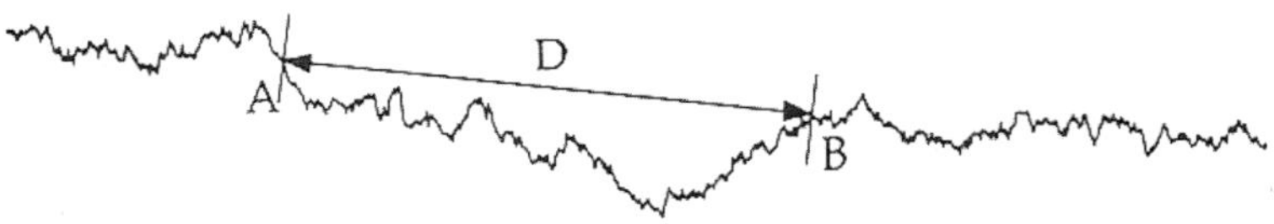

Figure 3.8. *Example of a fractal line (generated using the software Fractals)*

3.4.2. *Fractality of the fracture surface and fracture energy.*

Let us assume that the total work needed for the propagation of the crack is proportional to the conventional crack surface $A_0(D)$:

$$W_F(D) \propto A_0(D) \propto D^{d_f} \qquad [3.44]$$

where d=2. If the crack surface is fractal, i.e. its dimension is not [m^2], then the fracture energy needs to be modified [CAR 94a, b, c]. Instead of the classical relationship $W_F(D) \propto G_F(D).D^2$ where $G_F(D)$ is the fracture energy, we have:

$$G_F(D) = G_{F_0}\left(\frac{D}{u_0}\right)^{\phi_G} = \breve{G}_F D^{\phi_G}, \quad \phi_G = d_f - 2 \qquad [3.45]$$

G_{F0} is the conventional fracture energy (non fractal crack surface), $\breve{G}_F$ is the fractal fracture surface with dimension $FL^{-(1+\phi_G)}$ instead of FL^{-1}.

This simple theory is capable of describing size effect over a range from 1 to 4. Its generalization to structures of larger or smaller sizes has been proposed by Carpinteri *et al.* [CAR 96], it is called the multifractal size effect law and reduces to the formula:

$$G_F(D) = \frac{G_F^{\infty}}{\sqrt{1 + l_G / D}} \qquad [3.46]$$

where G_F^{∞} is the fracture energy for structure of infinite sizes, and l_G is a characteristic length. This size effect law is plotted in Figure 3.9. It is very different from the deterministic size effect law due to Bazant as it is founded on very different arguments. l_G and G_F^{∞} are fitted from size effect test data, as in Bazant's law.

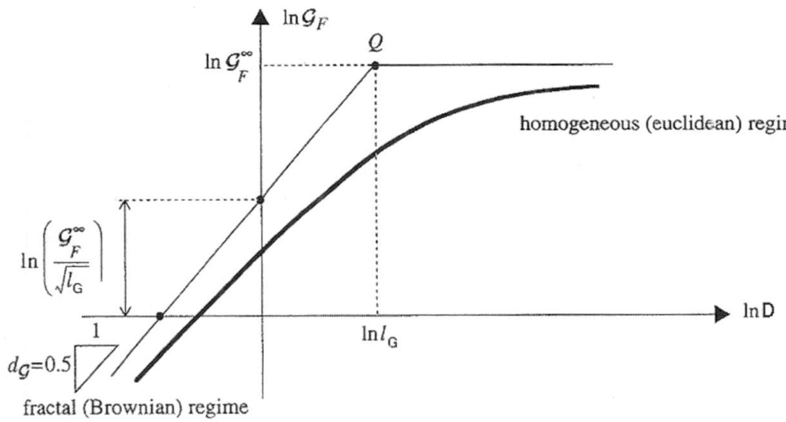

Figure 3.9. *Multifractal size effect law [CAR 96]*

3.4.3. *Lacunar fractality and multifractality of the nominal stress at failure*

Similar ideas may be applied to the nominal stress at failure σ_{Nu} instead of to the fracture energy. In this case, fractality is lacunar [CAR 94a, b, c]. If the crack surface is fractal, then the effective stress, which is the applied load divided by the same surface should be fractal too. For a cross-section of material defined as $D \times D$, subjected to the load F, the effective nominal stress is $\sigma_N = F/D^2$. If the cross-section is modified to $A_0 = cD^{2+\phi_\sigma}$, where c is a constant, then the effective nominal stress ought to be modified accordingly, so that the product of the effective nominal stress with the area of the cross-section is homogeneous to the applied load. Let σ_{R0} be the material stength. The maximum applied load reads $F_u = \sigma_{R0} c D^{2+\phi_\sigma}$ and the effective nominal stress is:

$$\sigma_{N_u} = \sigma_{R_0} c D^{\phi_\sigma} = \breve{\sigma}_R D^{\phi_\sigma} \qquad [3.47]$$

where $\breve{\sigma}_R$ is the fractal stress with dimension $FL^{-2+\phi_\sigma}$, and ϕ_σ is the fractal exponent. ϕ_σ is negative. The nominal stress decreases when the size of the structure increases and the corresponding size effect law is quite similar to Weibull size effect formula with a Weibull modulus equal $m = -3/\phi_\sigma$.

The theory possesses the same limitations as the above formulation with a fractal fracture energy, and has also been extended to a multifractal description of the effective nominal stress at failure [CAR 96]. The nominal stress is written as:

$$\sigma_{N_u} = \sigma_R^{\infty} \sqrt{1 + \frac{l_\sigma}{D}} \quad [3.48]$$

where σ_F^{∞} is the nominal failure stress for a structure of infinite size and l_σ is a characteristic transition length. Again, these two parameters are fitted from size effect test data. From such fits on concrete specimens, the decrease of the structure size $D = l_\sigma$ by a factor of 5 yields an increase of stress of 70% approximately and the increase of structure size from the same reference by a factor 5 yields an decrease of nominal stress by 20%.

3.5. Size effect and calibration of non-local models

Structural size effect, understood in the deterministic sense as proposed by Bazant, is the result of the interaction between the size of the fracture process zone and the size of the structure during the failure process. It can be regarded as a direct consequence of the introduction of an internal length in the description of the material response (see Chapter 2). In the presence of material strain softening, it is necessary to introduce an internal length in the constitutive model, which controls the size of the localization zone; in other words, which controls the size of the fracture process zone in the present case of quasi-brittle materials [BAZ 88]. Whenever this length is not negligible compared with the size of the structure, some deviation from size/scale invariance should be expected. Hence, some structural size effect may be captured from computations with non-local material models, which incorporate such an internal length.

In this section, we see that this size effect, described with a non-local model, is consistent with Bazant's size effect law for notched specimens. In return, the fit of size effect test data may provide a way to quantify the internal length from experiments. We use the non-local (integral) isotropic damage model described in Chapter 2 as it is quite simple, with a limited number of model parameters.

We will start from the three-point bend experiments that have been carried out by Le Bellégo *et al.* [LEB 03] on notched geometrically similar specimens of various height D = 80, 160, and 320 mm, of length L = 4D, and of thickness b = 40 mm kept constant. The length-to-height ratio is L/D = 4, the span-to-height ratio is *l*/D = 3, and the notch length is D/10 (Figure 3.10). These are mortar specimens with a water to cement ratio of 0.4 and a cement to sand ratio 0.46 (all by weight). The maximal sand grain size is $d_a = 3$ mm.

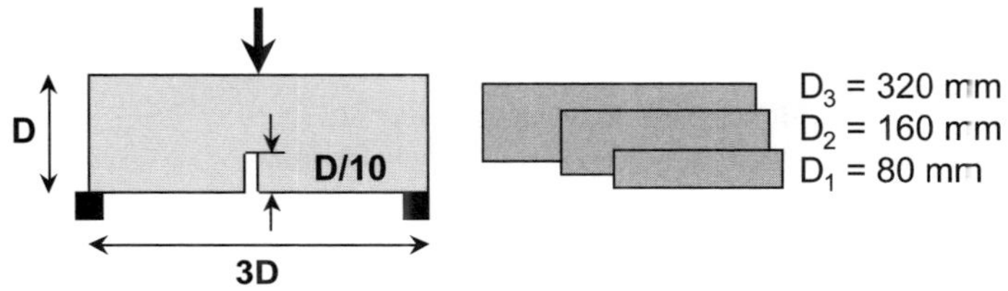

Figure 3.10. *Geometry of the three-point bend size effect tests*

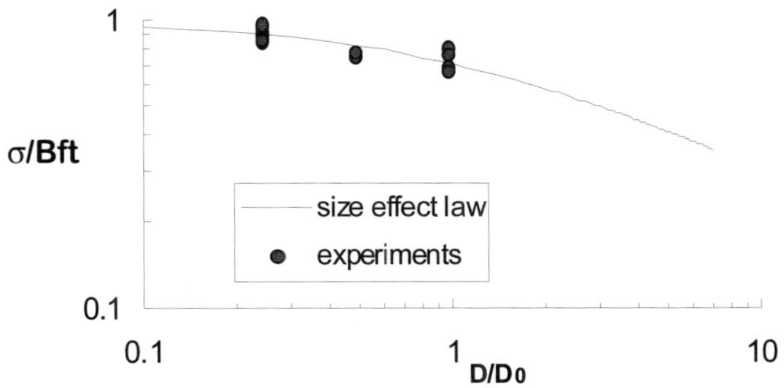

Figure 3.11. *Experimental results of size effect in three-point bend tests (after [LEB 03])*

Experimental and computational data will be interpreted with the help of Bazant's size effect law, in one of its simplest form (equation [3.31]). The nominal strength is obtained with the formula:

$$\sigma_N = \frac{3}{2}\frac{Fl}{b(0.9D)^2} \qquad [3.49]$$

where F is the peak load.

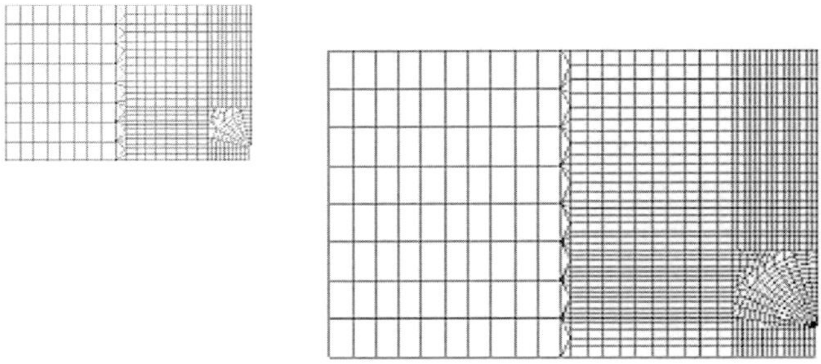

Figure 3.12. *Examples of finite element meshes for small and medium beams*

κ_0	At	Bt	E	ν
3×10^{-5}	0.95	9,000	38,500 MPa	0.24

Table 3.1. *Reference set of model parameters and optimal set from manual calibration*

Figure 3.11 shows the size effect law in a log-log plot. It can be seen that the nominal value of the material strength decreases as the size of the beam increases. This is typical of a damage process in which the process zone is size independent (controlled by the internal length) compared to the structure. For very small sizes, the process zone covers the entire specimen and a material strength criterion is expected, whereas for infinitely large specimens the process zone size is negligible and LEFM should apply.

The description of size effect with the non-local damage model devised in Chapter 2 [MA 84, PIJ 87], and the calibration of the model parameters from inverse analysis on size effect tests rely first on an accurate finite element representation of each size of specimen.

Experience shows that the accuracy of the finite element discretization in the fracture process zone depends on the ratio of the size of the finite element to the internal length. Therefore, geometrically similar finite element meshes should not be used for the calibration. In order to avoid some possible mesh bias, the finite element size in the process zone should be kept constant and small enough compared to the internal length (at least one-third of the internal length). Figure 3.12 shows two examples of finite element meshes. The following computations have been performed with the finite element code CASTEM 2000, in which a secant nonlinear solver is implemented.

3.5.1. *Description of size effect according to the non-local damage model*

The load deflection curves for the three sizes of specimen have been computed and then the parameters in the size effect law were obtained from the linear regression (equations [3.34] and [3.35]). Then, the nominal stress computed according to equation [3.49] has been plotted in the normalized size effect coordinates (σ/Bf_t versus D/D_0).

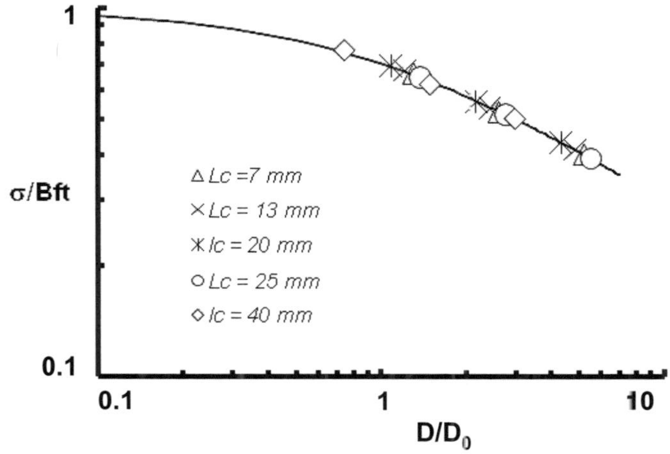

Figure 3.13. *Size effect described according to the non-local damage model for various internal lengths (after [LEB 03])*

Note that for each set of computations (on the three sizes), a different set of parameters Bf_t and D_0 is obtained and used in the normalized size effect plot. The difference between each fit is the location of the corresponding points on the size effect curve. For instance and for a given specimen size D, if D_0 changes from one set of parameter to another, the ratio D/D_0 will change too and the corresponding point in the normalized size effect plot will be shifted on the horizontal axis. The set of points shifts to the right when D_0 decreases, indicating that the response of the beams is more "brittle", i.e. closer to a LEFM criterion. Conversely, when D_0 increases, the set of points shifts to the left of the curve, which means that the response of the beams is more "ductile", i.e. closer to a strength of material criterion.

For the set of model parameters given in Table 3.1, Figure 3.13 shows the fit of the size effect law. In this figure, several values of the internal length have been considered. We may conclude that the computations in the non-local damage model are consistent with the deterministic size effect law. The same would be observed with a gradient damage model. Figure 3.13 also shows that the smaller the internal length, the closer the results to LEFM. The size of the fracture process zone decreases with the internal length and, therefore, the data on the size effect plot shift to the right.

3.5.2. *Calibration of internal length from size effect test data*

Starting from an initial (guessed) set of model parameters, it is possible to optimize this set in order to achieve a best fit of the load deflexion responses for the three different sizes of specimen. The optimization process uses the responses of the three specimen sizes simultaneously. A Levenberg-Marquardt algorithm is implemented. The algorithm performs the minimization of the following functional $\Im(\vec{P})$:

$$\Im(\vec{P}) = \frac{1}{2}\sum_{size=1}^{3}\left[\sum_{\text{measured data points}}\left\{\left(\frac{R^i(\vec{P}) - R^i_{\exp}}{Max(R^i_{\exp})}\right)^T\left(\frac{R^i(\vec{P}) - R^i_{\exp}}{Max(R^i_{\exp})}\right)\right\}\right] \quad [3.50]$$

where $R^i_{\exp}$ is the experimental response for the size i, $i \in [1,3]$, and $R^i(\vec{P})$ is the numerical response for size i corresponding to the vector of model parameters $\vec{P}$. Several definitions can be used for the responses. Here, we use a simple definition: the stress computed from the applied load according to equation [3.49] for 100 values of the deflection equally spaced and covering the experimental data range. Because the experimental and numerical responses have rather different values from one specimen size to another, it is necessary to express them in a non-dimensional way (or to use a weighted functional). This is the reason why the error between the numerical and experimental responses is divided by the maximal experimental response for each size.

When a single size of beam is used as the experimental data set, the non-local model cannot be fitted properly. In order to clearly demonstrate this result, Figure 3.14 shows the results of the automatic calibration on the middle-sized specimen. Three fixed different values of the internal length have been considered. For each one, the optimization has been performed yielding different values of the damage parameters. The computed responses are the same, in good agreement with the experimental data. Bending of notched beams is quite often used for parameter identification because it is relatively easy to perform compared with tensile tests that are very sensitive to boundary conditions [CAR 99]. This figure shows that the model cannot be calibrated from inverse analysis of a single load deflection curve.

Figure 3.15 shows the result of the optimization from the three sizes of specimens simultaneously. The model parameters, with the notations defined in Chapter 2, are given in Table 3.2.

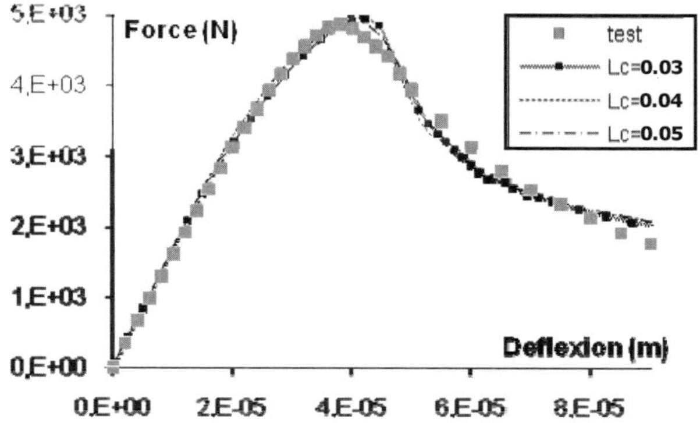

Figure 3.14. *Automatic calibration of the model parameters on the middle size experiment for different fixed internal lengths (after [LEB 03])*

l_c	κ_0	A_t	B_t	E	ν
45.9 mm	2.92×10^{-5}	0.72	10,000	38,500 MPa	0.24

Table 3.2. *Optimal set of model parameters from automatic calibration*

We have tried to change the initial set of model parameter in order to investigate its stability. Overall, the final sets did not differ very much. The largest difference was found on the internal length, which was between 40 and 50 mm. This fitting process provides an optimal set of model parameters, with a single internal length.

3.6. Conclusions

In the design of structural components, a stress is computed and most often compared to a limit strength. This strength is considered to be a characteristic that is not dependent on the structure size. In other words, size effect is neglected in most design codes or recommendations. Except in some special cases, such as diagonal shear failure (where size effect is taken into account in some design codes), the reduction of strength due to size effect between a standardized specimen and the real structure is not critical because safety factors are high enough.

Design regulations and specifications take into account the uncertainty of the material properties, applied loads, and structural characteristics due to the construction process. Better design tools, better material processing and innovative technologies, however, provide more and more confidence during the construction process. Specifications may become easier to reach and safety margins could become too large. As a consequence, one could be tempted to reduce safety factors. Typically, the dead weight of a structure is known, although the safety factor on dead loads is high (1.35). It could be reduced without danger, but if this is done, variations of the nominal strength of concrete due to size effect may become of importance for design. Simple calculations (see [BAZ 04]) show that the decrease of strength due to size effect may compensate the overestimation of the dead load in some structures such as bridges. Therefore, introducing size effect in the design codes may not be a meaningless issue if safety factors are to be reduced in the future.

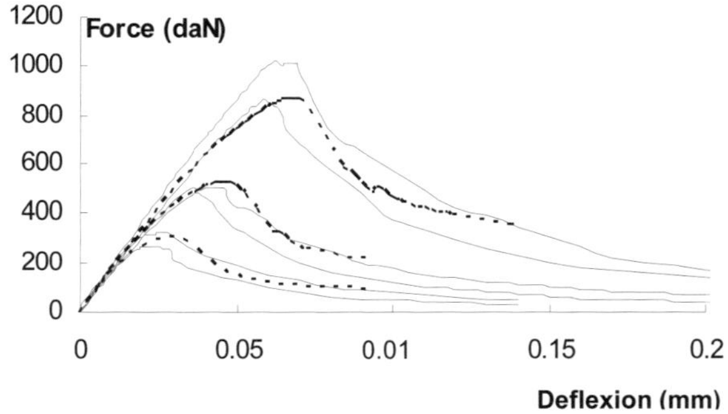

Figure 3.15. *Automatic calibration of the model parameters. The dotted lines are the model responses and the plain lines are, for each size, the minimum and maximum curves obtained from the experimental data (after [LEB 03])*

Size effect is also useful for the validation and the calibration of constitutive models that contain an internal length. Such models should be able to describe deterministic size effect and inverse analysis of the structural size effect test provides a consistent technique for the determination of the internal length of the material. It should be stressed that experiments on a single structure with data restricted to load-displacement records are not sufficient for the determination of the internal length. Some information related to the size of the fracture process zone is required and the size effect provides this information indirectly.

Finally, let us stress that deterministic and probabilistic size effect may be combined in order to achieve a more comprehensive modeling. This issue and others, related to size effect in compression, in reinforced concrete structures, in shear and punching shear failures, or related to the interaction with time dependent phenomena are discussed in detail in the book by Bazant and Planas [BAZ 98]. Some issues are still unresolved and form a very active research field today.

3.7. Acknowledgement

Partial financial support from the ERC-2008-AdG project Failflow (227769) is gratefully acknowledged.

3.8. Bibliography

[BAZ 84] BAZANT Z.P., "Size effect in blunt fracture: concrete, rock, metal", *J. Eng. Mech., ASCE*, vol. 110, pp. 518-535, 1984.

[BAZ 87] BAZANT Z.P., PFEIFFER P.A., "Determination of fracture energy from size effect and brittleness number", *ACI Mater. J.*, vol. 84, pp. 463-480, 1987.

[BAZ 88] BAZANT Z.P., PIJAUDIER-CABOT G., "Non local continuum damage, localisation instability and convergence", *J. Appl. Mech.*, vol. 55, pp. 287-293, 1988.

[BAZ 95] BAZANT Z.P., "Scaling of quasibrittle fracture and the fractal question", *ASME J. Mater. Technol.*, vol. 117, pp. 361-367, 1995.

[BAZ 97] BAZANT Z.P., "Scaling of quasibrittle fracture: asymptotic analysis", *Int. J. Fract.*, vol. 83, pp.1-40, 1997.

[BAZ 98] BAZANT Z.P., PLANAS J., *Fracture and Size Effect in Concrete and Other Quasibrittle Materials*, CRC press, 1998.

[BAZ 99a] BAZANT Z.P., "Size effect", *Int. J. Solids Struct.*, vol. 37, pp. 69-80, 1999.

[BAZ 99b] BAZANT Z.P., "Size effect on structural strength: a review", *Arch. Appl. Mech.*, vol. 69, pp. 703-725, 1999.

[BAZ 04] BAZANT Z.P., *Introduction aux Effets d'Echelle sur la Résistance des Structures*, Hermes Science, 2004.

[BRO 87] BROWN S. R., "A note on the description of surface roughness using fractal dimension", *Geophys. Res. Lett.*, vol. 14, pp. 1095-1098, and vol. 15, pp. 286, 1987.

[CAR 94a] CARPINTERI A., "Fractal nature of material microstructure and size effects on apparent mechanical properties", *Mech. Mater.*, vol. 18, pp. 89-101, 1994.

[CAR 94b] CARPINTERI A., "Scaling laws and renormalisation groups for strength and toughness of disordered materials", *Int J. Solids Struct.*, vol. 31, pp. 291-302, 1994.

[CAR 94c] CARPINTERI A., FERRO G., "Size effects on tensile fracture properties: a unified explanation based on disorder and fractality of concrete microstructure", *Mater. Struct.*, vol. 27, pp. 563-571, 1994.

[CAR 95a] CARPINTERI A., CHIAIA B., FERRO G., "Size effects on nominal tensile strength of concrete structures: multifractality of material liagments and dimensional transition from order to disorder", *Mater. Struct.*, vol. 28, pp. 311-317, 1995.

[CAR 95b] CARPINTERI A., FERRO G., INTERVENIZZI S., "Fracture mechanics of concrete structures", in E.T.H. ZURICH, F.H. WITTMAN, *Proceedings of FRAMCOS-2*, Aedificatio Publishers, pp. 557-578, 1995.

[CAR 95c] CARPINTERI A., CHIAIA B., "Fracture mechanics of concrete structures", in E.T.H. ZURICH, F.H. WITTMAN, *Proceedings of FRAMCOS-2*, Aedificatio Publishers, pp. 581-596, 1995.

[CAR 96] CARPINTERI A., CHIAIA B., "Size effects on concrete fracture energy: dimensional transition from order to disorder", *Mater. Struct.*, vol. 29, pp. 259-266, 1996.

[CAR 99] CARMELIET J., "Optimal estimation of gradient damage parameters from localisation phenomena in quasi-brittle materials", *Int. J. Mech. Cohesive Frict. Mater.*, vol. 4, pp. 1-16, 1999.

[FRE 68] FREUDENTHAL A. M., "Statistical approach to brittle fracture", Chapter 6 in H. LIEBOWITZ , *Fracture*, Academic Press, vol. 2, pp. 591-619, 1968.

[GAL 1638] GALILEO G. L., *Discorsi i Demostrazioni Matematiche intorno à due Nuove Scienze*, Elsevirii, Leiden; (English translation by T. Weston, London (1730)), pp. 178-181, 1638.

[GRI 21] GRIFFITH A. A., "The phenomena of rupture and flow in solids", *Phil. Trans.*, vol. 221A, pp. 179-180, 1921.

[HER 73] L'HERMITE R., "Influence de la dimension absolue sur la résistance à la flexion", *Ann. l'ITBTP*, no. 30-310, pp. 39-41, 1973.

[HIL 76] HILLERBORG A., MODÉER M., PETERSSON P.E., "Analysis of crack formation and crack growth in concrete by means of fracture mechanics and finite elements", *Cement Concrete Res.,* 6, pp. 773-781, 1976.

[LEB 03] LE BELLEGO C., DUBÉ J.F., PIJAUDIER-CABOT G., GÉRARD B., "Calibration of non local damage from size effect tests", *Eur. J. Mech.*, vol. 22, pp. 33-46, 2003.

[LON 91] LONG A. Y., SUQUIN L., LUNG C.W., "Studies of fractal dimension of a fracture surface formed by slow stable crack propagation", *J. Phys. D; Appl. Phys.*, vol. 24, pp. 602-607, 1991.

[MAN 84] MANDELBROT B. B., PASSOJA D. E., PAULLAY A., "Fractal character of fracture surface of metals", *Nature*, vol. 308, pp. 721-722, 1984.

[MAR 1686] MARIOTTE E., *Traité du Mouvement des Eaux et des autres Corps Fluides*, E. Michallet, Paris, 1686. Published posthumously, edited by M. DE LA HIRE (translated into English by J.T. DESVAGILIERS, London 1718).

[MAZ 84] MAZARS J., Application de la mécanique de l'endommagement au comportement non linéaire et à la rupture du béton de structure, PhD thesis, University Paris VI, 1984.

[MIH 77] MIHASHI H., IZUMI M., "Stochastic theory for concrete fracture", CEM *Concrete Res.*, vol. 7, pp. 411-422, 1977.

[MIH 81] MIHASHI H., ZAITSEV J. W., Statistical nature of crack propagation, Section 4-2 in WITTMANN F. H, report of RILEM TC 50 – FMC, 1981.

[PIJ 87] PIJAUDIER-CABOT G., BAZANT Z.P., "Nonlocal damage theory", *J. Eng. Mech. ASCE*, vol. 113, pp. 1512-1533, 1987.

[RIL 90] RILEM, "Size-effect method for determining fracture energy and process zone size of concrete", *Mater. Struct.* vol 23, pp. 461 - 465, 1990.

[SAO 94] SAOUMA V.C., BARTON C.C., "Fractals, fracture and size effect in concrete", *ASCE J. Eng. Mech.*, vol. 120, pp. 835-854, 1994.

[TAD 85] TADA H., PARIS P.C., IRWIN G.R., *The Stress Analysis of Cracks Handbook* 2nd Edition, Paris Production, St Louis, 1985.

[VIN 1500s] DA VINCI L., *The Notebooks of Leonardo da Vinci (1954)*, E. MCCURDY, London, pp. 546, 1500s.

[WEI 39] WEIBBUL W., "The phenomenon of rupture in solids", in *Proceedings of Royal Swedish Institute of Engineering Research*, Stockholm, vol. 153, pp. 1-55, 1939.

[WEI 49] WEIBBUL W., "A statistical representation of fatigue failures in solids", in *Proceeding of the Roy. Inst. Of Tech.*, no. 27, 1949.

[WEI 51] WEIBBUL W., "A statistical distribution function of wide applicability", *J. Appl. Mech. ASME*, vol. 18, pp. 293-297, 1951.

[WEI 56] WEIBBUL W., "Some observations of Leonardo, Galileo, Mariotte and others relative to size effect", *Ann. Sci.*, vol. 13, pp. 23-29, 1956.

[YOU 1807] YOUNG T., *A Course of Lectures on Natural Philosophy and the Mechanical Arts*, vol. I, London: William Sauvage [for Joseph Johnson], pp. 144, 1807.

[ZAI 74] ZAITSEV J.W., WITTMANN F.H., "A statistical approach to the study of the mechanical behavior of porous materials under multiaxial state of stress", in *Proceedings of the 1973 Symp. on Mechanical Behavior of Materials*, Kyoto, Japan, 1974.

[ZEC 77] ZECH B., WITTMANN F.H., "A complex study on the reliability assessment of the containment of a PWR, Part II. Probabilistic approach to describe the behavior of materials", in T.A. JAEGER, B.A. BOLEY *Proceedings of the 4th Int. Conf. on Structural Mechanics in Reactor Technology*, European Communities, Brussels, vol. H., J1/11, pp. 1-14, 1977.

PART 2

Concrete Under Cyclic and Dynamic Loading

Chapter 4

Cyclic Behavior of Concrete and Reinforced Concrete

4.1. Characterization tests of the cyclic behavior

Concrete resists compression stress well whereas it resists traction stress badly. This is why tensile strength is often ignored in order to proceed to the dimensioning of reinforced structures. This behavior difference generates a stress dissymmetry for a symmetric deformation as is the case for simple flexion. Usual structures are mainly stressed due to a monotone loading. Nevertheless, there are many cases where loading is alternated: wind action, earthquake action, stress of rotating machines, etc. Among these different types of loading, seismic action is probably the most important. The force levels reached are likely to damage structures because the first eigen frequency of a regular building ranges from 1 to 10 Hz. It corresponds to the fundamental frequency of earthquakes or wind approximately.

Both earthquakes and wind are loadings that are quite tough to reproduce, and demand complex experimental facilities. The random characteristics of seismic or wind signals are difficult to reproduce in the framework of a predictive test. This is why many studies have been carried out on the cyclic behavior of reinforced concrete structures. This type of experiment does not take into account the inertia effects but allows the identification of the phenomena that occur on structures submitted to alternated loadings, such as earthquakes. Load control systems are then quite simple, it is easy to repeat these tests and a parametric study can be carried out with controlled boundary conditions. Although experiments on the seismic behavior

---

Chapter written by Jean-François DUBÉ.

of structures can be envisaged, these require substantial means of servo control and acquisition [CAR 93; BRA 94].

In this chapter, we present some tests and results on the cyclic behavior of concrete in order to define the tension-compression relationship, and especially, the stiffness restitution due to crack closure, called the "unilateral characteristic" of the material. We will then focus on studies of reinforced beams and multi-storey buildings showing the overall effects of the local material response. Finally, we will discuss the consequences of an irregular seismic type of loading on the behavior of a reinforced concrete column.

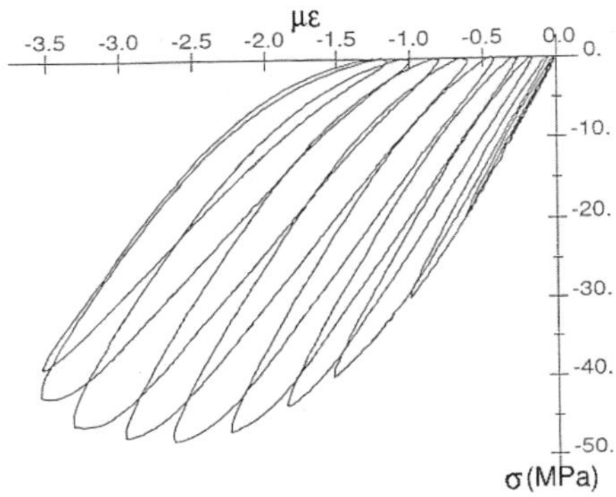

Figure 4.1. *Compressive cyclic behavior of concrete [RAM 90]*

4.1.1. *Uniaxial cyclic behavior*

Three more or less different complex types of tests characterizing the uniaxial behavior of concrete can be found: 1) cyclic compression, 2) cyclic tension, and 3) alternated tension-compression. The simplest studies consist of studying separately the tensile response and the compressive response of the material. These tests are displacement or strain controlled in order to obtain the post-peak behavior of concrete (Figures 4.1 and 4.2). Tensile and compressive responses show the same trends:

– the envelope curve is almost not affected by the cyclic loading;

– loading-unloading hysteresis loops are observed and do not change for small number of cycles;

– the apparent Young's modulus decreases when deformation increases;

– there is an increasing inelastic deformation with the total deformation.

The tensile response, however, is quasi-elastic until the peak is reached, whereas on the compressive response, inelastic deformations appear before the peak along with the loading-unloading hysteresis loops. We can then use some theoretical modeling, such as continuous damage, for the decrease of Young's modulus and plasticity for inelastic deformations.

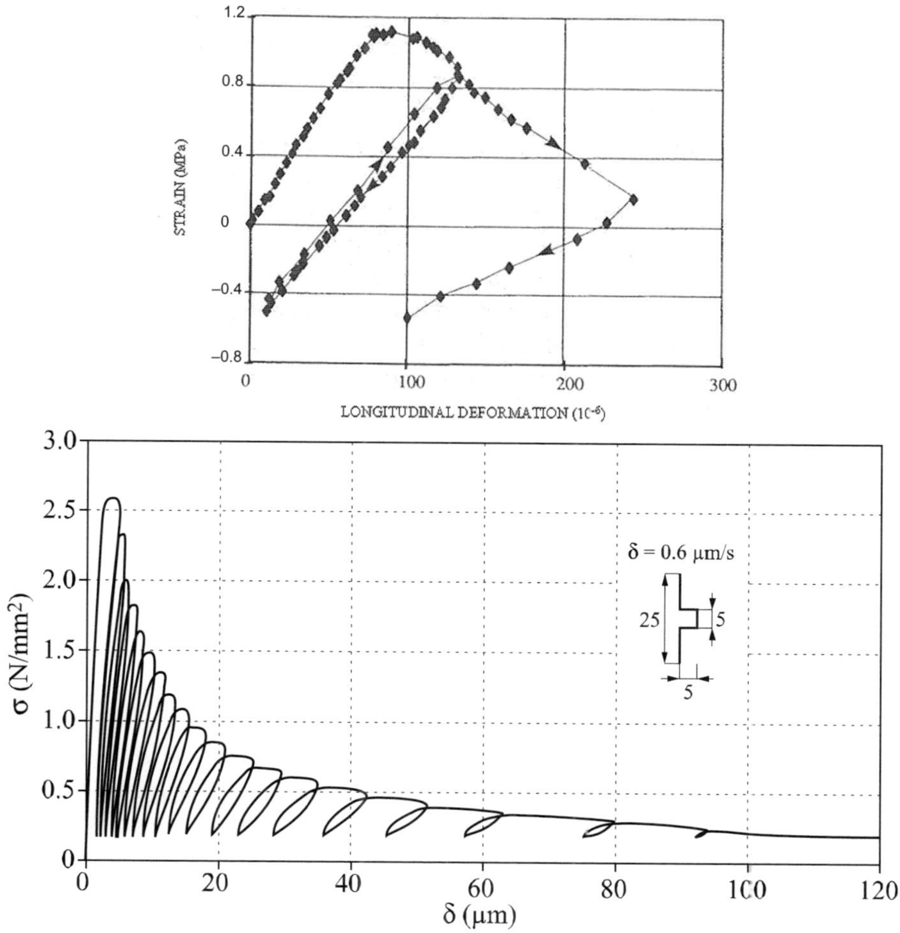

Figure 4.2. *Loading-unloading response of concrete in uniaxial tension [MAZ 89; REI 84]. Specimen without notch (top) and specimen with a notch (bottom). In the bottom experiment the test is notch opening controlled. The notch geometry is schematically shown in the figure*

These tests give valuable information on the mechanisms involved in the mechanical response of concrete but they have to be completed with alternated cyclic tests. These tests are more complex and more difficult to carry out because of the application of alternated loads. Strain localization and thus cracking, is either controlled and occurs at the tip of the notch in the specimen [REI 84], or it is avoided and cracking is distributed over the specimen (Figure 4.3) [MAZ 89].

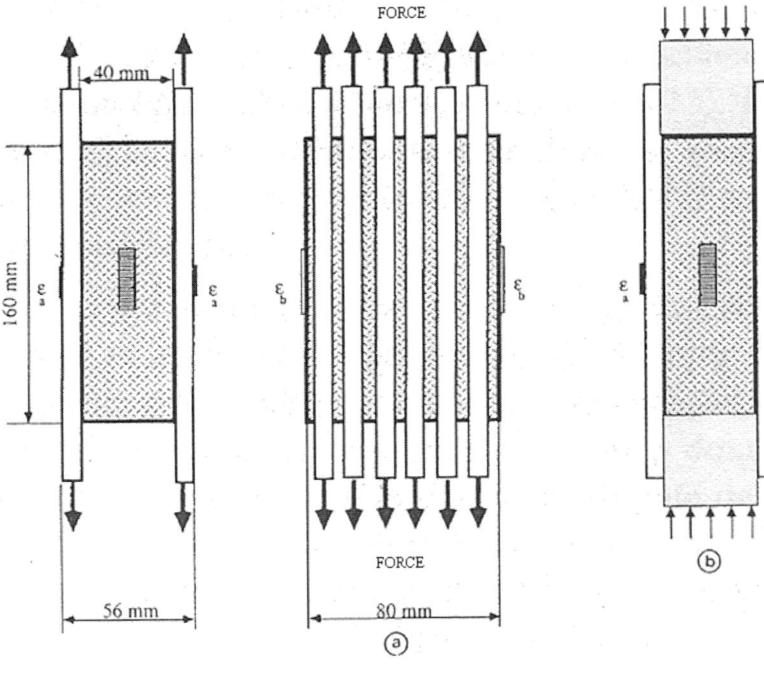

Figure 4.3. *PIED test specimen [MAZ 89]*

Monotonic and alternated tests give the same overall results (Figures 4.2 and 4.4). The decrease of tensile Young's modulus and the associated growth of continuum damage do not affect the compressive response. We also observe a zone where there is a progressive restitution on stiffness between tension and compression. The usual interpretation suggests that cracks close progressively (due to their roughness), and a compressive stress (closure stress) has to be applied in order to recover the stiffness completely and reach a full crack closure. These tests have validated the assumptions that tension and compression damage are not of the same nature (Figure 4.5). Tensile cracks are perpendicular to the direction of the stress and occur because of longitudinal deformation, whereas compression cracks are parallel to the direction of stress and occur due to the positive deformation induced by the Poisson effect.

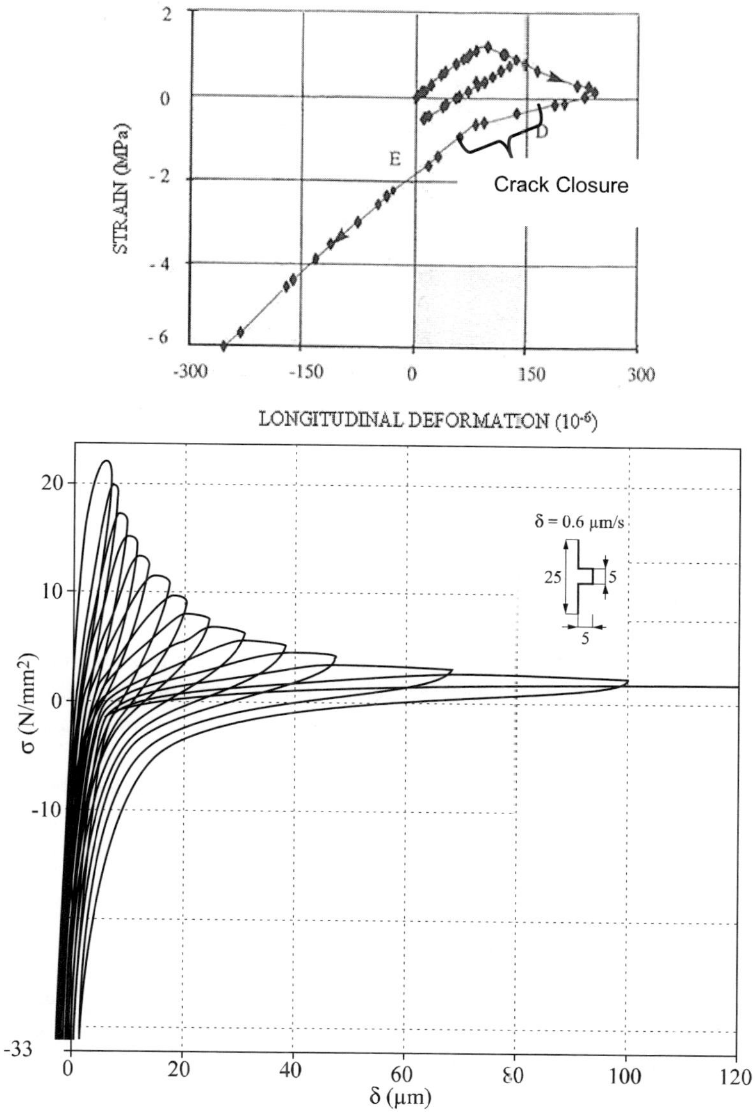

Figure 4.4. Tension-*compression curve of a damaged concrete [MAZ 89, REI 84]. Top: detail of the crack closure in the PIED test. Bottom: tension compression curve of a notched specimen*

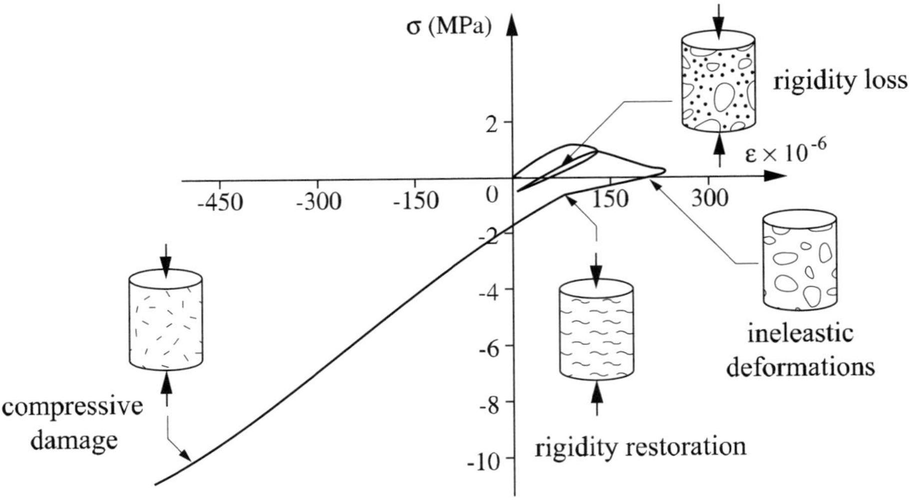

Figure 4.5. *Unilateral behavior of concrete [RAM 90]*

4.1.2. *Cyclic tests on structural elements*

Many studies deal with the cyclic behavior of beams, frames, and panels [GAU 78; GAU 79; DEL 88; FOU 90; MIN 95; GHO 01]. We will present a few examples of each type to illustrate that fact the different outputs from these experiments are similar. As the different tests are meant to represent the behavior of a structural element within its environment, compression is generally applied. Nevertheless, compression is not important enough to confine concrete. Tests provide the response of the structural elements subjected to bending or shear loads, i.e. with a tensile (stress or strain) component.

4.1.2.1. *Tests on beams and columns*

Many tests are concerned with the cyclic behavior of beams or beam-column joints [GAU 78; DEL 88; MIN 95; GHO 01]. La Borderie and co-workers performed a cyclic test on reinforced concrete beams along with the numerical modeling described in the next section [LAB 91]. It is actually a reinforced concrete beam of length 1.7 m. The loading is three-point bending with a span of 1.5 m. the bean cross-section is 0.2 m×0.15 m (Figure 4.6). Reinforcement is made of four HA12 bars. Plates are placed at the support level in order to overcome problems of localized rupture. An increasing cyclic displacement is applied at the center of the beam. The force-deflection response shows an elastic phase, a positive hardening, and an elastic unloading phase with a decrease of apparent stiffness (Figure 4.7). Stiffness variation is such that the global response shows a pinching of the cyclic response with evolving growing permanent displacement.

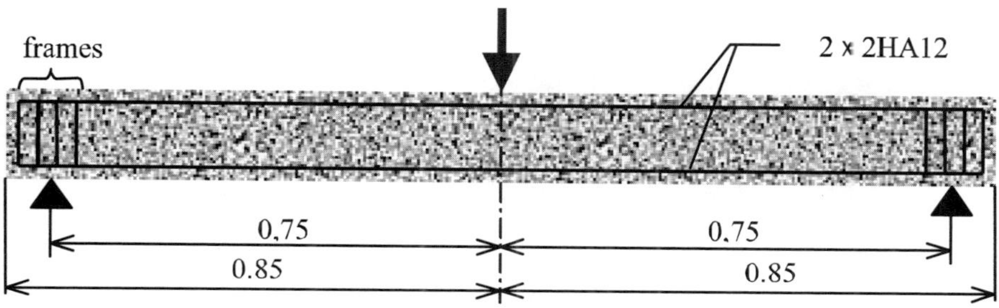

Figure 4.6. *Three-point bending beam [LAB 91]*

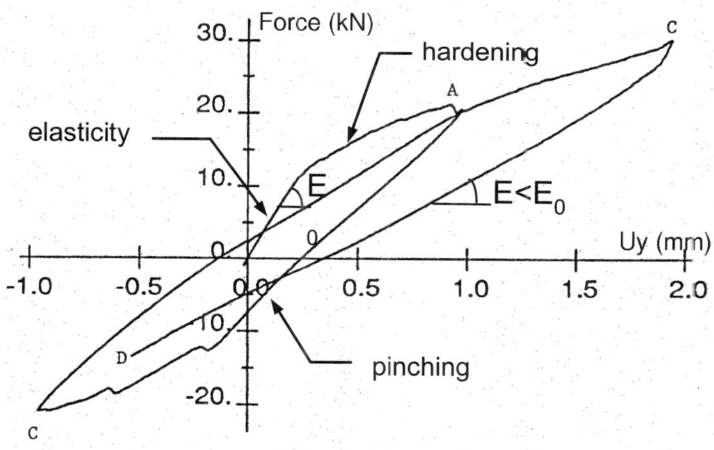

Figure 4.7. *Force-deflection (test) evolution [LAB 91]*

The test presented in Figure 4.8 illustrates the characterization of the cyclic response of a beam-column joint [GHO 01]. Figure 4.8a shows the test of a column loaded in compression, 3 m height and 250 mm × 400 mm, on which a beam with a 1.75 m length and the same section is embedded. The beam is loaded by an increasing cyclic displacement at its end. Reinforcement is made of 6HA20+2HA16 bars for the column and 4HA20 longitudinal bars at both the top and bottom of the beam. Shear reinforcements are made of frames and stirrups with a ∅10 mm spacing in compliance with the DTU regulation, the beam shear reinforcements are similar with a 150 mm spacing over 600 mm from the clamped end and then 200 mm spacing.

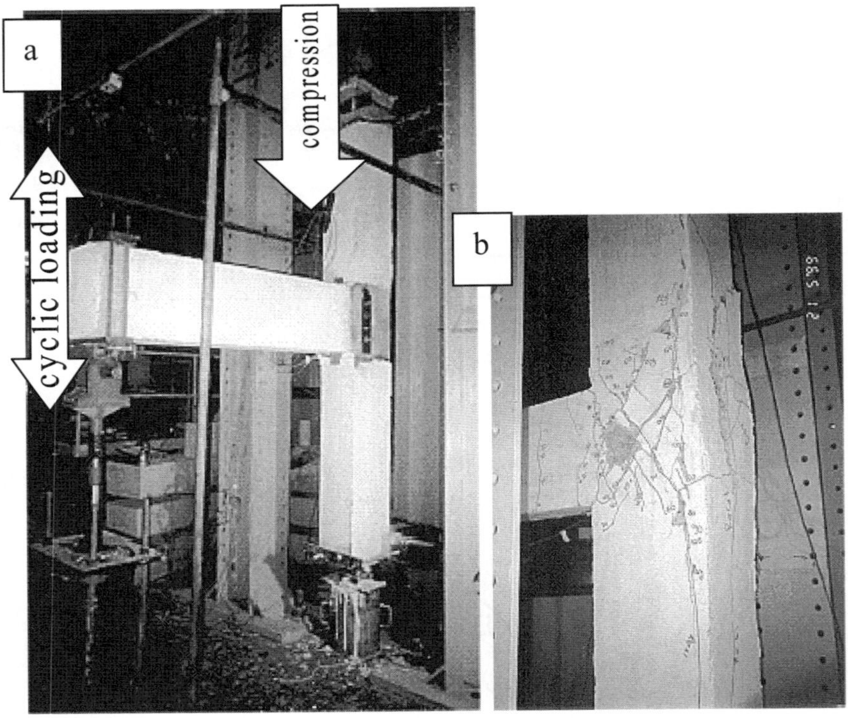

Figure 4.8. *Beam-column bond cyclic test [GHO 01]*

Figure 4.8b shows that bending vertical cracks develop, along with shear cracks and with the degradation of the steel-concrete bond. The force-displacement curve shows an elastic unloading with a small variation of slope at the beginning. When the force is low, the unloading slope decreases which gives a pinching to the curves (Figure 4.9). This variation of slope is associated with the compression-traction interaction: Tensile cracks are not totally closed, whereas the compressed part of the beam switches to tension. The global cyclic behavior is mainly ruled by the compressive behavior with inelastic displacements. These displacements result from the inelastic deformations of the material.

Cyclic statistic tests can be complemented with dynamic tests. The study presented and carried at the "Commissariat à l'Energie Atomique" (CEA) in 1978

[GAU 78] investigated the seismic behavior of columns and buildings clamped at their base (Figure 4.10).

Beams at a 1/3 scale have a 17 cm × 17 cm squared section for a 1.44 m height. Reinforcement is made of 12 HA8 bars fixed to the footing and of 29 Ø4 shear reinforcements associated to Ø2.4 pins. The fundamental frequency of the structure is 5 Hz. It is located within the most important zone of the used spectrum (San Francisco and Taft). Columns are then submitted to increasing earthquake levels resulting in the increase of relative deflections. Accelerations at the top and bottom of the column allow the forces to be determined. Temporal measurements of the displacements in four points of the column give information on its cyclic response. Force-displacement cyclic curves are complemented with analyses on the evolution of the frequencies and damping. We can observe that the natural frequency of the column decreases when deflection increases (Figure 4.11).

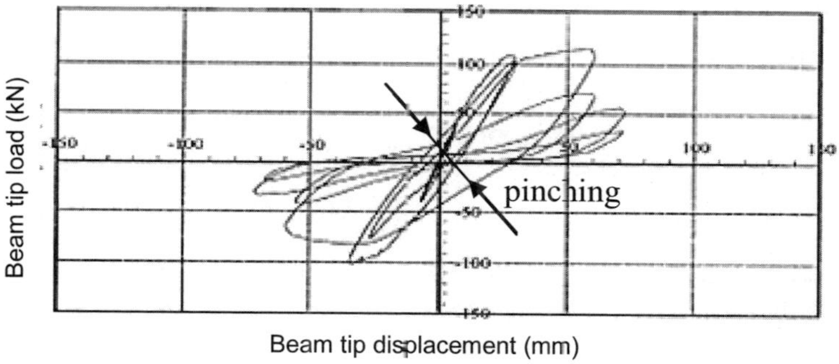

Figure 4.9. *Force-displacement evolution at the beam end [GHO 01]*

As the mass of the column remains constant, it is the stiffness of the column that decreases, and this is consistent with the previous results of the decrease in Young's modulus when deformation increases. Repetition of the same earthquake yields a small variation of the eigen frequency (Figure 4.12). The cycles remain approximately similar after the first one, as observed in the material testing.

4.1.2.2. *Tests on panels*

The tests performed on panels at the CEBTP between 1990 and 1992 show the various collapse modes for different configurations in order to elucidate the cyclic response of shear concrete walls [FOU 90]. The test program included tests on a panel with bending and shear reinforcements (PV1) and four tests on non reinforced panels or piers (except for compression reinforcements) denoted PV2 to 5. The base of the panel was fixed to the ground and a mobile metallic trimmer was fixed to its upper part. This trimmer was subjected to an imposed cyclic horizontal displacement. The imposed displacement was either centered at the middle or at the top of the panel, which generated bending moments (Figure 4.13).

The tests provide observations on the collapse modes of the panels as well as global and local measurements. The force-displacement curves show the dissipative response of the panels during the loading cycles (Figure 4.14). The cycles are slightly symmetric and do not change for an imposed displacement. Stiffness loss is also observed, same as in other experiments. Observations on the cracks and their propagation provide valuable information on the mechanical properties of the panel. Thus, initially, the strongly reinforced panel develops a bending horizontal crack at the base, which remains limited (Figure 4.15). Subsequently, an inclined shear crack suddenly spreads and leads to the plastic strain in the longitudinal reinforcements. Finally, horizontal cracks due to the reinforcements propagate at the height of the panel. Regarding the piers, the bending crack at the base spreads from the edges to the center of the panel. Finally, concrete is crushed at the corners of the panel (Figure 4.16).

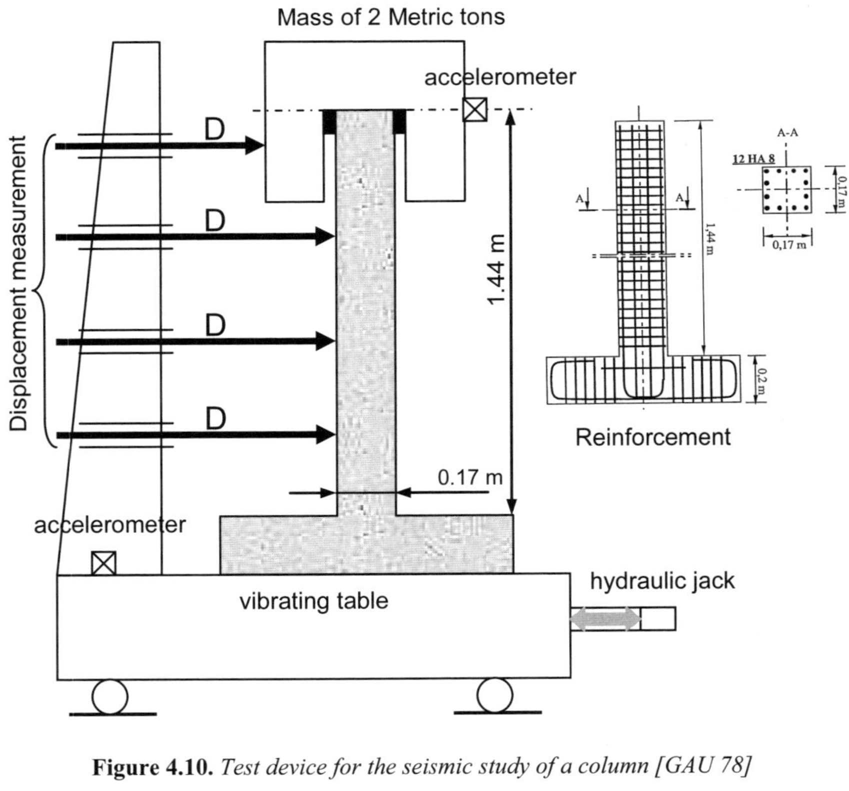

Figure 4.10. *Test device for the seismic study of a column [GAU 78]*

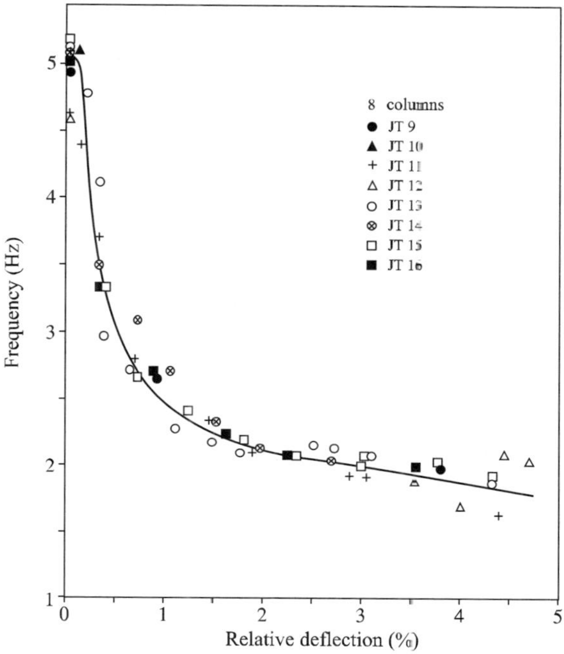

Figure 4.11. *Evolution of the basic frequency of the column as a function of the maximal deflection [GAU 78]*

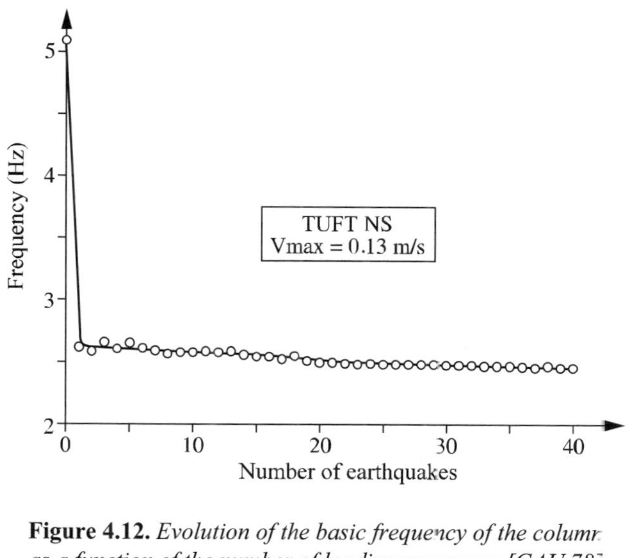

Figure 4.12. *Evolution of the basic frequency of the column as a function of the number of loading sequences [GAU 78]*

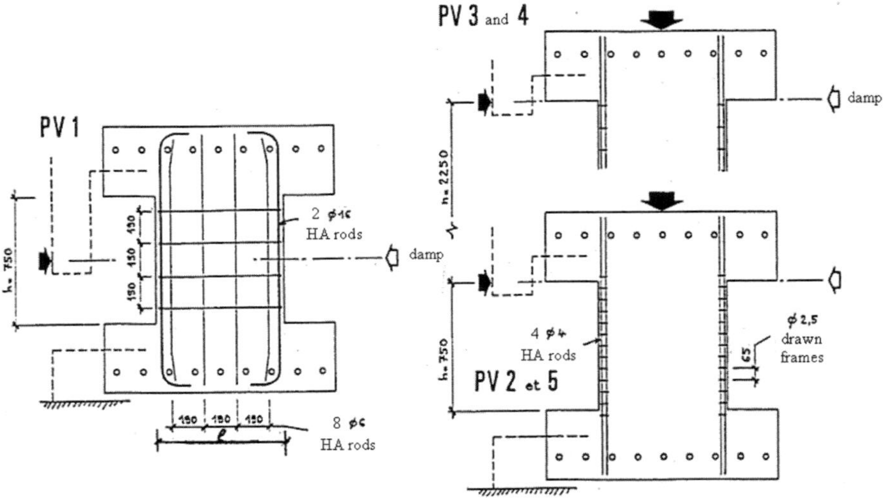

Figure 4.13. *Test device for the study of the cyclic behavior of the panel [FOU 90]*

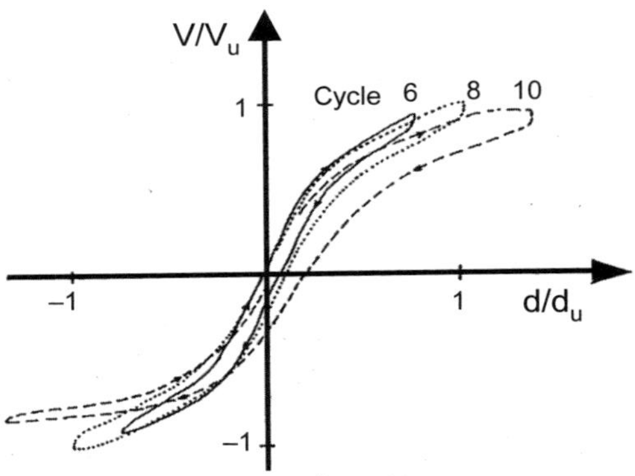

Figure 4.14. *Force-displacement curve of a reinforced panel (d_u is the displacement corresponding to the maximal loading V_u) [FOU 90]*

4.1.3. *Conclusions*

Material cyclic tests are complex to carry out and the deformation of the material may not be controlled due to strain localization. Tests on structures or on the elements of the structures demand quite significant means but are useful, in order to obtain the response of structural element and at the same time to provide some information on the local material response. Static tests are cheaper and can be more easily controlled. Dynamic tests, which provide a more realistic loading, complement static tests. Several observations can then be made:

– the significant inelastic compressive and tensile deformations;

– the degradation of stiffness of the material;

– the crack closure upon load reversal, the phenomenon induces the pinching of the load displacement curves;

– the loading-unloading hysteresis.

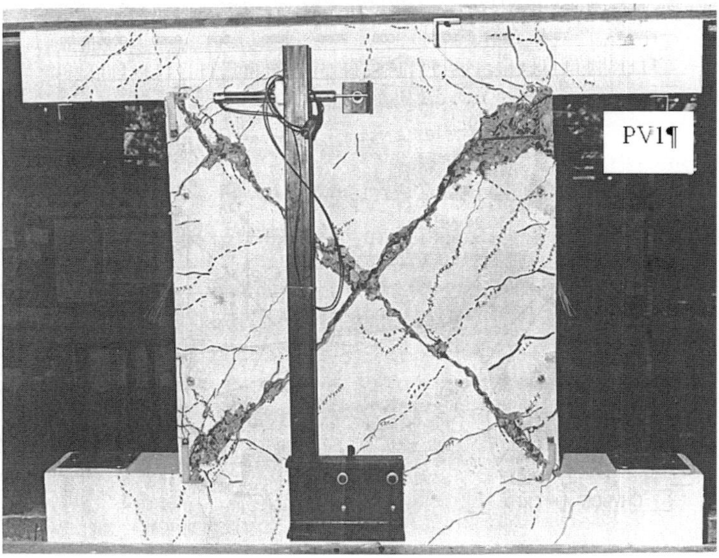

Figure 4.15. *Photograph of reinforced panel after the test [FOU 90]*

4.2. Modeling the reinforced concrete cyclic behavior

In this section, we will show how different authors have approached the modeling of the various phenomena enumerated above. We will use the basic model defined in Chapter 2, in particular damage and plasticity bases models. The three constitutive models discussed here are applicable within a well-defined range of loading conditions. These models do not always apply to all the situations. Nevertheless, they are accurate and simple enough. They are therefore good compromises.

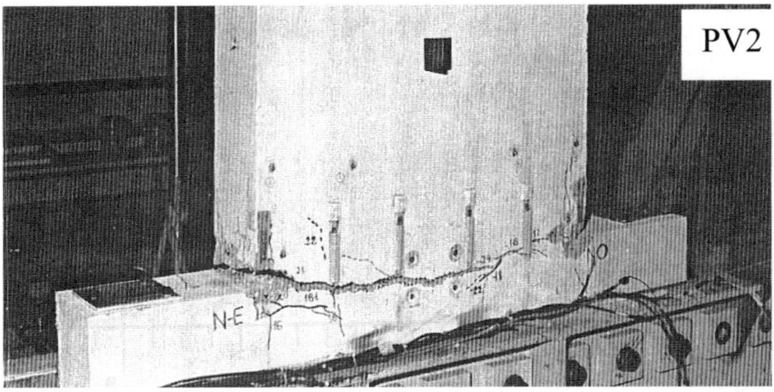

Figure 4.16. *Photograph of a unreinforced pier after the test [FOU 90]*

4.2.1. *Thermodynamic-based model with crack closure*

The model developed by La Borderie [LAB 91] is used in the simulations of section 4.3 at the end of this chapter. It describes:

– the envelope curve of the tensile and compressive behavior of concrete;

– stiffness loss;

– crack closure effect;

– inelastic deformations;

but it does not describe the elastic hysteresis: the tensile or compressive loading-unloading curves are linear. We can find from the same type of assumption in the works of many authors [NAH 86; SIM 88; MER 90].

This model is based on the thermodynamics of irreversible processes and uses the idea of continuous damage. It is written within the principal stress reference frame, which is here equivalent to the principal strain reference frame. The principal stress tensor is made of its positive part σ^+ (positive terms of the diagonal: traction) and its negative part σ^- (compression) [LAD 83].

The deformation tensor ε comes from a potential function χ (Gibbs free energy). Derivation of this potential with regard to the stress tensor σ gives a deformation split into two parts: ε^e representing the elastic deformation and ε^{an} representing the inelastic deformation.

$$\varepsilon = \varepsilon^e + \varepsilon^{an}$$

with:

$$\varepsilon^{e} = \frac{\sigma^{+}}{E_0(1-D_t)} + \frac{\sigma^{-}}{E_0(1-D_c)} + \frac{\nu}{E_0}(\sigma - (Tr\sigma)\mathbf{1})$$

the "elastic deformation" term of damaged material and:

$$\varepsilon^{an} = \frac{\beta_t D_t}{E_0(1-D_t)} F(\sigma) + \frac{\beta_c D_c}{E_0(1-D_c)} \mathbf{1} \qquad [4.1]$$

the "inelastic deformations" term.

E_0 is the initial Young's modulus, ν is Poisson coefficient. D_t and D_c are the tensile and compressive variables, respectively.

β_t and β_c are constants and F($\boldsymbol{\sigma}$) a function that describes progressive crack closure. The crack closure function F($\boldsymbol{\sigma}$) is piecewise, linear, and defined on three domains: tension (opened cracks), tension-compression (progressive opening-closure of the cracks), and compression (closed cracks).

Tension	Opened cracks	$Tr\sigma \in [0, +\infty[\Rightarrow F(\sigma) = 1$
Compression	Closure of cracks	$Tr\sigma \in [-\sigma_f, 0] \Rightarrow F(\sigma) = \left(1 + \frac{Tr\sigma}{\sigma_f}\right)\mathbf{1}$
	Closed cracks	$Tr\sigma \in]-\infty, -\sigma_f] \Rightarrow F(\sigma) = 0.1$

Table 4.1. *Definition of the crack closure function*

Evolutions of the tension and compression damages are controlled by the damage rates, Y_i, obtained using the derivation of the potential, χ, with regards to D_i (i=t or c). Evolution of damage is written as:

$$D_i = 1 - \frac{1}{1 + \left[A_i\left(Y_i - Y_{0_i}\right)\right]^{B_i}} \quad \text{with } i=t \text{ or } c \qquad [4.2]$$

Figure 4.17 shows the response of this model for two tension-compression cycles. This curve shows the tension damage phase with stiffness loss and strain softening. The crack closure function is triggered when the stress goes from zero, in order to recover the undamaged compression response starting at stress σ_f. Then concrete is damaged in compression. During the second cycle: tensile loading and unloading follow the same trend.

4.2.2. *Distributed cracking models*

Distributed cracking models [WEI 98] can be split into two families: fixed crack models [RAS 68] and rotating crack models [ROT 88], (see also Chapter 2).

In the first family, the direction of the cracks remains fixed after initiation. In the second family, the principal stress directions are the same as the principal strain directions and thus crack "rotate" during this process (see Chapter 2). Foster *et al.* [FOS 96] proposed, under plane strain, an incremental formulation with a uniaxial equation between stress and strain:

$$\varepsilon_{iu} = \sum_{j=1}^{n} \frac{\Delta\sigma_i^j}{E_i^j} \qquad [4.3]$$

with ε_{iu} the equivalent strain in the direction i, which is equal to the deformation when the stress in the other directions is zero. $\Delta\sigma_i^j$ is the stress increment in direction i with the loading increment j, and E_i^j is the associated tangential modulus. The material response is then controlled by the evolution laws of the tangent modulus. To do so, it uses the monotonic evolution under compression proposed by Saenz [SAE 64; DAR 77], connecting the principal stress in direction i to the uniaxial equivalent strain:

$$\sigma_i = \frac{\varepsilon_{iu} E_0}{1 + \left(\dfrac{E_0}{E_{sp}} - 2\right)\dfrac{\varepsilon_{iu}}{\varepsilon_{ip}} + \left(\dfrac{\varepsilon_{iu}}{\varepsilon_{ip}}\right)^2} \qquad [4.4]$$

E_0 is the initial modulus; E_{sp} is the secant modulus at the peak stress, also defined by its associated deformation ε_{sp}. Under tension, the response is linear up to the peak. The post-peak behavior is described by a bi-linear function (Figure 4.18).

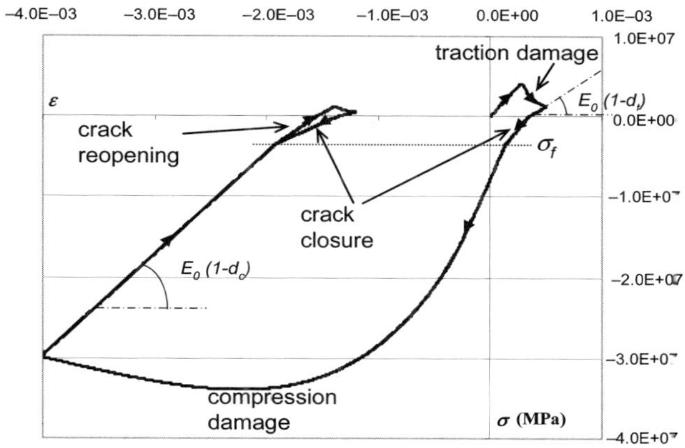

Figure 4.17. *Uniaxial response to the model with two damage variables and crack closure effect*

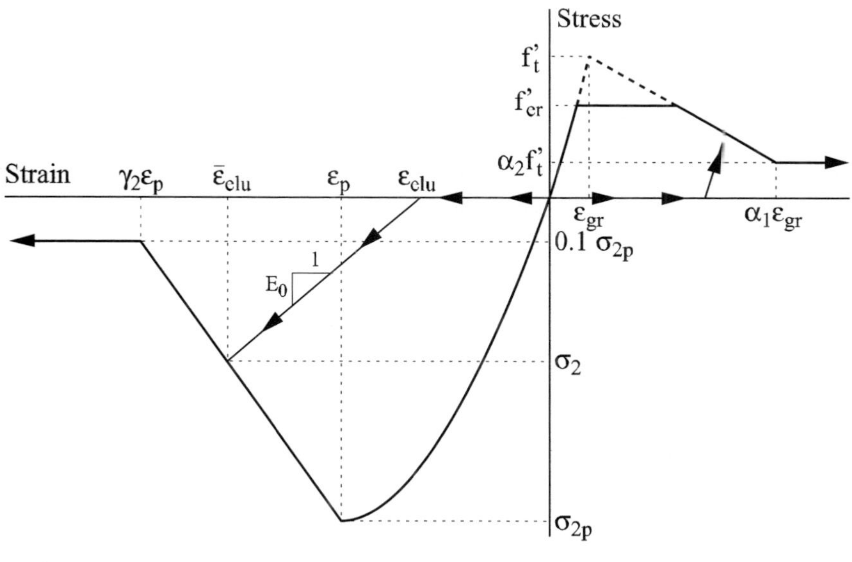

Figure 4.18. *Uniaxial stress-strain response of concrete [FOS 96]*

Figure 4.18 shows the uniaxial behavior of concrete with an unloading of tensile initial modulus E_0 and a variable compressive unloading modulus E_a function of the unloading point. We observe tensile and compressive inelastic deformations, the dissymmetry of the material response between tension and compression and the crack closure effect.

4.2.3. *Microplane model*

This approach is initially due to Taylor [TAY 38] and has been developed for cristal plasticity [BAT 49] and then has been applied to progressive cracking in concrete by Bazant [BAZ 84]. The model initially developed by [BAZ 88] has been generalized to cyclic behaviors [OZB 92] and works in three dimensions with a non-local strain driving damage growth. In this approach, the material response is derived from a set of constitutive relations defined on a given number of planes at each material point. The planes are defined by their normal vector $\vec{n}$, they are called microplanes.

Equations connecting the components of the deformation vector to the components of stress vector, for every plane are defined. In order to obtain a unique solution upon softening, a kinematic approach, where the deformations of the microplanes are the projections of the macroscopic deformation tensor, is implemented. The normal component of the strain vector, subscript N, is split into a volumetric part and a deviatoric part, the tangential component, subscript T, is split in two parts within the tangent plane [BAZ 88]:

$$\begin{aligned} \vec{\varepsilon}_N &= (\varepsilon_V + \varepsilon_D)\vec{n} \\ \vec{\varepsilon}_T &= \varepsilon_M \vec{m} + \varepsilon_K \vec{k} \end{aligned} \quad [4.5]$$

We thus have the following equations between the projections of the strain tensor and the stress vectors on each plane:

$$\begin{aligned} \dot{\sigma}_V &= E_V \dot{\varepsilon}_V, \dot{\sigma}_D = E_D \dot{\varepsilon}_D \\ \dot{\sigma}_M &= E_M \dot{\varepsilon}_M, \dot{\sigma}_K = E_K \dot{\varepsilon}_K \end{aligned} \quad [4.6]$$

where E_V, E_D, E_M, and E_K are tangent moduli. Determination of the macroscopic stress is carried out using the principle of virtual work:

$$\frac{2\pi}{3}\sigma_{ij}\varepsilon_{ij}^* = \int_S \left(\sigma_N \varepsilon_N^* + \sigma_M \varepsilon_M^* + \sigma_K \varepsilon_K^*\right)\Omega(\vec{n})dS \quad [4.7]$$

$\varepsilon_{ij}^*, \varepsilon_N^*, \varepsilon_M^*, \varepsilon_K^*$ are the virtual deformations. The left hand-side of equation [4.7] represents the macroscopic work within a material sphere of unit radius. The right hand-side part represents the work on the surface of the same sphere. $\Omega(\vec{n})$ is a weight function which may be introduced in order to describe the initial anisotropy of the material. For a material which is isotropic initially, we have $\Omega(\vec{n})=1$.

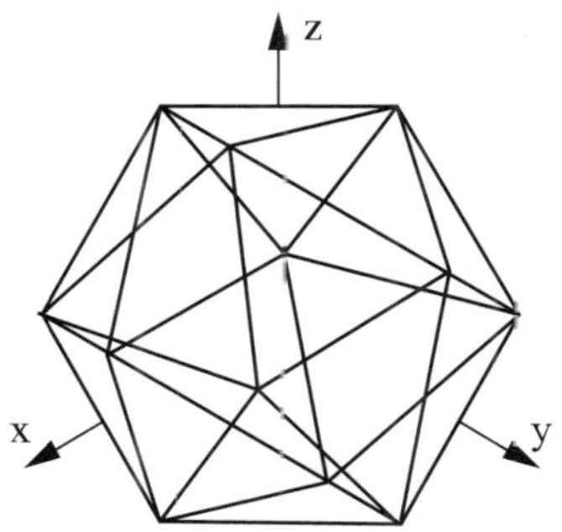

Figure 4.19. *Spatial discretization of the sphere using 21 integration points [OZB 01]*

The definition of the number of microplanes is still subject to controversy. Bazant and Oh considered that for a sphere, 21 integration points (Figure 4.19) are sufficient to describe the three-dimensional behavior correctly [BAZ 86]. Fichant *et al.* considered orthotropy was enough and reduced the scheme to a set of three microplanes [FIC 99].

The model developed by Ozbolt [OZB 01] captures the tensile and compressive material response with elastic hysteresis loops (Figure 4.20). We observe inelastic deformations and a slight difference between the monotonic response and the cyclic one.

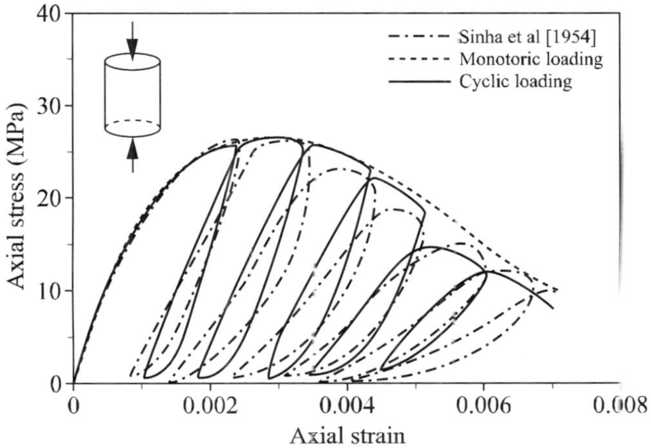

Figure 4.20. *Uniaxial compressive response: comparison of the experiments of Sinha et al. with three-dimensional simulations [OZB 01]*

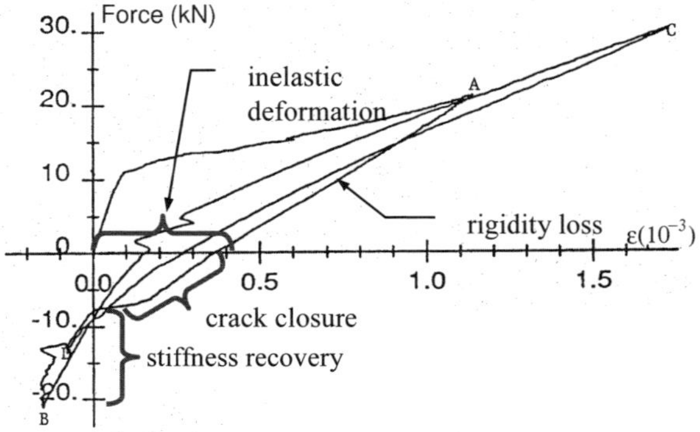

Figure 4.21. *Lower steel force-deformation evolution (test) [LAB 91]*

4.3. Modeling of the cyclic behavior of concrete

In this section we will illustrate how the characteristics of the cyclic behavior of concrete influence the response of structural elements. The constitutive models models defined in Chapter 2 and in section 4.2 will be used to perform the simulations using finite elements.

The examples are a reinforced concrete beam submitted to a three-point bending loading and the study of a building submitted to seismic excitation. The aim is to illustrate the impact of the characteristics observed in cyclic simple material tests. We show that the user can choose the sensitivity of the description of the material model as a function of the problem to be dealt with.

4.3.1. *Characteristics of the cyclic behavior and their influence*

Different characteristics of the behavior of concrete are considered:

– evolution (decrease) of Young's modulus during loading;

– inelastic deformations;

– crack closure;

– elastic hysteresis.

We will then look over the impact of these different phenomena on the response of a structural reinforced concrete element.

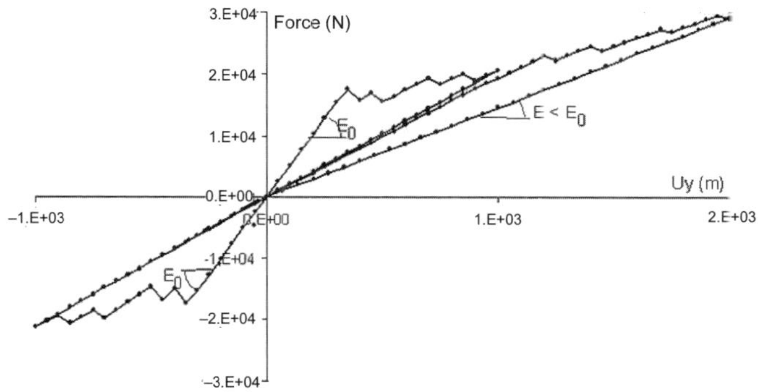

Figure 4.22. *Force-deflection evolution (model 1)*

The beam used in the simulations has been tested by La Borderie [LAB 91] (Figure 4.6). In order to compute the response of the beam, we use a finite element code with multilayered beams developed at LMT-Cachan called EFiCoS [LAB 91; DUB 94]. Each beam element is made of layers displaying a uniaxial behavior. The steel-concrete bond is assumed to remain perfect. The following sections will deal with the influence of the first three phenomena listed above only. Hysteresis will be considered later on. We will then discuss the effect of loading rate on the cyclic response.

We apply an increasing cyclic displacement at the center of the beam. The force-deflection response shows an elastic phase, a positive hardening, and an elastic unloading with an apparent loss of stiffness (Figure 4.7). The variation of stiffness due to crack closure occurs with a pinching of the cyclic response and there is a slightly evolving permanent displacement. The deformations within the lower reinforcement at the center of the beam remain lower than the elasticity limit of steel (Figure 4.21). The slope of force-displacement curve provides the initial stiffness. Local analysis shows the phenomenon of stiffness recovery.

The uniaxial response of the material is slightly "hidden" in the global response of reinforced concrete structure. Indeed, the crack closure effect is not very important on the global response. This is due to the reinforcements which remain elastic. The top and bottom longitudinal reinforcements are identical and the bending response is almost symmetric.

4.3.1.1. *Damage model-1*

We compute the response of the beam using damageable damage model with crack closure and without any inelastic deformation designated "model 1".

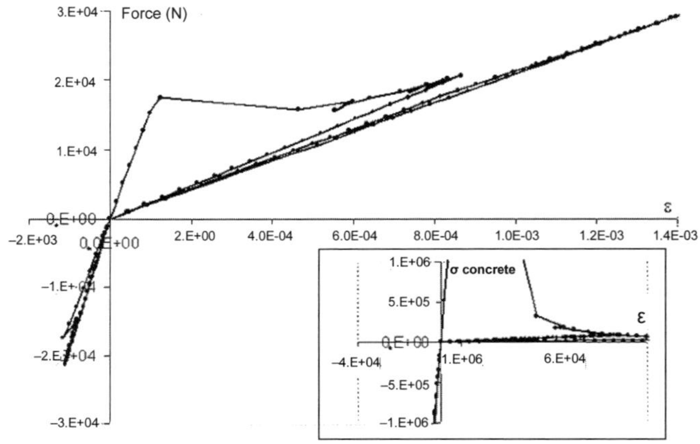

Figure 4.23. *Lower steel force-deformation evolution (model 1). Inset: strain-deformation evolution within concrete*

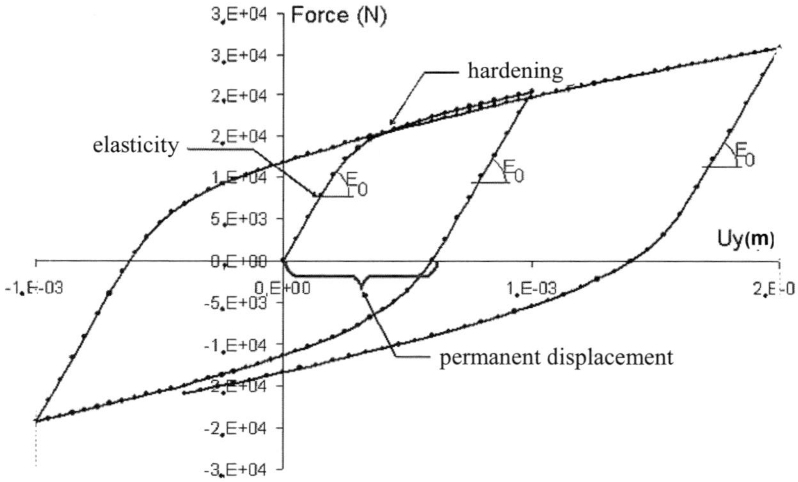

Figure 4.24. *Force-deflection evolution (model 2)*

In general, this type of model is easily implemented in a finite element code as it is explicit [MAZ84]. The other advantage is that the calculation times are usually small. We consider La Borderie's model with $\beta_t=\beta_c=\sigma_f=0$. We obtain responses similar to the test (Figure 4.22).

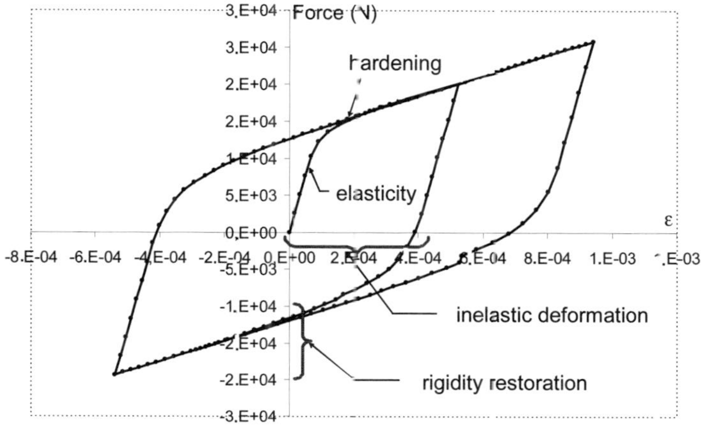

Figure 4.25. *Lower steel force-deformation evolution (model 2)*

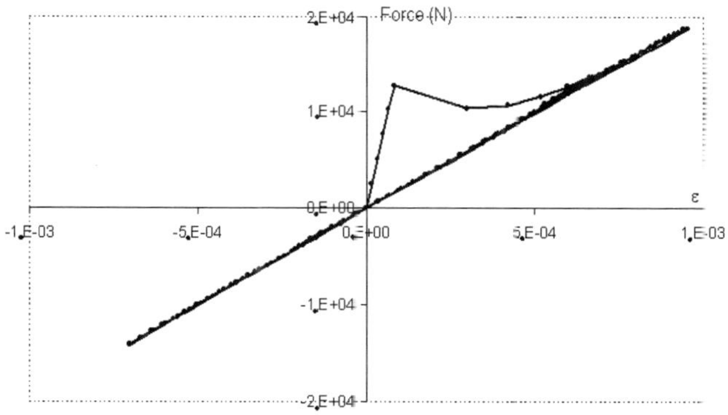

Figure 4.26. *Force-deflection evolution (model 3)*

The extreme force-deflection points are almost the same. Stiffness loss is slightly overestimated as there is no inelastic deformation. We observe pinching of the cycles at the origin. Local (stress-strain) analysis shows similar results without any inelastic deformation and a minimization of the maximal deformations (Figure 4.23).

The uncoupled tensile and compressive material responses allow us to describe the stiffness recovery when we get to the opposed half-cycle and the dissymmetry of the local behavior (see the inset in Figure 4.23). This type of model can suitably describe the global and local behaviors of the beam as the inelastic deformations are not very much relevant in this test.

4.3.1.2. *Inelastic deformations (model 2)*

Regarding this simulation, we use an elastoplastic model with a negative hardening called model 2.

This model is not supposed to represent the behavior of concrete accurately; it is presented in order to illustrate the effect of damage only. The global response shows high amplitude cycles with correct amplitudes but with permanent deflections that are too great compared at the reality (Figure 4.24).

The pinching phenomenon of the curves is not obtained. The cycles are overly large and dissipate far more than those of the test. The same result is found in the local (stress-strain) analysis whereby some cycles have an excessive surface (Figure 4.25).

There is no damage, no stiffness recovery and plastic deformations are excessive in this case. In general, considering the inelastic deformations without damage is not enough to describe the behavior of the beam correctly.

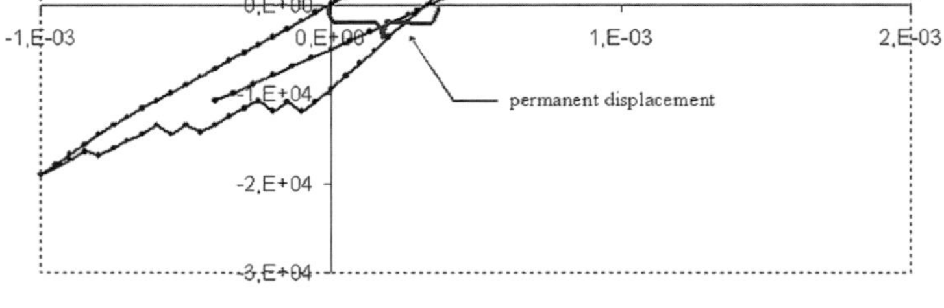

Figure 4.27. *Lower steel force-deformation evolution (model 3)*

4.3.1.3. *Crack closure and unilateral response*

The aim of this section is to show that considering crack closure may be important. We compare the response using a model without any unilateral response due to crack closure with the one obtained with a model that describes crack closure effect on the stiffness. Both models are based on La Borderie's model. The first (model 3) involves a modification such that damage in tension is equal to damage in compression (D_t=D_c=D). The second (model 4) is the complete model with damage in tension uncoupled from damage in compression, inelastic deformation, and crack closure. The global response obtained from the first model correctly describes the decrease of stiffness during the first half-cycle in tension, but not upon load reversal (Figure 4.26).

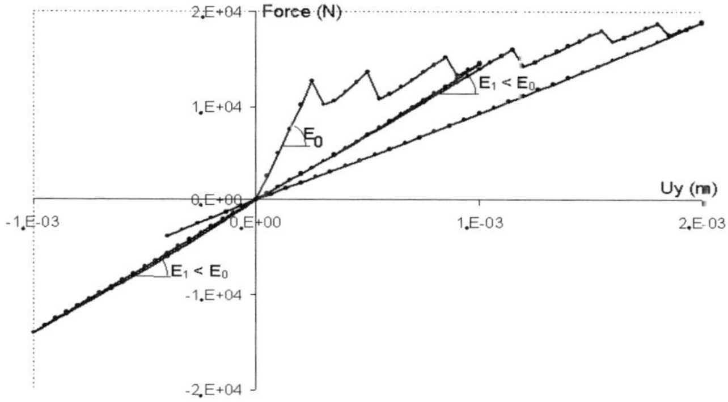

Figure 4.28. *Force-deflection evolution (model 4)*

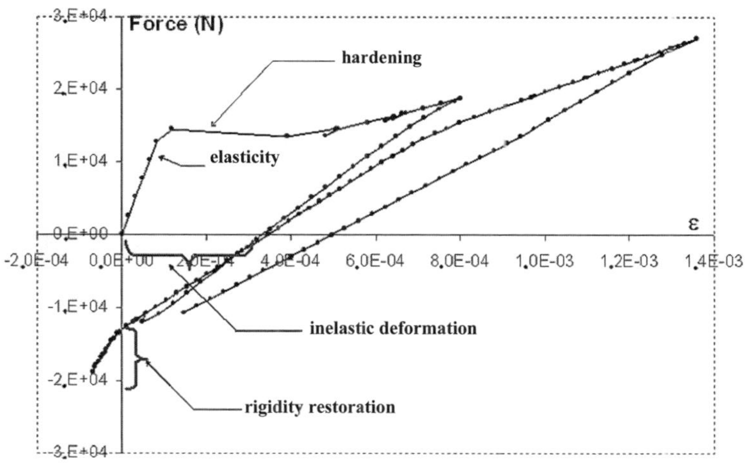

Figure 4.29. *Lower steel force-deformation evolution (model 4)*

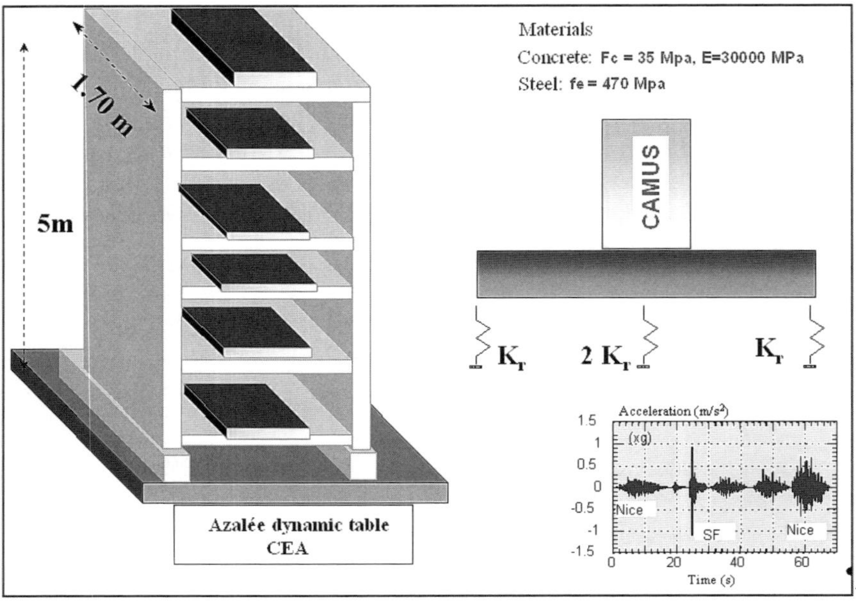

Figure 4.30. *CAMUS test principle*

Damage occurring during the second half-cycle, if there is any, is added to damage occurring in the first half cycle. The inelastic deformations are too small. Local (stress-strain) analysis shows the same results (Figure 4.27).

The second model provides a better description of the response of the beam. Indeed, damage is observed during each loading half-cycles (Figure 4.28). We can then observe the pinching phenomenon. The unilateral characteristic of the material induces the stiffness recovery during the load reversal. Local analysis confirms these results with a correct evolution of the inelastic deformations of the first and, especially, the second cycle (Figure 4.29). Locally, we also find the crack closure effect and the associated stiffness recovery.

4.3.1.4. *Elastic hysteresis*

The previous computations do not include the "elastic" hysteresis loops (loading-unloading hysteresis), which are negligible and do not provide any additional information for the study of the cyclic response of the reinforced concrete beam. Indeed, the previous results show that the description of the behavior of concrete without considering the elastic hysteresis is quite enough as far as accuracy is concerned. Nevertheless, the elastic hysteresis induces energy dissipation. It appears that this dissipation could largely explain the damping observed in the dynamic response of reinforced concrete structures. In most of the calculations, damping is described according to a fictitious matrix equal to a linear combination of both the weight and the rigidity matrixes. This formulation is arbitrary and it is not backed by any physical phenomenon. We can show, however, that the elastic hysteresis loops are sufficient to induce structural damping. Let us consider an example [RAG 00].

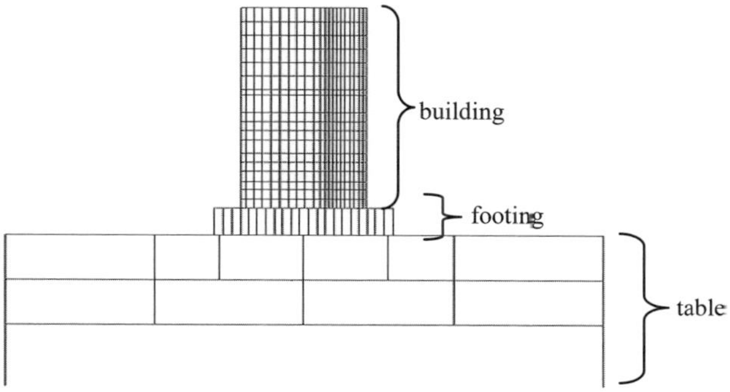

Figure 4.31. *Modeling of the Camus mock up (FE mesh resulting from the modeling with a multilayer beam)*

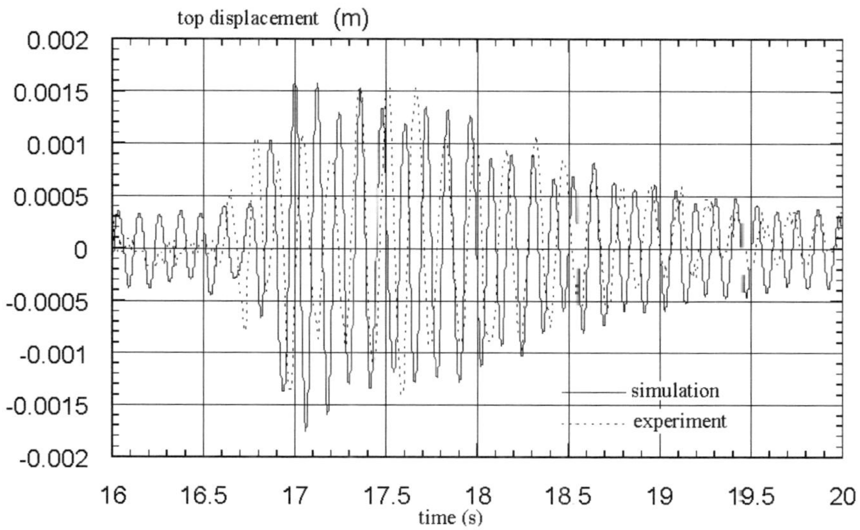

Figure 4.32. *Global response of the mock up*

It consists of the simulation of the response of the CAMUS building [MAZ 98] subjected to seismic loading. This mock up includes two slightly reinforced parallel walls of 5 m height and 1.7 m length, connected together with six floors with additional loads for similarity reasons. The scale model is placed on an Azalée seismic table (CEA) and is loaded by a series of earthquakes of increasing levels (Figure 4.30). In the computations, a multilayer model was used. The compartmentalization of the structure uses a 20 layer length and 16 elements with 20 layers each have been used in the FE model (Figure 4.31). The geometrical and mechanical characteristics of the model are those of the mock up.

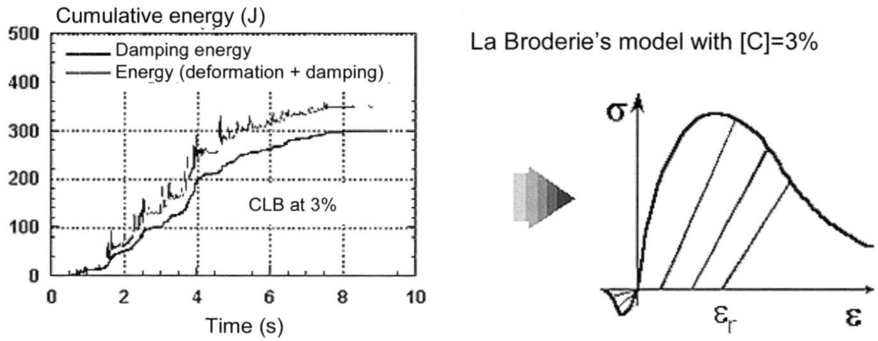

Figure 4.33. *Energy dissipation considering damping*

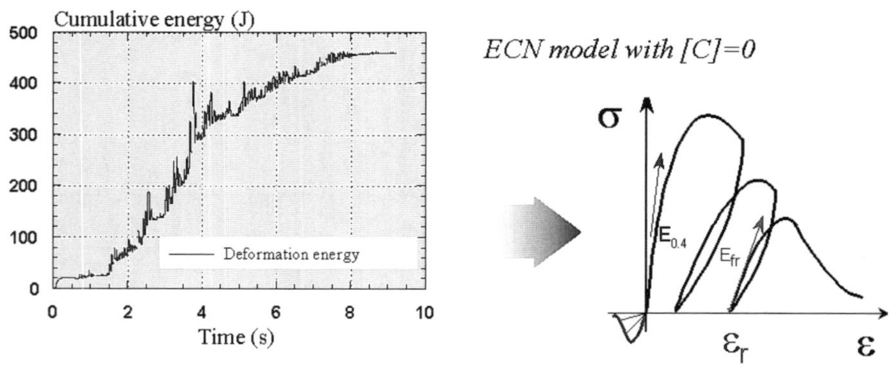

Figure 4.34. *Energy dissipation considering hysteresis*

Let us compare the response of a model including the elastic hysteresis loops (ECN model) with the previous model by La Borderie (CLB model). 3% of damping is introduced in the simulation with La Borderie's model. Figure 4.32 shows the comparison of the displacement at the top of the mock up between the experiment and the computations carried out with the model including the elastic hysteresis and without any additional damping [AMZ 94; AMZ 98]. It should be noted that the simulation correctly describes the global behavior during the earthquake. This is true during the loading phase, as well as during the unloading phase. This last decreasing phase can only be correctly captured when damping is introduced. Indeed, at the end of the earthquake, without any damping, a classical model vibrates with an amplitude which is slightly equal to the maximal amplitude.

Figures 4.33 and 4.34 explain the previous result. Figure 4.33 shows that the energy dissipated according to La Borderie's model is about five times less than the energy dissipated by damping. However, the model that considers hysteresis dissipates much more and this without any structural damping. It provides a dissipation that is equivalent to the damage model without any elastic hysteresis and with structural damping (Figure 4.34).

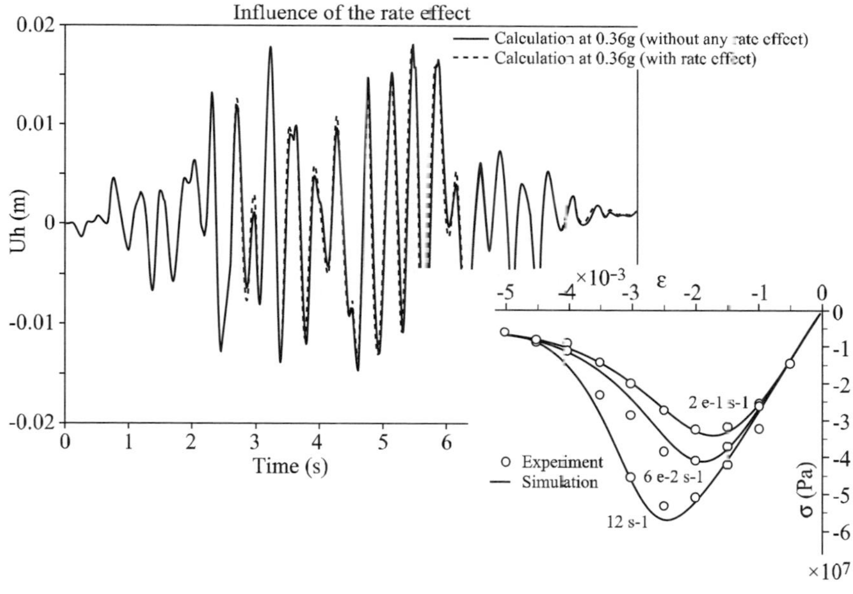

Figure 4.35. *Rate effect on the evolution of displacement at the top of the CASSBA structure. Inset: local response with rate effect*

4.3.2. *Influence of the loading rate*

Up to now, the loading rates during cyclic tests are low, with the most important rates obtained during shocks and earthquakes. Deformation rates are not higher than $10\ s^{-1}$. However, it is interesting to investigate whether the increase in strength due to the increase of strain rate modifies the global response of the structure. We have tested this effect on the simulation of a building subjected to an earthquake. We computed the response of the mock up structure of the CASSBA experiment [BRA 94], which is similar to the one of the CAMUS tests presented above. We use the unilateral damageable model with crack closure [LAB 91] and its rate-dependent damage formulation, which is likely to describe the effects of the strain rate [DUB 94]. Figure 4.35 shows that the effect of the strain rate on the global response of the structure is quasi-negligible. The strain rate remains low during loading, which explains why it has a negligible influence on the global response. Nevertheless, local analysis shows a slight difference between the two models (Figure 4.36). The rate dependent model limits the damage evolution which is restricted to the first level.

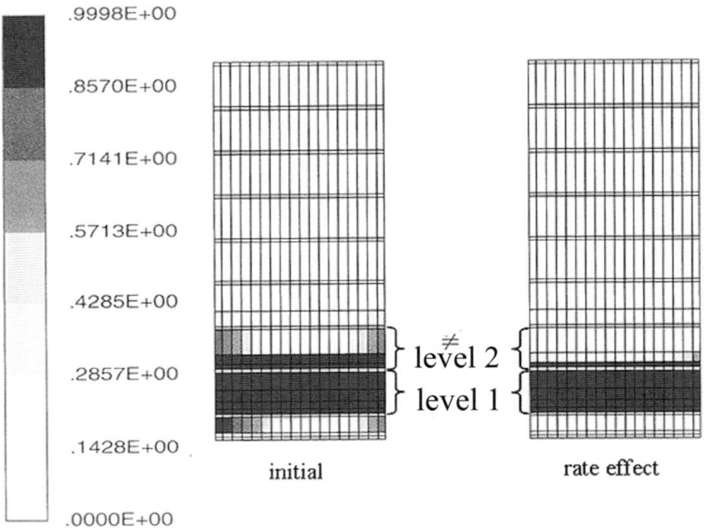

Figure 4.36. *Influence of the rate effect on the final damage distribution of the CASSBA structure*

4.4. Conclusions

Depending on the degree of global or local accuracy desired, it is possible to account for several characteristics of the cyclic response of concrete. Damage, inelastic deformations, crack closure effect, and hystereris upon loading-unloading cycles are the main ingredients. The most important one is the non-symmetric response of the material in tension and compression, which is captured with uncoupled damage variable due to tension and compression. Crack closure, e.g. stiffness recovery upon load sign reversal is the second. Inelastic deformations may be important but the computations presented in this chapter let us think that plastic strain in reinforcements and steel-concrete bond degradation are more important.

Most probably they are more important than crack closure too, at least when considering the global response of structural elements.

It can be important that the model describes elastic hysteresis loops in order to characterize structural damping. However, the classical and arbitrary use of damping, which is proportional to weight and rigidity, perceptibly gives the same results. Finally, the effects induced by the strain rate are small enough to be neglected in standard modeling. Rate effect will be dealt with in Chapter 6 of this volume dealing with blast loads and impacts. For a more comprehensive discussion on the modeling of reinforced concrete structures subjected to seismic and dynamic loads, the reader may also consider the volume edited by Mazars and Millard published in the same collection [MAZ 09].

4.5. Bibliography

[AMZ 98] S. AMZIANE, Comportement cyclique des bétons ordinaires et des BHP, PhD thesis, Ecole Centrale of Nantes, France, 1998.

[AMZ 96] S. AMZIANE, J. F. DUBE, "Analyse prévisionnelle non linéaire par éléments finis multicouches", GEO report, 1996.

[BAT 49] S. B. BATDORF, B. BUDIANSKI, *A Mathematical Theory of Plasticity Based on the Concept of Slip*, Technical Note n° 1871, National Advisory Committee for Aeronautics, Washington DC, 1949.

[BAZ 84] Z. P. BAZANT, P. GAMBAROVA, "Crack shear in concrete: crack band microplane model", *J. Eng. Mech.*, ASCE, vol. 110, pp. 2015-2035, 1984.

[BAZ 86] Z. P. BAZANT, B. H. OH, "Efficient numerical integration surface of sphere", *Z. Angew. Math. Mech.*, ZAMM 66, 1, pp. 37-49, 1986.

[BAZ 88] Z. P. BAZANT, P. C. PRAT, "Microplane model for brittle-plastic material, parts I and II", *J. Eng. Mech.*, ASCE, vol. 117, pp. 1070-1087, 1988.

[BRA 94] M. BRACHET-ROLLAND *et al.*, Opération CASSBA, *Conception et Analyse Sismique des Structures en Béton Armé: Modélisation du Comportement et Simulations du Fonctionnement de la Maquette CASSBA*, ITBTP, 1994.

[CAR 93] E. C. CARVALHO, *Cooperative Research Program on the Seismic Response of Reinforced Concrete Structures*, Commission Contractual Research Activities in the Area of the Response of Civil Engineering Structures to Severe Earthquake Loading, contract no. 4504-91-10 ED ISP P, final report, 1993.

[COP 80] R. J. COPE, P. V. RA., L. A. CLARCK, P. NORRIS, "Modelling of reinforced concrete behaviour for finite element analysis of bridges slabs", in C. TAYLOR *et al.*, *Numerical Methods for Nonlinear Problems I*, Pineridge Press, Swansea, pp. 457-470, 1980.

[DUB 94] J. F. DUBÉ, Modélisation simplifiée et comportement vico-endommageable des structures en béton, PhD thesis, ENS of Cachan, 1994.

[FIC 99] S. FICHANT, C. LA BORDERIE, G. PIJAUDIER-CABOT, "Isotropic and anisotropic descriptions of damage in concrete structures", *Int. J. Mech. Cohesive Frict. Mater.*, vol. 4, pp. 339-359, 1999.

[FOS 96] S.J. FOSTER, B. BUDIONO, R.I. GILBERT, Rotating crack finite element model for reinforced concrete structures, *Comp. Struct.*, vol. 58, pp. 43-50, 1996.

[FOU 90] B. FOURE, *Comportement sismique des murs en béton banché chaînés*, partial report no. 2, dossier CEBTP 22001 (FNB) and 2212-9-08, MNRT no. 89 F 0145, 1990.

[GAU 78] J. GAUVAIN *et al.*, *Essais sismiques de poteaux en béton armé*, Technical Note of CEA EMT/78/177, 1978.

[GHO 01] A. GHOBARAH, A. SAID, "Shear strengthening of beam-column joints", *Eng. Struct.*, vol. 24, pp. 881-888, 2001.

[LAB 91] C. LA BORDERIE, Phénomènes unilatéraux dans un matériau endommageable : modélisation et application à l'analyse de structures en béton, PhD thesis, University of Paris 6, 1991.

[LAD 83] P. LADEVÈZE, On an anisotropic damage theory, in J.P. BOEHLER, *Proc. of the CNRS Int. Coll.*, no. 351 Villars-de-Lans, *Failure Criteria of Structured Media*, pp. 355-363, 1983.

[MAZ 84] J. MAZARS, Application de la mécanique de l'endommagement au comportement non linéaire et à la rupture du béton de structure, PhD thesis, University of Paris 6, 1984.

[MAZ 89] J. MAZARS, Y. BERTHAUD, *Une technique expérimentale appliquée au béton pour créer un endommagement diffus et mettre en évidence son caractère unilatéral*, C.R. Acad. Sci. Paris, vol. 308, no. 7, pp. 579-584, 1989.

[MAZ 98] J. MAZARS, "French advanced research on structural walls: an overview on recent seismic programs", in P. BISCH, P. LABBE, A. PECKER, *Proc. 11th European Conf. on Earthquake Engineering,* Presse des Ponts Pubs, Paris, pp. 21-42, 1998.

[MAZ 09] J. MAZARS, A. MILLARD, *Dynamic Behavior of Concrete and Seismic Engineering*, ISTE, London and John Wiley, New York, 2009.

[MIN 95] C. MINOWA *et al.*, "Dynamic and static collapse test of reinforced-concrete columns", *Nucl. Eng. Des.*, vol. 156, pp. 269-276, 1995.

[NAH 86] G. NAHAS, Calcul à la ruine des structures en béton armé, PhD thesis, University of Paris 6, 1986.

[OZB 92] J. OZBOLT, Z. P. BAZANT, "Microplane model for cyclic triaxial behavior of concrete", *J. Eng. Mech.*, ASCE, vol. 118, pp. 1365-1386, 1992.

[OZB 01] J. OZBOLT, L. YIJUN, K. IVICA, "Microplane model for concrete with relaxed kinematic constraint", *Int J. Solids Struct.*, vol. 38, pp. 2683-2711, 2001.

[RAG 00] F. RAGUENEAU, C. LA BORDERIE, J. MAZARS, "Damage model for concrete – like coupling cracking and friction, contribution towards structural damping: first uniaxial applications", *Mech. Cohesive Frict. Mater.*, vol 5, pp. 607-625, 2000.

[RAM 90] S. RAMTANI, Contribution à la modélisation du comportement multiaxial du béton endommagé avec description du caractère unilatéral, PhD thesis, University of Paris 6, 1990.

[RAS 68] Y. R. RASHID, "Ultimate strength analysis of prestressed concrete pressure vessels", *Nucl. Eng. Des.*, vol. 7, pp. 334-344, 1968.

[REI 84] H. W. REINHARDT, H. A. W. CORNELISSEN, "Post-peak cyclic behaviour of concrete in uniaxial tensile and alternating tensile and compressive loading", *Cement Concrete Res.*, vol. 14, pp. 263-270, 1984.

[ROT 88] J. G. ROTS, Computational modeling of concrete fracture, PhD thesis, Delft University of Technology, 1988.

[SIM 88] J. C. SIMO, J. G. KENNEDY, S. GOVINDJEE, "Non-smooth multisurface plasticity and viscoplasticity. Loading/unloading conditions and numerical algorithms", *Int. J. Num. Meth. Eng.*, vol. 26, pp. 2161-2185, 1988.

[TAY 38] G. I. TAYLOR, "Plastic strain in metals", *J. Inst. Met.*, 62, pp. 307-324, 1938.

[WEI 98] S. WEIHE, B. KRÖPLIN, R. DE BORST, 'Classification of smeared crack models based on material and structural properties", *Int. J. Solids Struct.*, vol. 35, pp. 1289-1308, 1998.

Chapter 5

Cyclic and Dynamic Loading Fatigue of Structural Concrete

5.1. Introduction

The phenomenon of fatigue is a slow process of the microstructure's deterioration of a material when it is confronted with repeated loading. This deterioration generally corresponds to diffuse microvoids or microcracks which progress with the number of applied loading cycles. This irreversible process modifies the macroscopic mechanical properties of the material. It can lead to failure for a level of loading much lower than its static strength when the number of applied cycles is sufficient.

The first studies concerning the effect of repeated loadings on the strength of mechanical components were mentioned at the beginning of the 19th century. Industrial development following the creation of machines, in particular, the railway, necessitated these studies. The word fatigue itself seems to have appeared by 1939. Among the essential steps to understand this phenomenon, one may mention the introduction of the endurance curve by Wöhler around 1860, and subsequently, the investigations done by Palmgren in 1924 and Miner in 1945 concerning the prediction of failure under multilevel fatigue loadings. Finally, the concept of damage introduced by Kachanov in 1958 to quantify the progressive process of deterioration under sustained stress has subsequently become essential to understand the mechanism of fatigue deterioration.

Chapter written by Jean-François DESTREBECQ.

The first studies relating to concrete fatigue were attributed to Considere and De Joly, who, from 1898, carried out tests on mortar samples, with the first test on direct compression and the second on pure traction. Early research works mainly concerned fatigue of road or runway pavements made of concrete. The researchers concentrated their efforts mainly on concrete fatigue in direct compression or bending traction. Over time, the sphere of investigation got wider, due to the construction of off-shore structures, the increasing number of huge buildings, and the development of mass urban transports or the need for a better understanding of the behavior of constructions confronted with seismic risks [ACI 92; RIL 84; SHA 82]. They concentrated their interest on the influence of parameters such as the shape and frequency of the loading cycles, the effect of high level cyclic stresses, the repetition of a very large number of cycles, the environmental conditions, etc. More recently, improvements in experimental facilities and means of analysis have extended the investigation field, in particular, to the case of multiaxial fatigue loading.

In this chapter, a review of models for the fatigue behavior of concrete is presented including the most advanced features. A list of references given at the end of chapter gives the reader an opportunity to investigate the subject further, in particular, to find complementary details about the theoretical and experimental motivations of the models presented.

5.2. The mechanisms of concrete fatigue

5.2.1. *Nature of the phenomenon*

The process of fatigue of concrete is directly linked with the nucleation and the propagation of microcracks inside the material [BEN 69; SHA 70; ZHA 98a]. The presence of pre-existing defects (pores or shrinkage microcracks), as well as the discontinuity corresponding to the interface between the aggregates and the cement paste, stimulate their formation. The heterogeneous nature of concrete, i.e. a rigid aggregate skeleton inserted in a brittle mortar matrix, determines the evolution of the microcracks.

In general, the microcracks of fatigue form a similar network to the one that can be observed under the static loading of the same nature. However, they are known to be more numerous and better spread under the cyclic loading than under the static loading where they generally are fewer, but more developed. Fatigue induced microcracks propagate mainly around the aggregates and in the mortar matrix. Similar to the case in static loading, the favored orientation is parallel to the direction of loading if under compression and perpendicular to that direction under traction. In both cases the microcracks show an average orientation perpendicular to the direction of maximal extension in the mortar matrix

Fatigue microcracking constitutes an irreversible phenomenon that contributes to the slow, but progressive, deterioration of the inner structure of the material. The progression is quantifiable, recording of which allows an evaluation of the deterioration and its evolution up to the failure. The formation and the propagation of microcracks emit sounds that can be recorded in laboratory [SAK 97; WEI 79; YUY 00]. The detailed account of the acoustic events enables the process of deterioration to be monitored from the beginning of the loading. Compared with its final value at the time of failure, the number of acoustic events can be represented like a dimensionless index, ω, which varies from 0 for the intact material to 1 at failure. Taking into account the irreversibility of the microcracking, the index, ω, evolves in an increasing monotonous way during the life of the material: $\dot{\omega} \geq 0$.

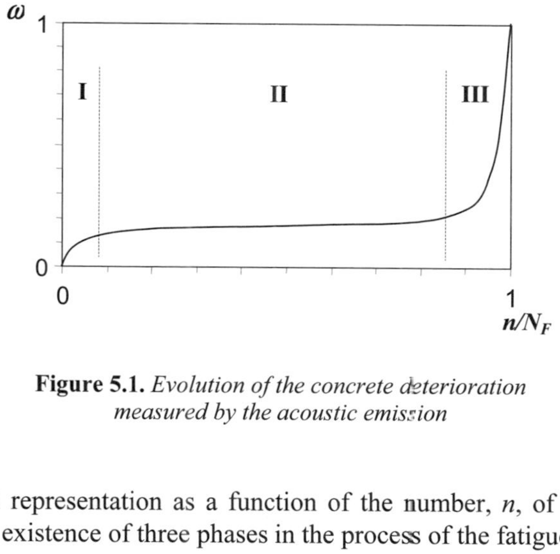

Figure 5.1. *Evolution of the concrete deterioration measured by the acoustic emission*

Its graphical representation as a function of the number, n, of applied loading cycles show the existence of three phases in the process of the fatigue of concrete:

– the initial phase of initiation corresponds to a rapid evolution in the deterioration of the concrete. This primary phase covers about 5 to 10% in the total number of loading cycles, N_F, that are necessary to cause the fatigue failure;

– the secondary phase during which the rate of the deterioration is stabilized. This accommodated phase constitutes the main part of the fatigue life of the material. It spans up to 80 to 90% of the total number of cycles to failure;

– the final phase, which corresponds to the progressive acceleration in the deterioration process and which leads finally to the failure.

Other physical effects can be measured in order to quantify the process of deterioration due to the fatigue loading:

– the dissipation of heat;

– the decrease in the speed of sound in the concrete;

– the volumetric strain and the increase in the deformation.

The first phenomenon corresponds to the release of energy due to the propagation of the microcracks. The last two result from a progressive weakening in the mechanical characteristics of the material, notably in its longitudinal modulus of elasticity. Regarding the acoustic emission, the evolutions of these phenomena follow *S*-shaped curves, which show a similar three-phase progression as the scheme of the material deterioration up to failure.

5.2.2. *Fatigue strength and number of cycles to failure*

A fatigue loading consists of identically repeated cycles that can be represented as the sum of a steady stress $\sigma_{mean}=(\sigma_{max}+\sigma_{min})/2$ and a cyclic stress $\sigma_a=(\sigma_{max}-\sigma_{min})/2$ in amplitude. According to the respective values of σ_{mean} and σ_a, the loading is said to be alternate, repeated, or undulated.

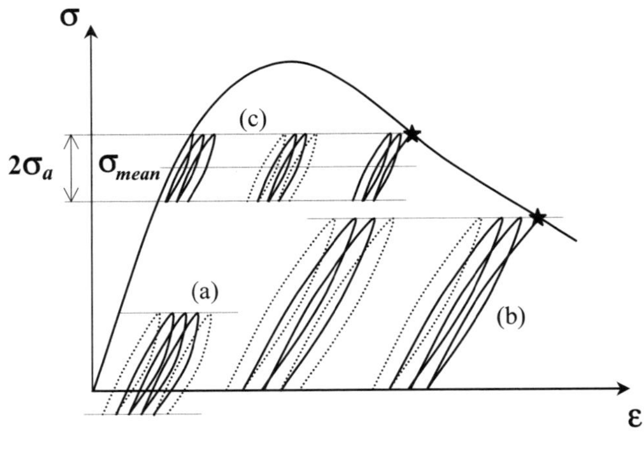

Figure 5.2. *Behavior of concrete under cyclic loading: a) alternate loading; b) repeated loading; c) undulated loading*

The response of concrete to cyclic compressive stress can be represented in a σ-ε diagram. The process of deterioration causes a decrease in the mechanical characteristics, particularly in the longitudinal modulus, and a development of a residual strain after unloading. When represented in a σ-ε diagram, this behavior takes the form of hysteretic buckles progressively moving along the strain axis with an increasing inclination [PLI 97]. The concept of an envelop curve is sometimes used to foresee the failure. Although not very reliable, this principle indicates the way in which the cycle parameters influence the fatigue life. According to this concept, the failure occurs when the point of loading meets the limit curve obtained for a static loading in the σ–ε diagram. The number of cycles to failure depends on the extreme levels of the cycle, and thus, upon the mean stress and the amplitude. When the maximum stress is fixed, a rise in the minimum stress causes a lower strain increase per cycle; thus, requiring more cycles to cause the failure. Conversely, an increase in the maximum stress to the given amplitude is associated with fewer cycles to reach the envelop curve, and so to initiate failure.

The relation between loading and the number of cycles to failure is classically presented under the form of a Wöhler curve in a σ-N_F diagram (stress *versus* the number of cycles to failure). In this representation, one of the parameters of the cycle is fixed (mean or maximal stress, amplitude). The stress, σ, is called the "endurance" or "fatigue strength". It corresponds to the level of loading that causes rupture for a given "number of cycles to failure", N_F.

Three different stages can be seen on the fatigue curves depending on the number of cycles to failure:

– low cycle fatigue (1): $N_F \leq 10^3$ c;

– large cycle fatigue (2): 10^3 c.$\leq N_F \leq 10^7$ c;

– very large cycle fatigue (3): 107 c.≤NF .

The portion of the curve corresponding to stage 2 can be approximated to a straight line in a semilogarithmic coordinate system, hence

$$\sigma = \alpha - \beta \, Log \, N_F \text{ (Wöhler's formula)}$$

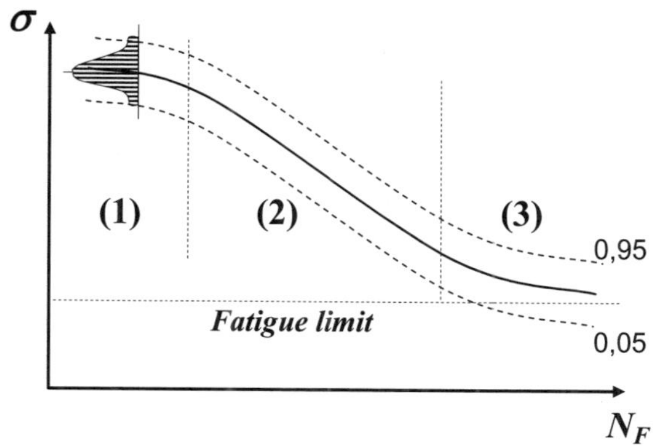

Figure 5.3. *The Wöhler's curve in the σ-N_F diagram*

Fatigue tests generally give very scattered results. In the case of concrete, the number of cycles to failure for the given level of loading may vary in the ratio of 1 to 100. Therefore, the fatigue strength is not defined by one single averaged value. A correct representation must include a notion of probability [HOL 82; SIE 82; YAN 94]. It is generally accepted that the fatigue strength of concrete follows a Gaussian distribution, equivalent to that which affects its static strength.

Owing to the nature and the particular behavior of concrete, the fatigue curves depend on several parameters:

– the characteristics of the material (composition and strength);

– the type and level of the loading cycle;

– the frequency and shape of the cycle;

– the environmental conditions.

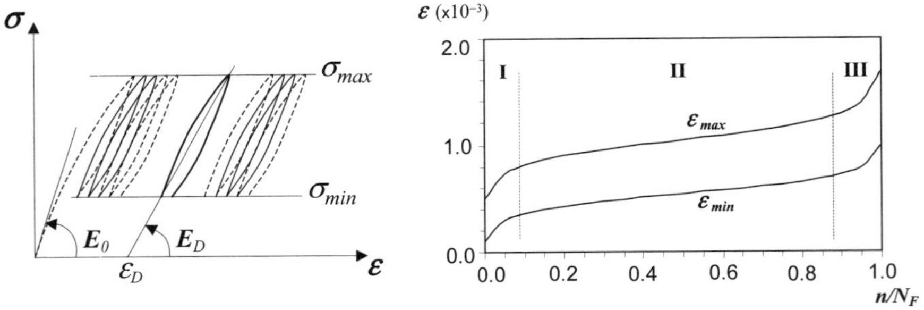

Figure 5.4. *Behavior of concrete subjected to fatigue under cyclic compression*

In fact, the fatigue strength of concrete depends on the same parameters that influence its static strength [AHM 99; KIM 96; KLA 82; RAI 74; STE 90]: nature and type of the aggregates, aggregate mixture grading, proportion of cement, water to cement ratio, porosity, and method of casting. This also depends on the size of the element and the curing and conservation conditions during the period of hardening. A supplementary parameter comes from the variation in the concrete strength over time. Therefore, the validity of a given σ-N_F conventional curve is restricted to concrete with a given age and characteristics. The influence of the static strength to the fatigue strength is neutralized if one traces Wöhler's curve in the *S-N* coordinate axis system, where *S* denotes the cyclic stress to the static strength ratio: $S=\sigma/f_c$. With this representation, the fatigue curve only depends on the loading parameters.

In the case of concrete, there is a difference in nature between low cycle fatigue and fatigue leading to failure for a higher number of loading cycles [AWA 72; CER 79; HSU 84]. For the elevated level of loading, the process of deterioration is governed by the duration of loading more than its cyclic nature. Conversely, it has not been clearly established whether a fatigue limit exists below which the endurance of the concrete becomes unlimited.

5.2.3. *Strains and damage*

The propagation of microcracking is accompanied by a decrease in the longitudinal modulus of elasticity as well as the development of an irrecoverable strain after unloading [ALL 92; HOL 82; GAO 98]. Figure 5.4 represents the evolution of strains (without creep effect) corresponding to the maximum and minimum levels of stress in cyclic compression. These curves present the form in *S*, previously mentioned, corresponding to the three phases of evolution in the fatigue deterioration of concrete.

It has been established [SPA 73] that the rate of evolution in phase II of the strain ε_{max} corresponding to the maximum level of the cyclic stress is proportional to the number of cycles to failure in the logarithmic coordinate system. This can be expressed as follows:

$$\frac{d\varepsilon_{\max}}{dn} = \left(\frac{N_F}{A}\right)^B$$

This expression is known under the name of Sparks and Menzies. It can be used to predict the residual service life of existing structures.

The progressive increase in the strain curves is due to the development of irrecoverable strain. The enlarging gap between the two curves shows a weakening in the longitudinal modulus of elasticity with the number of applied cycles. On this account, the uniaxial behavior of damaged concrete is given by the following relationship:

$$\sigma = E_D\left(\varepsilon - \varepsilon_D\right) \qquad \forall\, t$$

In this expression, E_D represents the actual value of the longitudinal modulus of elasticity. ε_D is the maximum value reached by the irrecoverable strain. The evolutions of E_D and ε_D are easily deduced from the curves of deformation:

$$E_D = \frac{\sigma_{\max} - \sigma_{\min}}{\varepsilon_{\max} - \varepsilon_{\min}}$$

$$\varepsilon_D = \frac{\sigma_{\max}\,\varepsilon_{\min} - \sigma_{\min}\,\varepsilon_{\max}}{\sigma_{\max} - \sigma_{\min}}$$

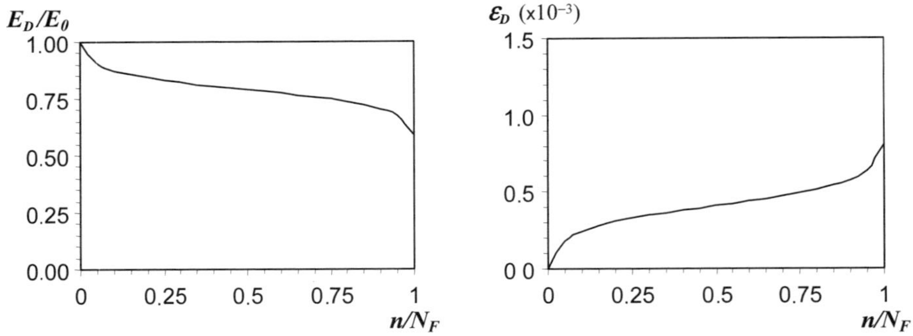

Figure 5.5. *Evolution of the longitudinal modulus and the irrecoverable strain*

Their graphical representations in terms of the number of applied cycles show a change in analogous S to that of strain. A progressive weakening of the longitudinal modulus is observed with the three phases discussed above. The shape of the curve depends on the characteristics of the cyclic loading. However, the modulus tends usually towards a unique final value about 60% of the initial value.

The theory of damage allows a link to be made between the physical deterioration undergone by the material and the progressive alteration of its mechanical properties [ALG 00; PAN 98; PAP 93; PAP 96; SUA 90. This is made possible by introducing the "ratio of damage", D, which is characteristically the state of deterioration of the material at a given moment:

– from the physical viewpoint, the ratio of damage, D, is assimilated to the accumulated total area, Ω_D, of microcracks caused by the cyclic loading on the initial area, Ω_0, of a elementary surface in the undamaged material:

$$D = \frac{\Omega_D}{\Omega_0}$$

– from the mechanical viewpoint, the change in the damage ratio is related to the evolution of the longitudinal modulus of elasticity. The elastic characteristics of the intact fraction of the area (Ω_0–Ω_D) are supposedly unchanged. The stress is supposedly supported only by the intact fraction. These hypothesis lead to the following relationship between the ratio of damage and the actual value of the modulus of elasticity:

$$D = 1 - \frac{E_D}{E_0}$$

This definition of the damage is equivalent to that previously introduced in Chapter 2. Figure 5.6 shows the evolution of the ratio of damage derived from the diagram of evolution of the modulus of elasticity in application of the previous relationship. The positive rate of evolution of the ratio of damage ($\dot{D} \geq 0$) results from the irreversible character of the microcracking.

The ratio of damage provides a link between the physical reality of the process of fatigue deterioration and its mechanical representation. It allows the expression of a law for the evolution of the actual behavior of the damaged material and introduces a criteria for the prediction of the failure that occurs when D=1. In practice, the failure occurs when the residual fraction of the surface area becomes insufficient to support the loading, which corresponds to the critical value of the ratio of damage D_{cr}<1. The acceleration in the damage process close to the failure (asymptotic shape of the curves) has only a slight influence on the number of cycles to the failure.

5.3. The fatigue strength under uniaxial compression or traction

5.3.1. *Aas-Jakobsen's formula*

A widely accepted expression for the fatigue strength of concrete under uniaxial cyclic compression or traction comes from observations made by Aas-Jakobsen and Lenshow [AAS 73] showing that the fatigue strength for a given number of identical cycles varies in a linear way with the loading ratio, R, between the minimum and the maximum intensity of the cyclic stress.

This linear relationship known as "Aas-Jakobsen's formula", is written:

$$\sigma_{max}/f_c = 1 - \beta\,(1 - R)\,Log\,N_F$$

where:

σ_{max}: maximum stress attained during the cycle;

σ_{min}: minimum stress attained during the cycle;

f_c: short-term resistance under static loading;

$R = \sigma_{min}/\sigma_{max}$: loading ratio;

β: material constant;

$Log\,N_F$: decimal logarithm of the number of cycles to failure.

The validity of this relationship has been established [TEP 79a; TEP 79b] by means of a statistical study carried out on a large number of test results concerning concretes with different strengths and characteristics for various frequencies and levels of loading. An average value of 0.0685 has been proposed for β (instead of 0.064 proposed in a preliminary work).

Aas-Jakobsen's formula (with β = 0.0685) applies to ordinary and to light-weight concrete. It gives an estimation of the fatigue strength under uniaxial cyclic compression or traction without the inversion in the stress sign during the loading cycle ($R \geq 0$). Its validity has been verified on a wide range of frequencies (0.1 to 150 Hz) for numbers of cycle to failure up to 2×10^6 and for levels of loading up to 80% of the static strength of concrete ($S_{max} \leq 0.80$). For higher levels of loading, the fatigue strength proves to be sensitive to the frequency, and the number of cycles to failure is generally less than the one predicted by this linear relationship.

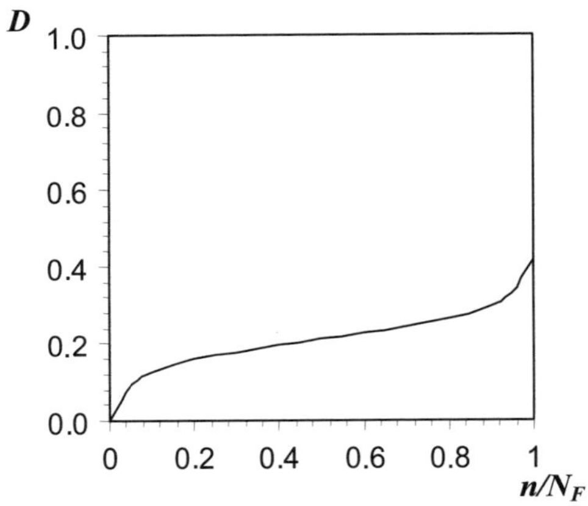

Figure 5.6. *Evolution in the ratio of damage*

5.3.2. *S-N-R curves*

Aas-Jakobsen's formula is usually represented in a semilogarithmic coordinate system, with the x-axis showing the number of cycles of rupture and the y-axis showing the maximum level of loading put in a dimensionless form: $S_{max}=\sigma_{max}/f_c$. The graph takes the form of straight lines converging towards a point (0, 1). Each line shows the existing relation between the maximum level of the loading cycle and the average number of cycles to failure corresponding to a particular value in the loading ratio $R=\sigma_{min}/\sigma_{max}$.

This representation has the advantage of simplicity. The dimensionless form adopted for the level of loading allows neutralizing the parameters that influence concrete strength. It is worth emphasizing that Figure 5.8 gives an average estimation of the fatigue strength of structural concrete. The dispersion is notably around these average values.

5.3.3. *Ros's diagram*

From Aas-Jakobsen's formula, it is possible to derive a relationship between S_{max} and $S_{min}=\sigma_{min}/f_c$, extreme levels of the loading cycle, i.e.:

$$S_{\max} = \frac{1-\beta\, Log\, N_F}{2}\left(1+\sqrt{1+\frac{4\,\beta\, Log\, N_F\, S_{\min}}{(1-\beta\, Log\, N_F)^2}}\right)$$

Figure 5.9 represents this expression in the Ros diagram, for the number of cycles to failure comprising between 10^3 and 10^7 cycles. This representation makes clear the influence of the minimal stress on the fatigue strength for the given numbers of cycles to failure.

5.3.4. *Haigh's diagram*

The influence of the amplitude of the loading cycle on the fatigue strength can be clearly shown representing fatigue curves in the Haigh diagram. This diagram has the maximum level S_{max} on the *x*-axis and amplitude ΔS of the loading cycles on the *y*-axis.

Derived from the linear formula of Aas-Jakobsen, the relationship between ΔS and S_{max} is written as follows:

$$\Delta S = \frac{S_{\max} - S_{\min}}{2} = \frac{S_{\max}\,(1 - S_{\max})}{2\,\beta\, Log\, N_F}$$

Considering the domain of validity of Aas-Jakobsen's formula, the expression below applies for:

$$\forall\ S_{\max} \geq S_{\min} \geq 0$$

that is to say for:

$$S_{\max} \geq 1-\beta\, Log\, N_F$$

or also for:

$$\Delta S \leq S_{\max}/2$$

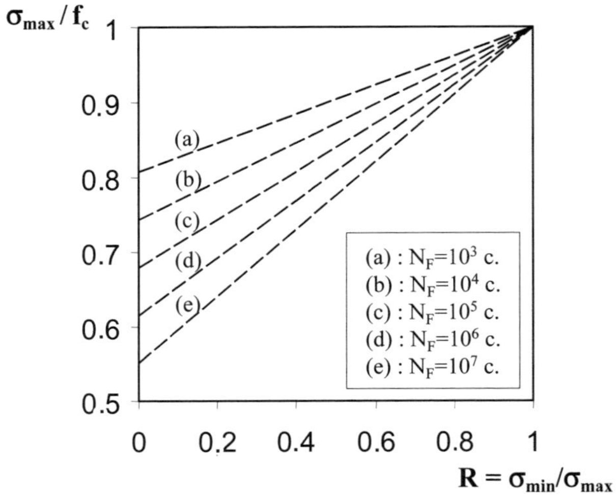

Figure 5.7. *Relationship between the fatigue strength and the loading ratio, R, for the given numbers of cycles to failure*

5.3.5. *Goodman-Smith's diagram*

The representation of fatigue curves in the Goodman-Smith (or modified Goodman-Smith) diagram shows the influence of the mean stress and the amplitude of the loading cycle on the fatigue strength for a given number of cycles to failure.

In the representation, the extreme levels, S_{max} and S_{min}, of the loading cycle are represented as a function of the mean level, S_{mean} with the fixed number of cycles to failure. The expressions derived from the Aas-Jakobsen formula are given below:

$$S_{\max} = \frac{1-2\,\beta\,Log\,N_F}{2}\left(1+\sqrt{1+\frac{8\,\beta\,Log\,N_F\,S_{mean}}{\left(1-2\,\beta\,Log\,N_F\right)^2}}\right)$$

$$S_{\min} = 2\,S_{mean} - S_{\max}$$

As for the preceding representation, these expressions apply:

$$\forall\ S_{\max} \geq S_{\min} \geq 0$$

that is to say for:

$$S_{mean} \geq \frac{1-\beta \; Log \, N_F}{2}$$

Their graphical representation is given in Figure 5.11.

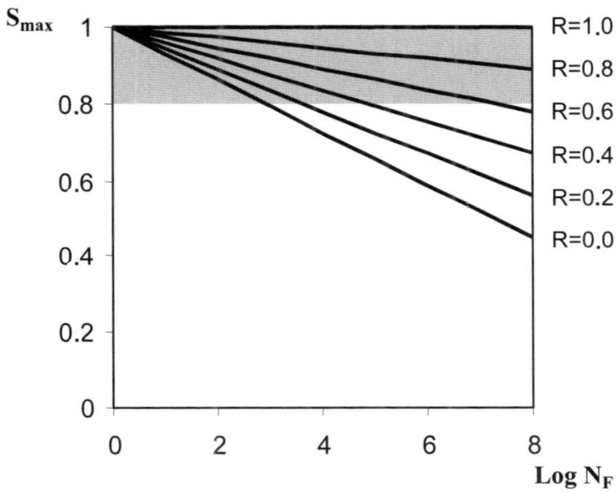

Figure 5.8. *Representation of Aas-Jakenson's formula in the form of S-N-R curves*

5.4. Extension to Aas-Jakobsen's formula

5.4.1. *Effect of the frequency and the cycle shape*

The fatigue strength is influenced by the shape and frequency of the loading cycle. This influence is mainly sensitive at the higher levels loading, i.e. when the maximum level of the loading cycle attains or exceeds the long-term static strength under the sustained load ($\sigma_{max} \geq 0.8 f_c$).

Therefore, considering the influence of the cycle shape, comparative tests under a given frequency and loading level have shown that triangle-shaped cycles produce higher numbers of cycles to failure rather than sinusoidal cycles, whereas the rectangular cycles, lead to lower numbers of cycles to failure [RIL 84]. This effect is more obvious if the level of loading is higher. Compared with sinusoidal cycles, the number of cycles to failure for $S_{max} \geq 0.8$ can be doubled for triangular cycles or, conversely, can be divided by six for rectangular cycles.

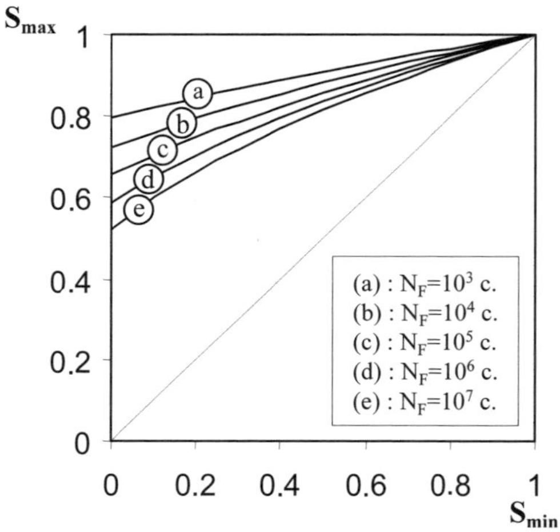

Figure 5.9. *Aas-Jakobsen's formula shown in the Ros diagram*

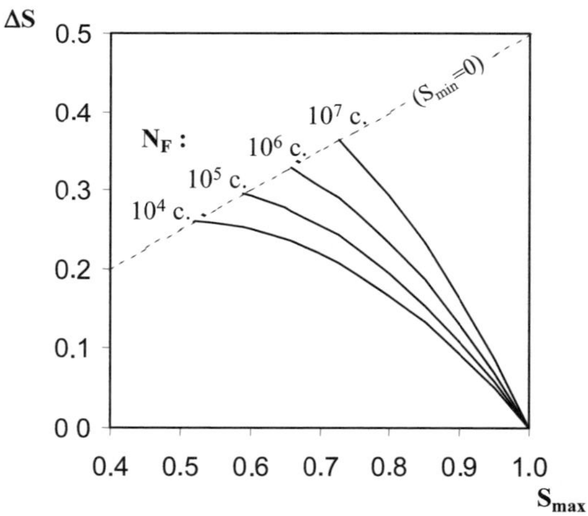

Figure 5.10. *Influence of the amplitude*

The frequency of the cyclic loading influences directly on the number of cycles to failure. As a general rule, the number of cycles to failure increases with the frequency. This effect is not really sensitive for the low or moderate levels of loading. It increases when the maximum level of the cyclic stress exceeds the static strength under sustained load. The increase of the frequency by a factor 10 leads to an increase of the number of cycles to failure approximately by a factor of 2. For the cycles of higher levels ($\sigma_{max} \geq 0.8f_c$) the loading frequency becomes a dominating parameter of the number of cycles to failure.

The influence of the frequency can be taken into account by expressing the loading level as the ratio between the cyclic stress and the static strength, measured when a compression test is done at the same rate of loading as the average stress rate of the ascending branch of the cycle [SPA 82]. In case of a cycle different from a triangular one, the average stress rate can be estimated using $\dot{\sigma} = 4\, f\, \sigma_a$ (slope of the ascending branch of a triangular cycle with the same frequency *f* and the same amplitude σ_a as the real cycle). Taking this correction into account, the fatigue curves obtained experimentally for different frequencies are almost identical.

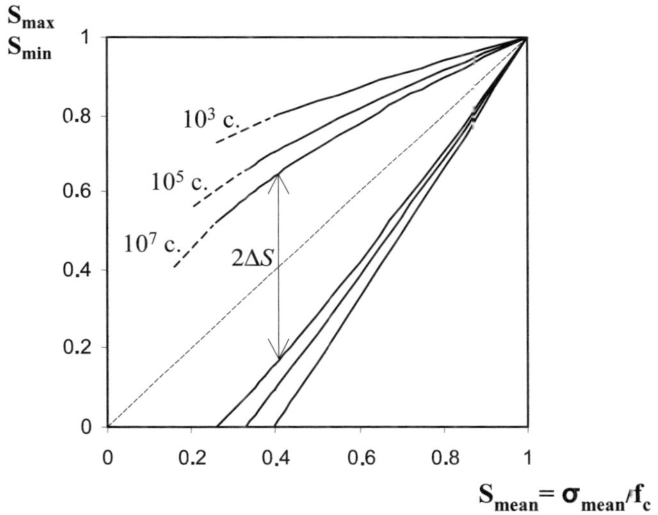

Figure 5.11. *Influence of the mean stress and the amplitude on the fatigue strength*

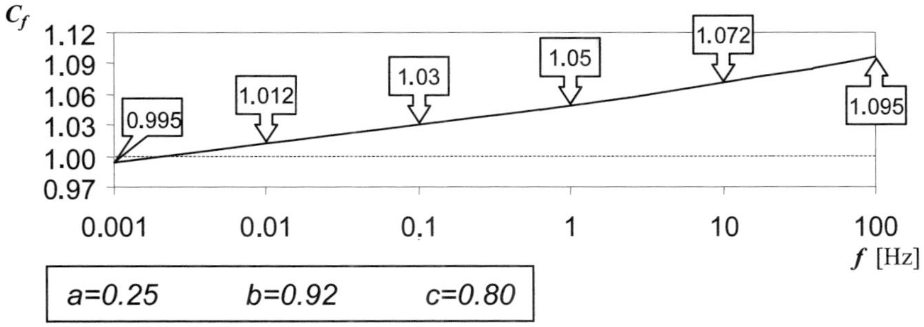

Figure 5.12. *Value of the corrective coefficient C_f in terms of the frequency*

This correction can be applied to Aas-Jakobsen's formula. As mentioned above, this formula has been established and calibrated based on tests performed with frequencies between 0.1 and 150 Hz. Its validity is established "on average" upon this range of frequency. Nevertheless, it tends to overestimate the number of cycles to failure for the lower frequency or underestimate them when the frequency is higher. According to the previous observations, the change to Aas-Jakobsen's formula assumes that the static strength of concrete and its fatigue strength are affected in a similar proportion by the rate of loading [ZHA 96]:

$$S_{\max} = \sigma_{\max}/f_c = C_f \left[1 - \beta(1 - R) Log N_F\right]$$

where:

$$C_f = f_{cf}/f_c = a\, b^{-Log f} + c$$

f_c: static strength at the standardized stress rate;

f_{cf}: static strength at the average stress rate of the ascending branch of the cycle;

f: frequency of the loading cycle (in Hz).

The parameters a, b, and c are derived from compressive tests results performed with various loading rates. The introduction of the coefficient, C_f, requires recalibrating the coefficient, β (the new value is 0.0807).

Figure 5.12 represents the expression of C_f calculated for the given values of the parameters a, b, and c (NB: cf_c corresponds to the ultimate static strength under sustained load).

The modified Aas-Jakobsen formula elucidates the influence of frequency on the fatigue strength. Its validity has been verified for frequencies between 0.014 and 150 Hz. It can be applied for any loading level up to $0.75f_c$. Beyond this level, it tends to overestimate the number of cycles to failure.

5.4.2. *Fatigue under alternate uniaxial loading*

The mechanisms of deterioration of concrete subjected to uniaxial cyclic compression or traction can be simply described as follows:

– the diffuse microcracking caused by a compressive loading presents a general orientation more or less parallel to the direction of loading;

– conversely, the microcracking induced by a tensile loading develops perpendicularly to the direction of loading.

A priori, both mechanisms do not seem to be cumulative. Conversely, an interaction cannot be excluded between both forms of deterioration when the material is alternatively loaded in compression and traction during the fatigue cycle. The existing studies show contrasting opinions regarding this matter [COR 82; TEP 82]. Some conclude that there is little influence, almost negligible, of the inversion in the stress sign during the loading cycle. Others show, on the contrary, that the alternate character of the loading has a significant influence on the number of cycles to failure. In fact, the behavior differs according to the predominant sense of loading. When the cycle is dominated by compression, the incursion on the traction side has very limited influence on the fatigue strength. Conversely, the alternate nature of loading has much more influence when the cyclic stress concerns mainly the traction.

For the discussion, the following notations are introduced:

– $\sigma^+ > 0$ denotes a compressive stress and $\sigma^- < 0$ a tensile stress;

– $f_c > 0$ and $f_t < 0$ are the static strength in compression and traction, respectively;

– $\sigma^+/f_c > 0$ and $\sigma^-/f_t > 0$ represent the ratios of the positive stress (compression) or the negative stress (traction), respectively, against the corresponding static strength;

– the indices *min* and *max* refer to the extreme levels, maximum and minimum, of the loading cycle ($f_t < \sigma_{\min} < \sigma_{\max} < f_c$).

Taking into account the dissymmetry of the behavior of concrete, the mode of loading, as well as the predominant sense if the loading is alternated, can be determined by comparing the levels achieved in the course of the loading cycle by the extreme loading levels: σ^+_{max}/f_c or σ^-_{max}/f_t for the maximum level, σ^+_{min}/f_c or σ^-_{min}/f_t for the minimum level. The different possible cases are presented in Figure 5.13.

Recent studies have made possible to extend the validity of Aas-Jakobsen's formula in case of alternated loadings, by modifying it as follows:

$$S_{\max} = C_f\ \left[1 - \beta\left(1 - R'\right) Log N_F\right]$$

The dimensionless value S_{max} represents the loading that governs the process of fatigue ($S_{max} > 0$ in any cases). C_f is the corrective coefficient introduced previously to allow for the influence of frequency, and the parameter β keeps its adjusted value given at 0.0807. The loading ratio R is replaced with R' defined as follows:

compression only:

$$0 \le \sigma^+_{\min} < \sigma^+_{\max} < f_c \qquad S_{\max} = \frac{\sigma^+_{\max}}{f_c} \qquad 0 \le R' = R = \frac{\sigma^+_{\min}}{\sigma^+_{\max}} < 1$$

predominant compression:

$$0 \leq \frac{\sigma^-_{\min}}{f_t} < \frac{\sigma^+_{\max}}{f_c} < 1 \qquad S_{\max} = \frac{\sigma^+_{\max}}{f_c} \qquad -1 < R' = \alpha_{tc}^{-1} \frac{\sigma^-_{\min}}{\sigma^+_{\max}} \leq 0$$

predominant traction:

$$0 \leq \frac{\sigma^+_{\max}}{f_c} < \frac{\sigma^-_{\min}}{f_t} < 1 \qquad S_{\max} = \frac{\sigma^-_{\min}}{f_t} \qquad -1 < R' = \alpha_{tc} \frac{\sigma^+_{\max}}{\sigma^-_{\min}} \leq 0$$

traction only:

$$f_t < \sigma^-_{\min} < \sigma^-_{\max} \leq 0 \qquad S_{\max} = \frac{\sigma^-_{\min}}{f_t} \qquad 0 \leq R' = \frac{\sigma^-_{\max}}{\sigma^-_{\min}} < 1$$

where:

$$\alpha_{tc} = {|f_t|}/{f_c} > 0 .$$

5.5. Fatigue under multiaxial loading

The knowledge of the fatigue behavior of concrete under multiaxial loading is still limited. The two-dimensional case of loading is generally treated by combining normal and shear stress. The three-dimensional case is generally restricted to the effect of a transversal confinement combined with a loading of cyclic compression applied in the longitudinal direction.

All in all, the existence of a confinement, when it is applied transversely to the direction of the cyclic compressive loading, contributes to the improvement in the fatigue strength of the concrete. However, this effect may vary, depending on whether the confinement is active (lateral pressure) or passive (transverse reinforcement).

In the case of an active confinement, the lateral pressure exists continuously during the cycle, independently from the level of applied cyclic compression. Due to its influence on the state of stresses all over the material, and especially at the microcrack tips, it slows down the nucleation and propagation of the microcracks which, as stated above, tend to develop parallel to the direction of the cyclic compressive loading. In contrast, the concrete remains free to expand transversely.

In the presence of a transverse reinforcement, the confining effect results from the traction in the reinforcement induced by the transverse deformation of the concrete. This effect is actually active in the phase of microcracking of the concrete. The confining effect depends, therefore, on the level of damage of the concrete and varies with the compressive stress during the loading cycle. The confining effect is limited by the plastic yielding of the steel reinforcement. However, the presence of transverse reinforcement delays the disaggregation of the concrete and, therefore, the failure.

5.5.1. *Fatigue in the presence of passive confinement*

The fatigue strength under uniaxial compression is greatly improved by the presence of confining reinforcement [TAN 00]. The reinforcement consists of enclosing links or cross-ties arranged in such a way to cause an isotropic confinement effect transverse to the direction of loading. The positive effect of the confinement is more significant for concretes of moderate strength. The improvement in the fatigue strength is accompanied by a decreased dispersion in the number of cycles to failure.

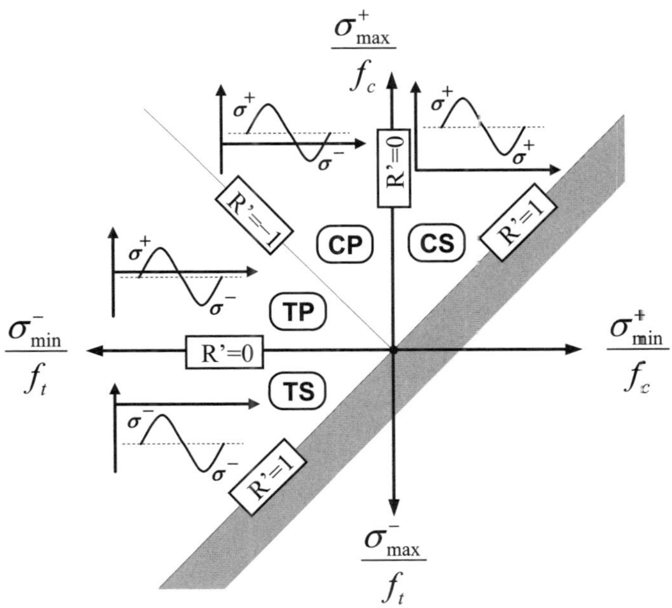

Figure 5.13. *Diagram showing the determination of the mode of loading. CS=compression only; CP=predominant compression; TP=predominant traction; TS=traction only*

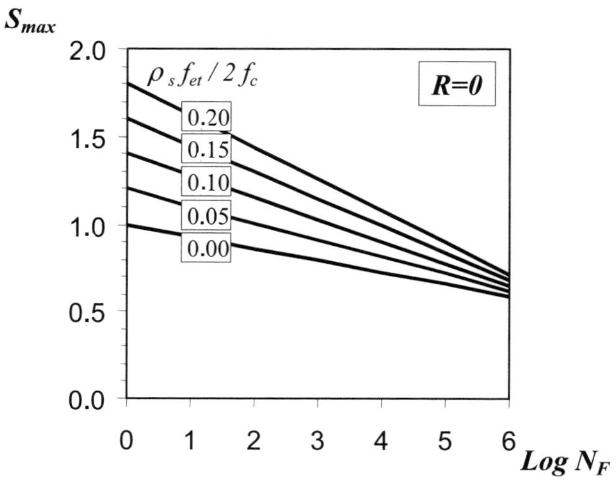

Figure 5.14. *S-N curves with passive confinement*

In a first approach, it may be assumed that the fatigue strength is modified by the effect of passive confinement in the same proportion as the static strength. On the basis of this hypothesis, it is possible to predict the fatigue strength in the presence of transverse reinforcement by means of Aas-Jakobsen's formula, considering S_{max} as the ratio of the maximum compressive stress, σ_{max}, to the static strength of the confined concrete, f_{cc}:

$$S_{\max} = \sigma_{\max} / f_{cc}$$

The strength f_{cc} may be estimated by the following expression:

$$f_{cc} = \alpha\ f_c$$

where

$$\alpha = 1 + k_0\ \rho_s\ \frac{f_{et}}{2 f_c}$$

with:

– f_c: compressive static strength for unconfined concrete;

– k_o: coefficient of adjustment comprising between 4 and 6 (recommended value: 4);

– ρ_s: confining steel ratio (by volume unit of the confined concrete);

– f_e: yield stress of the confining reinforcement.

NB: $\rho_s f_{et}/2$ represents the confining pressure when the yield stress is attained in the transverse reinforcement.

This simple approach slightly underestimates the fatigue strength in the presence of transverse confining steel. A more precise estimation is possible, considering a modified expression of Aas-Jakobsen's formula

$$S_{\max} = \sigma_{\max}/f_c = \alpha - \beta'(1-R)\, Log\, N_F$$

where $\alpha = 1 + k_0\, \rho_s \dfrac{f_{et}}{2 f_c}$

and $\beta' = k_1 + k_2\, \rho_s \dfrac{f_{et}}{2 f_c}$

This expression has been established and validated (considering k_o=4, k_1=0.080 and k_2=0.50) for concretes having a static strength comprising between 25 and 75 MPa with a confinement ratio ($\rho_s f_{et}/2f_c$) up to 0.155. It is worth noting that $\beta'=k_1$ in the absence of confinement. Therefore, the parameter k_1 corresponds to the constant β in the non-modified Aas-Jakobsen formula. The coherence between these two formulas may be preserved choosing adjusted values for the parameters k_1 and k_2:

$k_1=\beta=0.0685$,

$k_2=0.56$.

The modified Aas-Jakobsen's formula is represented in Figure 5.14 in the form of *S-N* curves for R=0 and for various confinement ratios.

5.5.2. *Fatigue under biaxial loading*

The way concrete behaves under biaxial fatigue loading has been investigated by means of alternate torsion tests on hollow cylinders [SUB 98; SUB 99]. With this mode of loading, the material being supposed homogeneous in its initial stage, the principal stresses are found to be inclined at ±45° to the longitudinal axis of the cylinder. The principal stresses are constantly equal in intensity but opposed in sign $\sigma_t=-\sigma_c$. They remain proportional over time to the cyclic effort of torsion applied to the cylinder, and they vary in phase with this effort. If the effort of torsion varies between two extreme values, T_{min} and T_{max}, with the same sign, the material is subjected to a cyclic stress of compression along one principal direction and a cyclical stress of traction along the orthogonal direction:

$$\sigma_1 \leq \sigma_c \leq \sigma_2$$

$$-\sigma_1 \geq \sigma_t \geq -\sigma_2$$

Cylinders tested under repeated torsion, with the intensity of the torsion effort being varied between $T_{max}=0.75T_{rupt}$ and $0.85T_{rupt}$ (T_{rupt} being the ultimate strength in static torsion) and $T_{min}=0.05T_{max}$, have shown a similar behavior as in the case of cyclic uniaxial compression. So the torsional stiffness, K, derived from the curves of torsional deformation recorded during the tests, evolves with the cycle ratio to failure in an analog manner to the longitudinal modulus in the case of uniaxial fatigue:

– phase I ($n/N_F \leq 0.05$) corresponds to the rapid development of primary damage;

– during phase II ($0.05 \leq n/N_F \leq 0.90$), a progressive and quasilinear reduction in the torsional stiffness occurs, which suggests a process of damage accumulation at a constant rate;

– phase III ($0.90 \leq n/N_F$) corresponds to an acceleration in the damaging process, which is shown by a rapid and important decrease in the torsional stiffness.

The slope of phase II is strongly correlated with the number of cycles to failure. The rate of decrease per cycle of the torsional stiffness is linear in terms of the logarithm of the number of cycles to failure. This feature follows that found by Sparks and Menzies for uniaxial compression.

Complementary information is obtained by measuring the natural frequencies of the damaged cylinder at different stages of fatigue. The frequency related to the first fundamental natural modes evolves in a similar manner to the torsional stiffness, which confirms the nature of the damaging process described previously. As for the torsional stiffness, the rate of decrease per loading cycle of the natural frequencies during phase II is proportional to the logarithm of the number of cycles to failure. This property may be useful for the prediction of the residual number of cycles to failure.

The damage process leads to rupture. As in the static case, the rupture in biaxial fatigue is localized under the form of a crack inclined at 45°, perpendicular to the principal direction of traction.

5.5.3. *Fatigue under triaxial cyclic loading*

Tests carried out on cylinders subjected to a longitudinal cyclic compression combined with a cyclic transverse pressure provide information about the influence of triaxial modes of loading on the fatigue behavior of concrete [TAL 96; TAL 98]. The conclusions reported below concern the case where the mean intensity and amplitude of the radial transverse component of the cyclic loading are much lower than that of the longitudinal component. In this case, the transverse pressure acts as an active confinement against the longitudinal loading. Generally speaking, the observed effects are as follows.

As in the uniaxial case, the number of cycles to failure depends on the mean level and amplitude of the cyclic stress. At the fixed mean level of the cyclic loading, the number of cycles to failure increases when the amplitudes of longitudinal and transverse components of the loading are diminished.

The existence of confining stresses favorably influences the fatigue damage process. This effect is shown by a lower decrease in the longitudinal modulus of elasticity, and increases as the rate of confinement is raised. Similarly, the number of cycles to failure increases with the intensity of the stress of confinement.

When the two loading components are in phase, the confinement effect is optimal during the duration of the loading cycle. As a result, there is an increase in the number of cycles to failure. Conversely, the effect of confinement is far less efficient when the two components are in opposite phase. The confining stress is then minimal when the longitudinal compressive stress is at a maximum and *vice versa*. The favorable effect of confinement persists, but the increase in the number of cycles to failure is much more limited.

As for uniaxial loading, the rate per cycle of the longitudinal strain is strongly correlated with the number of cycles to failure. At a given strain rate, the number of cycles to failure is increased in the presence of a confining stress as in the case of uniaxial cyclic loading. This effect is more significant when the two longitudinal and transverse components of the cyclic loading evolve in phase. As in the uniaxial case, this property can help in forecasting the residual number of cycles to failure.

5.6. Fatigue under high-level cyclic loading

The description of the fatigue loading includes two time-related parameters, namely the frequency and the shape of the cycle. The influence of these two parameters is moderate when the maximum level of the cyclic stress is less than 80% of the static strength of the concrete. This threshold corresponds slightly to the long-term static strength of concrete under sustained stress. As seen previously, under this limit the effect of frequency can be taken into account by multiplying the fatigue strength by a corrective coefficient C_f (Aas-Jakobsen's adjusted formula). Above this limit, this adjustment turns out to be insufficient and leads generally to overestimate the number of cycles to failure. The gap is all the more obvious when the minimum level of the cyclic stress reaches, or even exceeds, this limit.

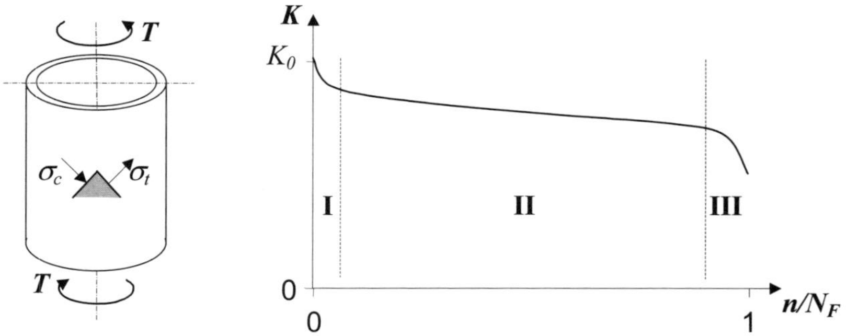

Figure 5.15. *Evolution of the torsional stiffness under the effects of fatigue*

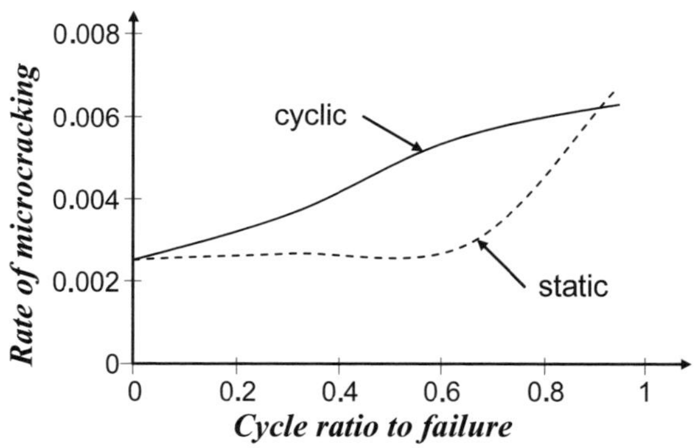

Figure 5.16. *Propagation of the microcracking under cyclic or constant stress*

Judicious observations have provided a comparison of the process of damage in concrete loaded at high-level cyclic compression with that in the same concrete subjected to sustained compression equal to the maximum of the cyclic stress [SHA 70]. The process of damage corresponds in both cases to the propagation of a diffuse microcracking:

– In the static case, the propagation of the microcracks reveals the existence of a creep damage governed by time, which leads finally to a delayed failure under constant stress. The time to failure depends on the stress intensity.

– In the cyclic case, this mode of damage is added to the pure fatigue damage. The propagation of the microcracking is accelerated, which leads to failure for a lower number of cycles than the one that would result from a single process of fatigue.

This combined damage process elucidates why the number of cycles to failure decreases when the level, frequency, and shape of the loading cycle improve the duration of incursion of the cyclic stress in the high-stress domain. In particular, rectangular cycles are more damaging and lead to lower numbers of cycles to failure than the ones obtained for sinusoidal or triangular cycles. Similarly, the creep damage accumulated during every cycle increases with the cycle period, which finally leads to a lower number of cycles to failure when the frequency is diminished.

5.6.1. *General principles for the analysis*

Various methods have been suggested to predict the number of cycles to failure under high-level fatigue loading. Some consist of an adaptation of formulas applicable to moderate-level loading in order to consider the effect of time-dependent parameters. Other approaches suggest appropriate shapes for the fatigue curves, which are derived from the damage theory. Two methods are presented below:

– the first consists of modifying Aas-Jakobsen's linear formula in order to extend its validity to high-level cyclic loading [ZHA 98b];

– the second analyses the damage process that leads to fatigue failure and proposes suitable expressions for the *S-N* curves [DES 93; DYD 94; SZE 95].

Both methods take into account the static strength under sustained stress as the limit above which the time-dependent parameters become significant:

$$\sigma_S = A\, f_c\, t_c^{1/\alpha} \qquad \Leftrightarrow \qquad t_c = \left(\frac{\sigma_S}{A\, f_c}\right)^{\alpha}$$

σ_s is the intensity of the sustained stress, which causes the delayed failure at time t_c. A and α are material parameters. The damage is supposed to evolve in a linear manner between 0 and 1 from the time of loading until the failure:

$$D_t(t) = \frac{t}{t_{c\,(\sigma_S)}} \qquad \Leftrightarrow \qquad \dot{D}_t(t) = \frac{1}{t_{c\,(\sigma_S)}} \qquad \forall\ 0 \le t \le t_e$$

The application of this relationship to a cyclic stress varying between σ_{min} and σ_{max} leads to the calculation of the damage, D_{t1}, accumulated during one loading cycle of period T. This enables the derivation of the duration, T_1, that would be necessary for a static stress, σ_{max}, in intensity to produce the same damage, D_{t1}:

– cyclic loading:

$$D_{t1} = \int_0^T \frac{dt}{t_c(\sigma)}$$

– static loading:

$$D_{t1} = \frac{T_1}{t_c(\sigma\ \max)}$$

Hence:

$$T_1 = \zeta\ T \text{ where } \zeta = \frac{1}{T}\left(\frac{\sigma_{\max}}{A\ f_c}\right)^{\alpha} D_{t1} \qquad (0 < \zeta < 1).$$

The dimensionless coefficient ζ only depends on the constant α, the loading ratio $R=\sigma_{min}/\sigma_{max}$ and the cycle shape. Figure 5.17 gives values of ζ calculated for α=–82.6 in the case of a trapezoidal-shaped cycle defined by a parameter λ. A sinusoidal cycle corresponds to λ=2–4/π (trapezoidal cycle with the same area).

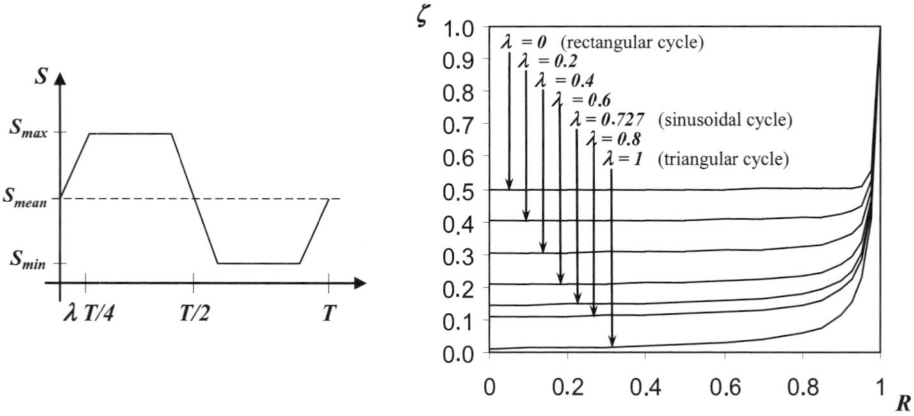

Figure 5.17. *Coefficient ζ as a function of the level and the shape of the loading cycle*

5.6.2. *Modified Aas-Jakobsen's formula*

Two modified versions of Aas-Jakobsen's formula take into account the effects of time-dependent parameters on the fatigue strength [ZHA 98b]:

– the first expression is obtained in analogy with the static case, by multiplying the fatigue strength by a reduction function of the duration of the equivalent sustained stress:

$$S_{\max} = C_f \left[1 - \beta_1 (1 - R) \, Log \, N_F\right] (\zeta \, N_F T)^{\gamma_1}$$

– the second expression is obtained by introducing a complementary term that takes into account the reduced number of failure cycles due to the additional damage induced by the high level of loading:

$$S_{\max} = C_f \left[1 - \beta_2 (1 - R) \, Log N_F - \gamma_2 \, Log(\zeta \, N_F T)\right]$$

where:

C_f: corrective coefficient, function of the frequency (defined above)

β_1, γ_1: parameters with the following values

β_1=0.0691–0.0260R

γ_1=–1.16×10^{-2}

β_2, γ_2: parameters with the following values

β_2=0.0661–0.0226R

γ_2=2.47×10^{-2}

These two modified versions of Aas-Jakobsen's formula are roughly equivalent. They apply to cyclic compression or traction ($R \geq 0$) so that alternate loadings ($R<0$) when compression predominates. In the last case, R must be replaced by $R'=R/\alpha_{tc} \leq 0$ (note that $\alpha_{tc} = -f_t / f_c$). The second version presents the advantage that it can be inversed in order to provide an expression of the number of cycles to failure:

$$Log \, N_F = \frac{1 - (S_{\max} / C_f) - \gamma_2 \, Log(\zeta \, T)}{\beta_2 (1 - R) + \gamma_2}$$

5.6.3. *Formulas based upon the theories of damage*

Damage to concrete caused by high level cyclic loadings can be expressed as a linear combination of pure fatigue damage D_n and creep or 'static' damage D_t (damage over time), hence

$$\dot{D}(t) = \phi_n \, \dot{D}_n(t) + \phi_t \, \dot{D}_t(t) \qquad \forall \ t \geq 0$$

where ϕ_n and ϕ_t are two coupling coefficients. The law of evolution for static damage can be represented by Rabotnov's formula:

$$\dot{D}_t(t) = \frac{\beta_0}{r+1}\left(\frac{\sigma}{f_c}\right)^{-\alpha}(1 - D_t)^{-r} \qquad \forall \ t \geq 0$$

where α, β_o, and r are material parameters.

Integrating this relationship for a constant stress leads to an expression for the time to failure under sustained stress (the rupture is supposed to occur when $D_t = 1$):

$$t_c(\sigma) = \frac{1}{\beta_0}\left(\frac{\sigma}{f_c}\right)^{\alpha}$$

Fatigue damage supposedly evolved in a linear manner up to the failure, therefore:

$$\dot{D}_n(t) = \frac{1}{N_F T} \qquad \forall \ t \geq 0$$

N_F being the number of cycles to failure in pure fatigue.

The first method [DES 93; DYD 94] assumes that static damage evolves at a constant rate, taking into account r=0 in Rabotnov's formula. In the case of moderate levels of loading, the process of fatigue is reduced to pure fatigue damage (D_t=0) which forces consideration of ϕ_n=1. Finally the law of evolution for the total damage is written:

$$\dot{D}(t) = \frac{1}{N_F T} + \frac{\phi_t}{t_c(\sigma)} \qquad \forall \ t \geq 0$$

Hence, after integration:

$$D(t) = \frac{n}{N_F} + n\,\phi_t\, D_{t1} \qquad \forall\ t \geq 0$$

$n=t/T$ being the number of loading cycles of period T applied until time t. The failure occurs at $n=N_R$ cycles, when the damage becomes equal to 1. Hence:

$$N_R = \frac{N_F}{1+\phi_t\, N_F\, D_{t1}} \quad \text{where } D_{t1} = \zeta\,\frac{T}{t_c(\sigma_{\max})}$$

The validity of this expression has been verified considering $N_F = 10^{\frac{1,2 - S_{\max}}{0,116\,(1-R)}}$ and $\phi_t = 0.367$. It applies to cyclic loading at moderate or high stress levels. It is worth noting that the term D_{t1} depends on the period, the level, and the shape of the loading cycle. In accordance with experimental observations, this expression predicts a decrease in the number of cycles to failure when the level or the period of the cycle is augmented. Moreover, considering the influence of the cycle shape on the value of coefficient ζ, the number of cycles to failure predicted by this relationship is much lower for a rectangular-shaped cycle than for a triangular one, a sinusoidal cycle leading to an intermediate number of cycles to failure.

Another method [SZE 95] consists of assuming that the phenomenon of fatigue under high-level loading is dominated by creep damage. This means setting ϕ_n=0 and ϕ_t=1 in the law of damage, which is reduced to Rabotnov's formula. Its integration gives the law of evolution of damage as a function of the number of applied cycles:

$$D(t) = 1 - \left(1 - n\,D_{t1}\right)^{\frac{1}{r+1}} \qquad \forall\ t \geq 0$$

The failure occurs when D=1, leading to the following expression for the number of cycles to failure:

$$N_R = \frac{1}{D_{t1}} = \frac{t_c(\sigma_{\max})}{\zeta\, T}$$

This simple expression only applies to high-level cyclic loadings It can not be applied to other cyclic loadings with moderate stress levels, where the fatigue damage predominates.

5.7. Fatigue strength under variable level cyclic loadings

5.7.1. *Rule of linear accumulation*

The fatigue curves (*S-N* curves) are established for identical loading cycles. In reality, the fatigue loadings generally consist of cycles with variable levels and amplitude. For the prediction of the number of cycles to failure, a solution could consist of establishing the fatigue curves for the cyclic loading of variable characteristics, according to a representative spectrum of real loading conditions of the structure. This is a costly solution, rarely used, that leads to specific fatigue curves to be used for which they have been created.

Another solution consists of quantifying the evolution of the fatigue process by means of a rule of damage accumulation linked with a criterion of failure. The most frequently used method is the linear cumulative Palmgren-Miner law [MIN 45]. Assuming the loading constituted of several sequences of cycle series, each series having n_i identical cycles of level S_i, for series corresponds to a number of cycles to failure N_{Fi} deduced from the conventional fatigue curves (*S-N* curves). The ratio $\delta_i = n_i / N_{Fi}$ connected with the series of level S_i is called the "cycle ratio to failure". According to the linear cumulative Palmgren-Miner law, failure occurs when the sum of the cycle ratios to failure by the successive series becomes equal to unity:

$$\sum_i \delta_i = \sum_i \frac{n_i}{N_{Fi}} = 1$$

This simple rule provides a method for predicting the number of cycles to failure under variable-level cyclic loading from the conventional *S-N* curves. It also gives an estimation of the residual number of cycles to failure. Thus, the number n_2 of level S_2 cycles, which can be applied to a concrete of which the cycle ratio to failure by previous loadings is worth δ_c, is given by:

$$n_2 = (1 - \delta_c)\, N_{F2}$$

N_{F2} being the number of cycles to failure given by the *S-N* curves for a level of loading S_2. This simple example can be easily generalized.

The linear Palmgren-Miner law is, at times, compared to a rule of damage accumulation [LEM 85]. The damage is also a cumulative variable whose value evolves from 0 for the intact material, to 1 at the time of failure.

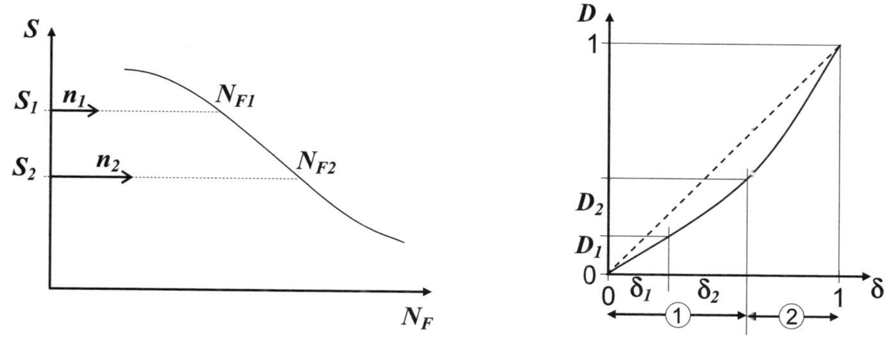

Figure 5.18. *Illustration of the linear Palmgren-Miner law: a) cycle ratio to failure; b) residual number of cycles to failure*

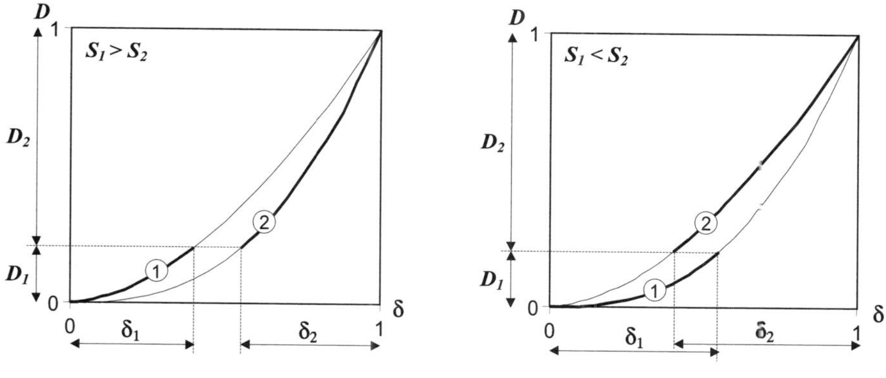

Figure 5.19. *Non-linear accumulation of the cycle ratio to failure*

When the damage evolves in a linear way with the number of applied cycles, the increase of damage caused by a given sequence of loading is equal to the cycle ratio to failure related to this sequence. So, the damage evolves like the cumulated cycle ratio to failure and its evolution is represented by the bisecting axis in a D-δ diagram. In this case, the linear Miner law is equivalent to a rule of accumulation of damage.

Conversely, if the damage evolves in a non-linear manner with the applied number of cycles, its increase for a given sequence of loading cannot be identified with the cycle ratio to failure. The Palmgren-Miner formula then becomes a simple rule of counting that enables prediction of failure. However, it remains compatible with the property of linear accumulation of damage, as far as the latter follows a unique law in terms of the cycle ratio to failure (unique curve in a D-δ diagram).

Initially established for metallic materials, Palmgren-Miner's law has been the subject of numerous checks for various materials, one of which is concrete. Its simplicity is the reason why it is widely used, although its validity is at times debated [KLA 79; OH 91b; SHA 84; SIE 82; TEP 77; VAN 79]. Its applicability can be questionable when it is applied to sequential loadings. It appears more reliable when the characteristics of the cyclic loading vary in a random manner. These various aspects are discussed in the following sections.

5.7.2. *Rule of non-linear accumulation*

The rule of linear accumulation based upon Palmgren-Miner's law is taken as the default case when the loading of fatigue consists of two successive sequences at different levels. According to this rule, the failure is expected to occur when the sum of the cycle ratios to failure becomes equal to unity. Actually, the experimental observations show that this sum is usually greater than unity when the loading sequences are applied in increasing stress levels. They are generally lower than unity if this is not the case:

$$S_2 > S_1 \qquad \Rightarrow \qquad \frac{n_1}{N_{F1}} + \frac{n_2}{N_{F2}} \geq 1$$

$$S_2 < S_1 \qquad \Rightarrow \qquad \frac{n_1}{N_{F1}} + \frac{n_2}{N_{F2}} \leq 1$$

The rate of fatigue is found to increase when the level of loading decreases; conversely, it decreases when the cyclic loading level increases. This phenomenon has been observed for cyclic compression or traction. It is explained by the necessary delay in the nucleation of microcracks and by the irreversible nature of the microcracking process. This delay diminishes when the level of loading is increased. The existence of higher cyclic levels at the beginning of the fatigue process contributes to reduce the global number of cycles to failure when the level of loading decreases. Contrary to this, the process of damage proves more progressive when the level of loading increases over the course of time.

It is possible to represent this phenomenon in supposing that the rate of evolution of the damage obeys a nonlinear law as a function of the loading level and of the cycle ratio to failure:

$$\dot{D} = \frac{\alpha}{T\, N_F}\left(\frac{n}{N_F}\right)^{\alpha-1} \qquad \forall\, n \leq N_F$$

$\alpha \geq 1$ being a parameter which decreases when the level of the loading cycle increases. The integration of this expression gives the law of evolution of damage as a function of the cycle ratio to failure for identically repeated cycles:

$$D(\delta) = \left(\frac{n}{N_F}\right)^{\alpha} \qquad \forall\, \delta = {n}/{N_F} \leq 1$$

The graph of this expression in a D-δ diagram is presented in the form of a series of curves, each corresponding to a given level of cyclic loading. From these relationships it is possible to determine the evolution of damage in the case of a fatigue loading constituted of loading sequences at different levels:

$$D = \sum_i \left(\frac{n_i}{N_{Fi}}\right)^{\alpha_i}$$

The failure occurs when the damage becomes equal to unity. Hence:

$$\sum_i \left(\frac{n_i}{N_{Fi}}\right)^{\alpha_i} = 1$$

This relationship is shown in Figure 5.19. It can be interpreted as a rule of non-linear accumulation of the cycle ratios to failure. When the loading sequences are applied by levels of increasing stresses, the sum of the cycles ratio to failure calculated according to Palmgren-Miner's law is greater than unity. In a contrary case, this sum is smaller than unity (NB: the rule of linear accumulation is evident if the α_i are equal, that is to say if the rate of evolution of damage does not depend on the level of the loading cycle).

The effect of loading sequences is also observed when the cyclic loading constitutes lots of sequences. However, this effect vanishes when the successive sequences occur in a random manner. In this case, the observed dispersion on the sum of the cycle ratios to failure at the moment of failure becomes comparable to that which affects the number of cycles to failure under identically repeated cycles [SIE 82].

5.7.3. *Fatigue under random loading*

In real situations, the loadings are susceptible to important variations over the course of time, with random distributions for the level, amplitude, and frequency. Real loading conditions of materials are generally very different to those applied when sequential or constant level tests are carried out.

This question has been the subject of experimental studies [HOL 82; PAR 90; VAN 75] in order to determine the validity of the linear cumulative rule under realistic loading conditions. Three types of random loading have been considered:

– loading by sequences made of identical cycles, the sequences being repeated in a random order up to failure;

– loading where the amplitude is varied in an random manner at each cycle, the other cycle characteristics remaining unchanged;

– loading where the mean stress and the amplitude are varied in a random manner at each cycle or each half cycle.

In each case, the frequency may be constant or variable.

Generally, these studies show that for random loadings of fixed characteristics, the dispersion in the number of cycles to failure is comparable to that which is usually obtained for fatigue tests at constant levels. The random character of the loading does not appear to be a source of supplementary uncertainty. The value attained by the sum of the cycle ratio to failure at the moment of rupture is itself a random value. When fatigue tests under random loading are repeated for identical concrete samples, the sum of the cycle ratios to failure, according to the linear Palmgren-Miner law, presents a comparable dispersion to that usually observed in conventional fatigue tests.

In the case of loading by sequences, the damage accumulated is calculated according to the linear rule and is, generally, inferior to unity at the time of failure. Therefore, this mode of loading is more damaging than a loading consisting of identically repeated cycles. Therefore, all random loadings are not equivalent. When only the amplitude is varied, the random loading with a constant maximum stress proves more unfavorable than those with a constant minimum stress. Tests carried out with a constant mean stress and variable amplitude hold an intermediary position.

In fact, the fatigue strength under random loading proves very sensitive to the presence of small amplitude cycles in the loading spectrum [HOL 82]. When the program of loading contains few cycles of small amplitude, the rupture occurs when the sum of the cycle ratio to failure is slightly superior to 0.60. However, if the proportion of small amplitude cycles is increased, the sum of the cycle ratio to failure accumulated according to the linear rule attains a value close to 1 at the moment of failure. The existence of periods of rest is also beneficial and contributes to increase the number of cycles to failure.

Finally, studies concerning random loadings confirm the principle of linear accumulation of the cycle ratio to failure, the order of the sequences or the cycles having no significant influence on the number of cycles to failure. In contrast, the random character of the loading increases the rate of the damage process, in particular, when the proportion of small amplitude cycles is low. The Palmgren-Miner rule of linear accumulation can be used to predict the number of cycles to failure under random loading, considering a modified failure criteria as a function of the loading characteristics:

$$\sum_i \delta_i = \sum_i \frac{n_i}{N_{Fi}} = \omega(S_{mean}, R) \leq 1$$

If there is no more precise indication, the limit $\omega(S_{mean}, R)$ can be taken as 0.6.

5.8. Bibliography

[AAS 73] AAS-JAKOBSEN K., LENSHOW R., "Behavior of reinforced columns subjected to fatigue loading", *ACI J.*, vol. 70, pp. 199-206, 1973.

[ACI 92] ACI COMMITTEE 215, *Considerations for Design of Concrete Structures Subjected to Fatigue Loading*, publication 215R-74 (revised), ACI, 1992.

[AHM 99] AHMED T.M.A., BURLEY E., RIGDEN S.R., "The effect of alkali-silica reaction on the fatigue behaviour of plain concrete tested in compression, indirect tension and flexure", *Mag. Concrete Res.*, vol. 51, pp. 375-390, 1999.

[ALG 00] AL-GADHIB A.H., BALUCH M.H., SHAALAN A., KHAN A.R., "Damage model for monotonic and fatigue response of high strength concrete", *Int. J. Dam. Mech.*, vol. 9, pp. 57-78, 2000.

[ALL 92] ALLICHE A., FRANÇOIS D., "Damage of concrete in fatigue", *J. Eng. Mech.*, vol. 118, pp. 2176-2190, 1992.

[AWA 72] AWARD M.E., HILSDORF H.K., "Strength and deformation characteristics of plain concrete subjected to high repeated and sustained loads", in *ACI Symp. "Fatigue of Concrete"*, publication SP41, ACI, pp. 1-13, 1972.

[BEN 69] BENNETT E.W., RAJU N.K., "Cumulative fatigue damage of plain concrete in compression", in *Conf. Civil Eng. Mater.*, Southampton, pp. 1089-1102, 1969.

[BUT 90] BUTLER J.E., "The flexural fatigue performance of plain and fibrous concretes", *Strain*, pp. 135-139, 1990.

[CER 79] CERVENKA V., "Behaviour of concrete under low cycle repeated loadings", in *AICAP/CEB Symp., "Structural Concrete Under Seismic Actions"*, Rome, pp. 15-20, 1979.

[COR 82] CORNELISSEN H.A.W., REINHARDT H.W., "Fatigue of plain concrete in uniaxial tension and in alternating tension-compression loading", in *IABSE Symp. "Fatigue of Steel and Concrete Structures"*, Lausanne, pp. 273-282, 1982.

[COR 84] CORNELISSEN H.A.W., "Fatigue failure of concrete in tension", *Heron*, vol. 29, pp. 1-68, 1984.

[DES 89] DESTREBECQ J.F., GHRICI M., "Prédiction de la résistance du béton sous compression répétée de niveau élevé", in *Greco-Cnrs "Rhéologie des Géomatériaux"*, scientific report, pp. 130-137, 1989.

[DES 93] DESTREBECQ J.F., SZERSZEN M., "A model of prediction for plain concrete subjected to high level compressive cyclic loading", in *Int. Conf. "Concrete and Structures"*, Hong Kong, pp. 45-49, 1993.

[DO 93] DO M.T., CHAALLAL O., AITCIN P.C., "Fatigue behavior of high-performance concrete", *J. Mater. Civ. Eng.*, vol. 5, pp. 96-111, 1993.

[DYD 94] DYDUCH K., SZERSZEN M., DESTREBECQ J.F., "Experimental investigation of the fatigue strength of plain concrete under high compressive loading", *Mater. Struct.*, vol. 27, pp. 505-509, 1994.

[GAO 98] GAO L., HSU C.T.T., "Fatigue of concrete under uniaxial compression cyclic loading", *ACI Mater., J.* vol. 95, pp. 575-581, 1998.

[HOL 82] HOLMEN J.O. , "Fatigue of concrete by constant and variable amplitude loading", in *Fatigue of Concrete Structures*, publication SP75, ACI, pp. 71-110, 1982.

[HOR 93] HORDIJK D.A., REINHARDT H.W., "Numerical and experimental investigation into the fatigue behavior of plain concrete", *Exp. Mech.*, vol. 33, pp. 278-285, 1993.

[HSU 84] HSU T.T.C., "Fatigue and microcracking of concrete", *Mater. Constr.*, vol. 17, pp. 51-54, 1984.

[KIM 96] KIM J.K., KIM Y.Y., "Experimental study of the fatigue behavior of high strength concrete", *Cement Concrete Res.*, vol. 26, pp. 1513-1523, 1996.

[KLA 79] KLAUSEN D., WEIGLER H., "Strength of concrete under constant and variable pulsating fatigue load", *Betonwerk Fertigteil-Tech.*, pp. 158-163, 1979.

[KLA 82] KLAIBER F.W., LEE D.Y., "The effects of air content, water-cement ratio, and aggregate type on the flexural fatigue strength of plain concrete", in *Fatigue of Concrete Structures*, publication SP75, ACI, pp. 111-131, 1982.

[LEM 85] LEMAÎTRE J., CHABOCHE J.L., "Mécanique des matériaux solides", Chapter 7 in *Endommagement*, Dunod, pp. 343-442, 1985.

[MEY 94] MEYER C., PASKOVA T., "Optimum number of specimens for low-cycle fatigue tests of concrete", *J. Struct. Eng.*, vol. 120, pp. 2242-2247, 1994.

[MIN 45] MINER M.A., "Cumulative damage in fatigue", *ASME J. Appl. Mech.*, vol. 67, pp. 159-164, 1945.

[OH 91a] OH B.H., "Fatigue life distributions of concrete for various stress levels", *ACI Mater. J.*, vol. 88, pp. 122-128, 1991.

[OH 91b] OH B.H., "Cumulative damage theory of concrete under variable-amplitude fatigue loadings", *ACI Mater. J.*, vol. 88, pp. 41-48, 1991.

[PAN 98] PANDOLFI A., TALIERCIO A., "Bounding surface models applied to fatigue of plain concrete", *J. Eng. Mech.*, vol. 124, pp. 556-564, 1998.

[PAP 93] PAPA E., "A damage model for concrete subjected to fatigue loading", *Eur. J. Mech., A/Solids.*, vol. 12, pp. 429-440, 1993.

[PAP 96] PAPA E., TALIERCIO A., "Anisotropic damage model for the multiaxial static and fatigue behaviour of plain concrete", *Eng. Fract. Mech.*, vol. 55, pp. 163-179, 1996.

[PAR 90] PARK Y.J., "Fatigue of concrete under random loadings", *J. Struct. Eng.*, vol. 116, pp. 3228-3235, 1990.

[PAS 97] PASKOVA T., MEYER C., "Low-cycle fatigue of plain and fiber-reinforced concrete", *ACI Mater. J.*, vol. 94, pp. 273-285, 1997.

[PIN 99] PINDADO M.A., AGUADO A., JOSA A., "Fatigue behavior of polymer-modified porous concretes", *Cement Concrete Res.*, vol. 29, pp. 1077-1083, 1999.

[PLI 97] PLIZZARI G.A., CANGIANO S., ALLERUZZO S., "The fatigue behaviour of cracked concrete", *Fatigue Fract. Eng. Mater. Struct.*, vol. 20, pp. 1195-1206, 1997.

[RAI 74] RAITHBY K.D., GALLOWAY J.W., "Effect of moisture condition, age, and rate of loading on fatigue of plain concrete", in *ACI Symp. "Fatigue of concrete"*, publication SP41, ACI, pp. 15-34, 1974.

[RED 95] REDJEL B., "Etude expérimentale de la fatigue du béton en flexion 3 points", *Cement Concrete Res.*, vol. 25, pp. 1655-1666, 1995.

[RIL 84] RILEM COMMITTEE 36-RDL, "Long term random dynamic loading of concrete structures", *Mater. Struct.* vol. 17, pp. 1-28, 1984.

[ROE 99] ROESLER J.R., BARENBERG E.J., "Effect of static and fatigue cracking on concrete strain measurements", *Transp. Res. Rec.*, no. 1684, pp. 51-60, 1999.

[SAK 97] SAKATA Y., OHTSU M., "Spectral response and acoustic emission of reinforced concrete members under fatigue bending", *J Acoust. Emission*, vol. 15, pp. 570-579, 1997.

[SHA 70] SHAH S.P., CHANDRA S., "Fracture of concrete subjected to cyclic and sustained loading", *ACI J.*, vol. 61, pp. 816-824, 1970.

[SHA 82] SHAH S.P., ACI COMMITTEE 215, *Fatigue of Concrete Structures*, publication SP-75, ACI, 1982.

[SHA 84] SHAH S.P., "Predictions of cumulative damage for concrete and reinforced concrete". *Mater. Constr.*, vol. 17, pp. 65-68, 1984.

[SHI 93] SHI X.P., FWA T.F., TAN S.A., "Flexural fatigue strength of plain concrete", *ACI Mater. J.*, vol. 90, pp. 435-440, 1993.

[SIE 82] SIEMES A.J.M., "Miner's rule with respect to plain concrete variable amplitude tests", in *Fatigue of Concrete Structures*, publication SP75, ACI, pp. 343-372, 1982.

[SPA 73] SPARKS P.R., MENZIES J.B., "The effect of rate of loading upon the static and fatigue strengths of plain concrete in compression", *Mag Concrete Res.*, vol. 25, pp. 73-80, 1973.

[SPA 82] SPARKS P.R., "The influence of rate of loading and material variability on the fatigue characteristics of concrete", in *Fatigue of Concrete Structures*, publication SP75, ACI, pp. 331-341, 1982.

[STE 90] STEMLAND H., PETKOVIC G., ROSSELAND S., LENSCHOW R., "Fatigue of high strength concrete", *Nord. Concrete Res.*, no. 9, pp. 172-196, 1990.

[SUA 90] SUARIS W., OUYANG C., FERNANDO V.M., "Damage model for cyclic loading of concrete", *J. Eng. Mech.* vol. 116, pp. 1020-1035, 1990.

[SUB 98] SUBRAMANIAM K.V., POPOVICS J.S., SHAH S.P., "Monitoring fatigue damage in concrete", in *Materials Research Society, Symp. Proc.*, vol. 503, pp. 151-157, 1998.

[SUB 99] SUBRAMANIAM K.V., POPOVICS J.S., SHAH S.P., "Fatigue response of concrete subjected to biaxial stresses in the compression-tension zone", *ACI Mater. J.*, vol. 96, pp. 663-669, 1999.

[SZE 95] SZERSZEN M., DESTREBECQ J.F., DYDUCH K., "Behaviour and durability of concrete with regards to cycle and time dependent effects", In *IABSE Symp.* "*Extending the Lifespan of Structures*", San Francisco, pp. 1265-1270, 1995.

[TAL 96] TALIERCIO A.L.F., GOBBI E., "Experimental investigation on the triaxial fatigue behaviour of plain concrete", *Mag. Concrete Res.*, vol. 48, pp. 157-172, 1996.

[TAL 98] TALIERCIO A.L.F., GOBBI E., "Fatigue life and change in mechanical properties of plain concrete under triaxial deviatoric cyclic stresses", *Mag Concrete Res.*, vol. 50, pp. 247-255, 1998.

[TAN 00] TAN T.H., "Effects of passive confinement on fatigue properties of concrete", *Mag Concrete Res.*, vol. 52, pp. 7-15, 2000.

[TEP 77] TEPFERS R., FRIDEN C., GEORGSSON L., "A study of the applicability to the fatigue of concrete of the Palmgren-Miner partial damage hypothesis", *Mag. Concrete Res.*, vol. 29, pp. 123-130, 1977.

[TEP 79a] TEPFERS R., "Tensile fatigue strength of plain concrete", *ACI J.*, vol. 76, pp. 919-933, 1979.

[TEP 79b] TEPFERS R., KUTTI T., "Fatigue strength of plain, ordinary, and lightweight concrete", *ACI J.*, vol. 76, pp. 635-652, 1979.

[TEP 82] TEPFERS R., "Fatigue of plain concrete subjected to stress reversals", in *Fatigue of Concrete Structures*, publication SP75, ACI, pp. 195-215, 1982.

[VAN 79] VAN LEEUWEN J., SIEMES A.J.M., "Miner's rule with respect to plain concrete", *Heron*, vol. 24, pp. 2-34, 1979.

[WEI 79] WEIGLER H., KLAUSEN D., "Fatigue behaviour of concrete – effect of loading in the fatigue strength range", *Betonwerk Fertigteil-Tech.*, pp. 214-220, 1979.

[YAN 94] YANG D., "A distribution function for the fatigue life of concrete", *Mag. Concrete Res.*, vol. 46, pp. 215-221, 1994.

[YUY 00] YUYAMA S., LI Z.W., YOSHIZAWA M., TOMOKIYO T., UOMOTO T., "Evaluation of fatigue damage in reinforced concrete slabs by acoustic emission", *Insight*, vol. 42, pp. 439-443, 2000.

[ZHA 96] ZHANG B., PHILLIPS D.V., WU K., "Effects of loading frequency and stress reversal on fatigue life of plain concrete", *Mag. Concrete Res.*, vol. 48, pp. 361-375, 1996.

[ZHA 97a] ZHANG B., WU K., "Residual fatigue strength and stiffness of ordinary concrete under bending", *Cement Concrete Res.*, vol. 27, pp. 115-126, 1997.

[ZHA 97b] ZHANG B., PHILLIPS D.V., WU K., "Further research on fatigue properties of plain concrete", *Mag. Concrete Res.*, vol. 49, pp. 241-252, 1997.

[ZHA 98a] ZHANG B., "Relationship between pore structure and mechanical properties of ordinary concrete under bending fatigue", *Cement Concrete Res.*, vol. 28, pp. 699-711, 1998.

[ZHA 98b] ZHANG B., PHILLIPS D.V., GREEN D.R., "Sustained loading effect on the fatigue life of plain concrete", *Mag. Concrete Res.*, vol. 50, pp. 263-278, 1998.

Chapter 6

Rate-Dependent Behavior and Modeling for Transient Analyses

6.1. Introduction

Concrete is a rate-dependent material, i.e. its properties – ultimate strength, fracture energy, and to a lesser extend Young's modulus – are highly dependent on the loading rate. For this reason, knowledge of the constitutive relationship for a wide range of strain rates is required to develop a realistic material law.

Understanding the behavior of concrete at very high strain rates is of critical importance in a range of applications. Some structures have to be designed to withstand impacts or impulsive loading: shelters, ammunition bunkers, structures to protect mountain roads from falling rocks, structures to protect bridges against impact of ships, accidental explosions. Under highly dynamic conditions, the strain rate dependence of material response and high levels of hydrostatic pressure cause the material behavior to be significantly different from what is observed under quasi-static conditions.

6.2. Experimental behavior

Many issues remain ambiguous in dynamic experiments as opposed to static ones and, in particular, the influence of time. For more frequent structural testing (contrary to material testing) the effects of time typically appear in two ways: by inertial forces induced by non-zero acceleration to which the structure is subjected

---

Chapter written by Fabrice GATUINGT.

and by a dependence of the mechanical characteristics of each elementary volume of material with respect to the time evolutions, which is described in terms of viscosity. The effects of viscosity are due to microscopic inertial phenomena. This distinction is strictly related to the traditional concept of representative volume elements used to define the behavior of the material. This remark is particularly pertinent to the case of concrete-like materials, where the size of the representative volume element is in the order of centimeters.

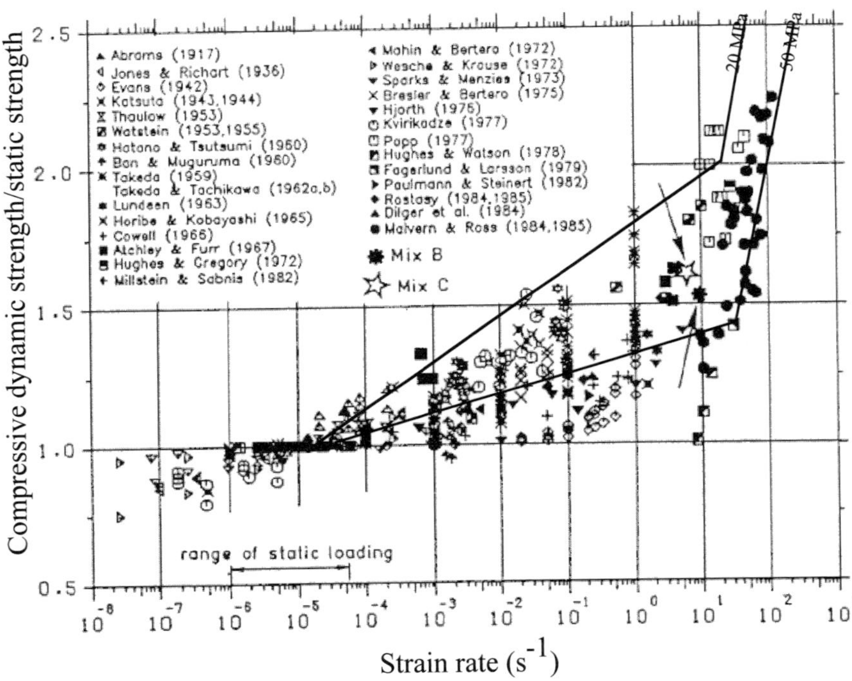

Figure 6.1. *Compressive strength increase versus strain rate*

6.2.1. *Dynamic behavior in uniaxial compression*

A number of studies has been undertaken to elucidate the behavior of concrete in dynamic compression and to study the influence of the loading rate. For example, Bischoff and Perry [BIS 91] show, as corroborated by a significant bibliographical synthesis in Figure 6.1, that there is an increase of the compressive strength of concrete at higher deformation rates. Moreover they distinguish two different modes producing the relative increase of the resistance in compression: the first mode corresponds to a range of variations of strain rate ranging between the quasi-static

$\dot{\varepsilon}=10^{-6}/s$ and $\dot{\varepsilon}<10/s$ leading to a 1.5 times increase in the resistance in compression. The majority of the authors agree that the explanation for such a strength increase pertains to the viscosity of the material, which is related to the presence of water in the pores of the concrete [ROS 91; TOU 94]. The free water in the micropores is assumed to exhibit the so-called Stefan effect causing a strengthening effect in concrete with increasing loading rate. This induces a delayed appearance of the microscopic cracks.

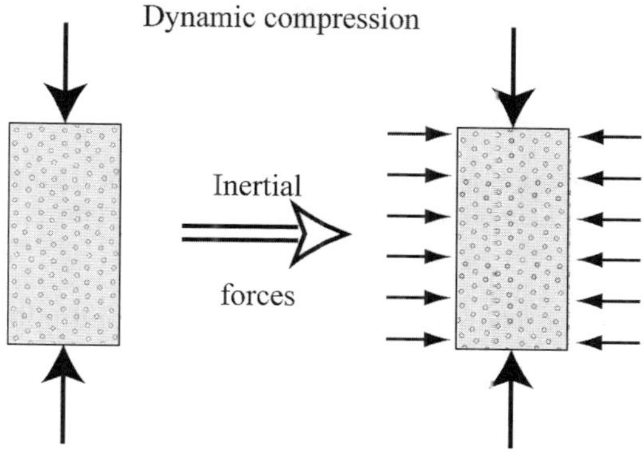

Figure 6.2. *Inertia effect in dynamic compression*

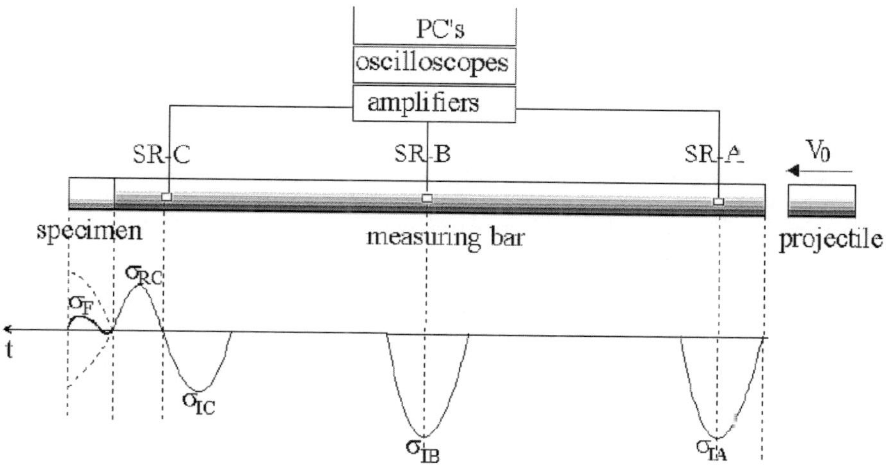

Figure 6.3. *Principle of the experiment*

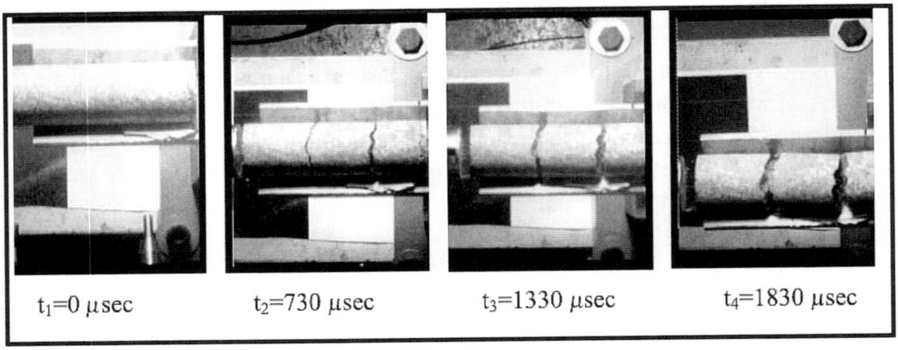

Figure 6.4. *Captured fracture process with two spalls, the load is applied on the left side of specimen*

The second mode, for which the strain rate $\dot{\varepsilon} > 10/s$, multiplies the resistance of the concrete by four, thus $\dot{\varepsilon} = 10^3/s$. According to various studies, this rate effect is considered to be of structural origin [BAI 94; LEN 96; DUB 97; CAM 00]. This origin seems to be a state of non-homogeneous stress in the specimen brought about by the inertia, which generates considerable radial constraints acting like a confining pressure [KOT 83] (see Figure 6.2).

Moreover, studies carried out on ceramics [DEN 00] show that the higher the rate of deformation, the more cracking remains diffuse (no localization of cracking), which also increases resistance.

6.2.2. *Dynamic behavior in uniaxial tension*

6.2.2.1. *Rate effect in tension*

As in compression, the simple tension test must be adapted to dynamic conditions. The experimental set up presented hereafter has been developed at LPMM-Metz [KLE 01], based on the Split Hopkinson Pressure Bar (SHPB), which enables very high loading rates to be attained. This setup consists of a concrete specimen in the form of round bar (40 mm diameter and 120 mm long), which is loaded by a compressive incident wave through the SHPB (40 mm diameter and 1000 mm long), made of a hard aluminum alloy, see Figure 6.3.

After projectile impact, the incident wave generated in the SHPB is transmitted into the concrete specimen, and a small part is reflected back to the bar due to the difference of impedances. The compressive wave transmitted into the specimen is reflected by the free face as a tensile wave.

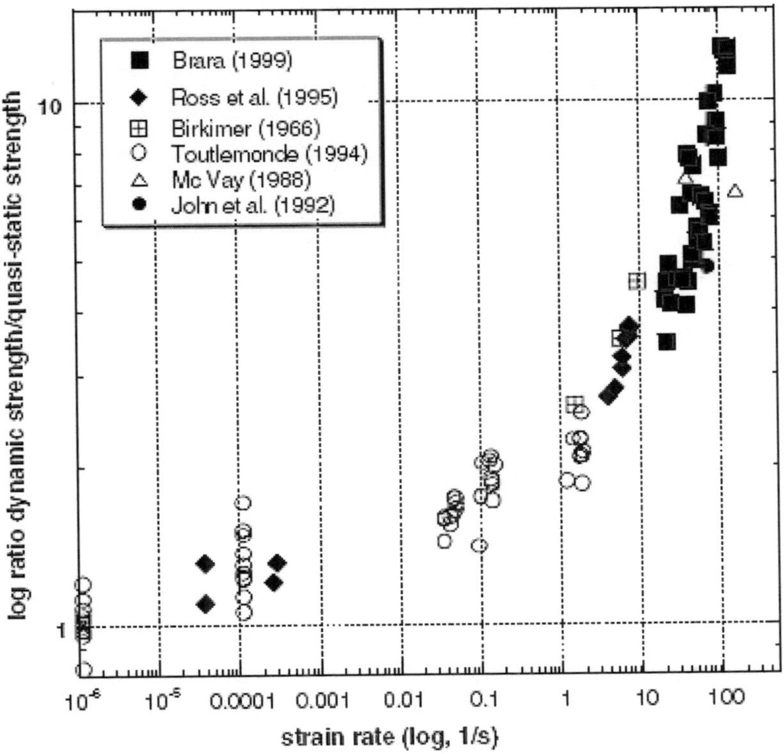

Figure 6.5. *Tensile strength increase versus strain rate [BRA 06]*

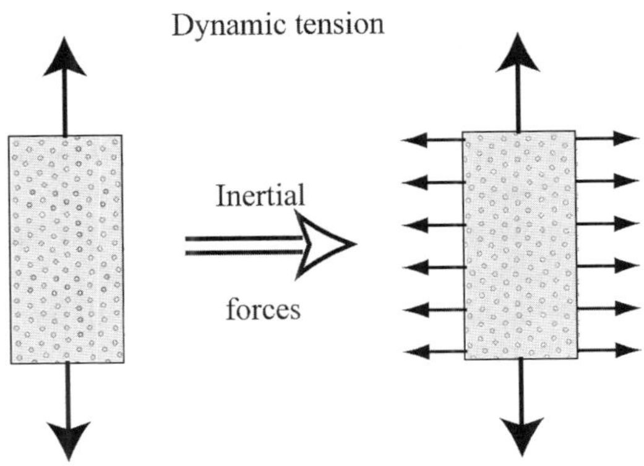

Figure 6.6. *Effect of inertial confinement in a) uniaxial compression and b) uniaxial tension*

Figure 6.7. *The Split Hopkinson Bar at Delft University of Technology and a detail of the glued specimen*

The superposition of the incident compressive wave and the reflected tensile wave generates a tensile stress that grows rapidly in time along the concrete specimen. Due to wave superposition, the tensile wave leads to spalling (tensile fracture) of the specimen at a certain distance from the free end, where the tensile stress reaches the critical value (see Figure 6.4). The whole process of wave propagation is recorded via the three strain gages glued on the Hopkinson bar surface at three specified distances. This arrangement allows for determination of the fracture stress caused by spalling, the stress history in the specimen, the critical time of loading, and the loading rate or strain rate.

The experimental results show a high rate sensitivity of tensile strength. Indeed, the tensile strength increases from four up to about 12 times when compared to quasi-static values, as shown in Figure 6.5a. Figure 6.5b presents results obtained from the literature. We can see from these figures that as in compression, two distinct modes appear in the increase of the tensile strength due to the high rate of strain. Moreover, for both modes, the increase seems to be higher than in compression.

In the first mode the dynamic resistance increase is probably due to the presence of water in material, but in the second mode, the Stefan effect is not sufficient. Numerical simulations [BRA 01] of this test obtained with the discrete elements method without accounting for viscous effects, showed that we were not able to reproduce the increase in tension resistance, successfully achieved by the same method in compression. It thus seems that the "inertial" effects do not have the same influence in traction (Figure 6.6) and that it might be necessary to account for other phenomena (e.g. crack speed). Some authors think that micro-inertia effects in the fracture process zone mainly cause the rate dependency of concrete in the high regime [WEE 01; WEE 03; BRA 06]. Nevertheless, with a probabilistic fragmentation analysis it is possible to introduce all microscopic aspects of fracture to describe both static and dynamic loading [DEN 00].

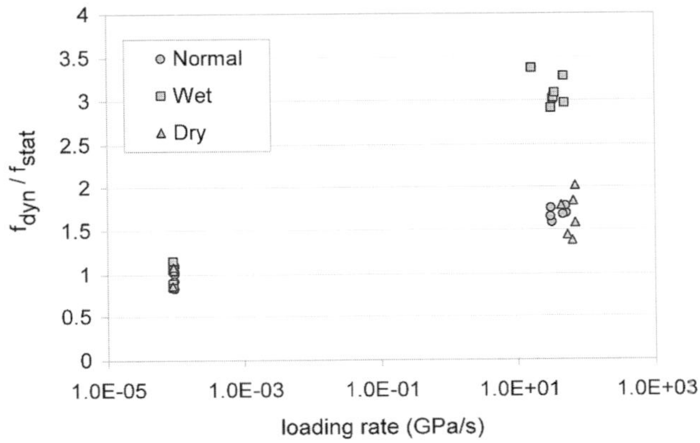

Figure 6.8. *The tensile strength ratio for dry, wet, and normal concrete as a function of the strain rate. (f_{stat} is the average value per test series)*

6.2.2.2. *Influence of moisture on the rate dependency of concrete*

If micro-inertia effects in the fracture process zone can explain the rate dependency of concrete in the high regime, the moisture content is assumed not to be dominant in this regime. However, in the moderate regime the moisture content is believed to play an important role in the strength increase of concrete [ROS 92; ROS 97; CAD 01; ROS 96]. The free water in the micropores is assumed to exhibit the so-called Stefan effect causing a strengthening effect in concrete with increasing loading rate [ROS 97]. This Stefan effect is the phenomenon that occurs when a viscous liquid is trapped between two plates that are separated quickly, causing a reaction force on the plates that is proportional to the velocity of separation. Cadoni *et al.* [CAD 01] give a different explanation for the influence of the moisture content. This interpretation is based on the principle of wave propagation in concrete. When a pore is not filled with water, it will locally reflect the incoming stress wave. The multiple reflections of all pores together can cause a considerable increase in stress. When a stress wave meets a pore that is filled with liquid the reflected stress is not big enough to cause the increase in stress that locally provokes the damage of the material. Therefore, the wet concrete will exhibit a more pronounced rate effect than the dry concrete. This interpretation only gives an explanation of the difference between wet and dry concrete and does not explain the increase in strength between static and dynamic loading.

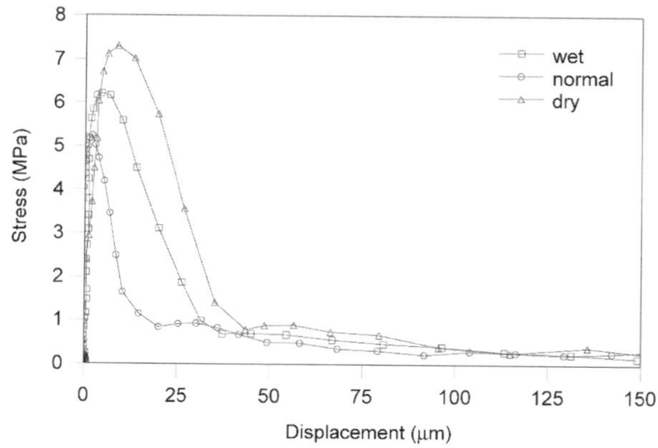

Figure 6.9. *Three stress-displacement curves for dry, normal, and wet concrete*

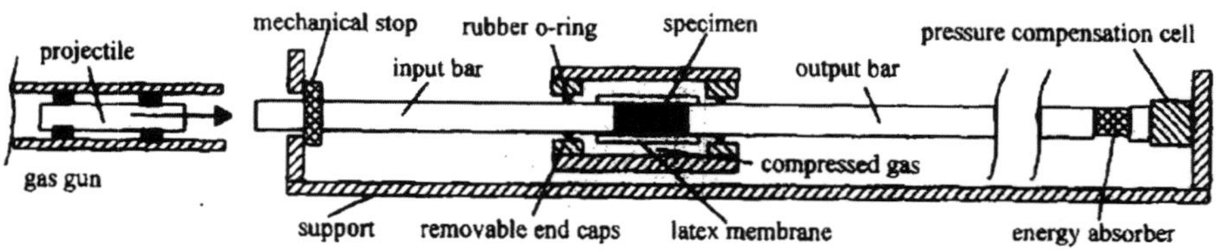

Figure 6.10. *Set-up for confined pressure tests*

To determine the dynamic tensile properties of concrete and to study the behavior under moderate loading rates the Stevin laboratory of Delft University of Technology developed a gravity driven Split Hopkinson Pressure Bar (SHPB). This experimental set up consists of two cylindrical aluminum bars (Ø 74mm) between which the concrete specimen (length 100 mm) is glued (Figure 6.7). The tensile stress wave is generated with a drop weight, which slides along the lower bar and hits an anvil at the end. In the applied set-up the strain-time relation is measured with strain gages at the upper bar, while the deformations are measured directly on the notched specimens. Notches are applied to avoid multiple fractures and limit data to a single fracture.

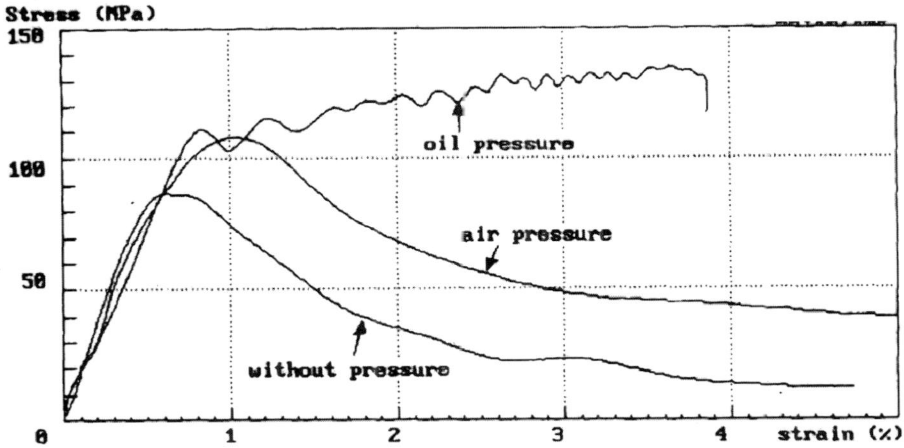

Figure 6.11. *Comparison between different lateral loading conditions (from [GAR 98])*

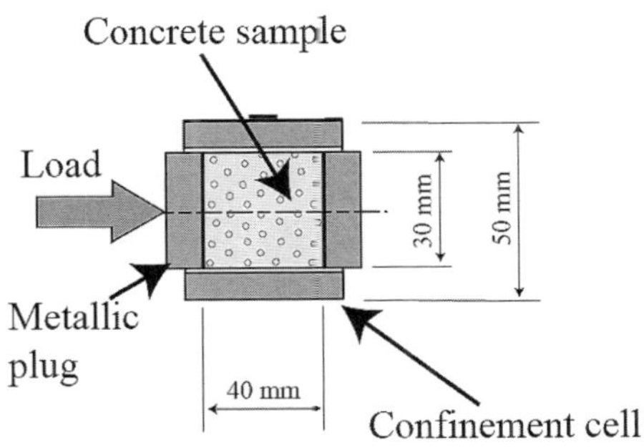

Figure 6.12. *Specimen used for quasi-oedometric tests (from [GAT 99])*

Figure 6.8 shows the results for three different curing conditions. From the figure it can be seen that the wet concrete exhibits a more pronounced rate effect than the normal and dry concrete. The absolute results and the dynamic-static strength ratio for the moderate loading rates of the wet and normal concrete are similar to results found by other researchers [CAD 01; ROS 96; ROS 92].

Figure 6.9 shows three stress-displacement diagrams obtained in the same tests. The stress displacement curves exhibit a pronounced kink shape in the descending branch. The tails of the curves seem not to be influenced by the higher loading rate. For all three moisture contents the rate effect seems to be most pronounced for the ascending branch and the first part of the descending branch. The peaks of the wet and dry concrete are much wider than the normal concrete. Also the peak value is reached at different displacement values. This indicates that the existence of microcracks, the initial crack extension, and final bridging of cracks play an important role in the rate dependency of concrete. The presence of many propagating microcracks causes an increase in damage of the material, and therefore, in the energy absorption, until the formation of a macrocrack takes place.

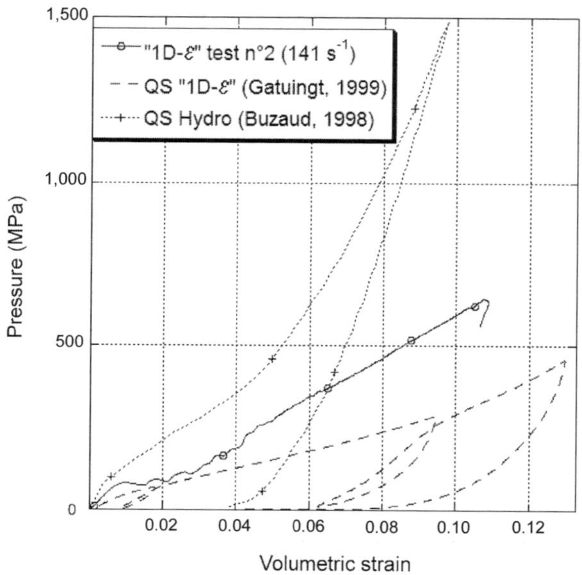

Figure 6.13. *Comparison of behaviors of a concrete based on measurements made with three different experimental devices, dynamic and static quasi-oedometric tests (from [GAT 99; FOR 08]) and purely hydrostatic compression tests (from [BUZ 98])*

6.2.3. *Dynamic behavior in multiaxial loading*

6.2.3.1. *Confined test with quasi-static pressure and split Hopkinson pressure bar [GAR 98]*

In this test developed at LMS-X a concrete specimen is introduced into a cylindrical quasi-static pressure cell [GAR 98]. The bars act as pistons and are introduced into the cell through sealing rings. A scheme of the complete set-up is shown in Figure 6.10. The lateral pressure can be applied with oil (up to 50 MPa), or with air (up to 10 MPa).

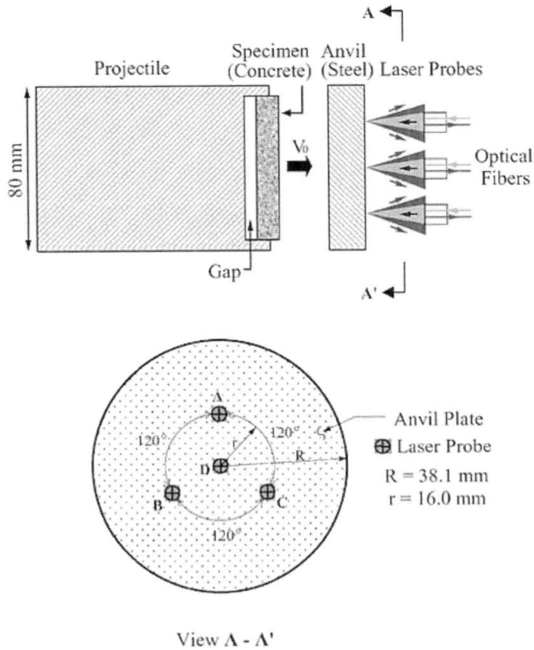

Figure 6.14. *Configuration of the plate impact experiment [GRO 01]*

Gary and Bailly [GAR 98] assumed that a static confined pressure applied with air is not significantly affected by the increase in the diameter of the specimen induced by the deformation. The situation is not so clear with oil, and it is unknown, because of transient effects in the fluid, whether a measurement of the oil pressure during the test at a point in the chamber would give an exact measure of the pressure applied to the specimen. To evaluate this point, they performed tests with oil and with air, using the same initial confinement pressure (Figure 6.11). Results under oil pressure are similar to the ones presented in [MAL 91] using a very similar device in which water lateral pressure was used.

When lateral pressure is applied with air, the stress-strain relation shows a much lower apparent strain hardening, as presented in Figure 6.11. The same kind of results has been obtained in tests on sand specimens [SEM 94] showing that oil pressure increases during the test. In that case, for similar tests (same specimen, same speed of loading, same initial lateral pressure), the stress associated with a given strain was minimum for air pressure tests, maximum for edometric (zero lateral strain) tests, and between the two for oil pressure tests.

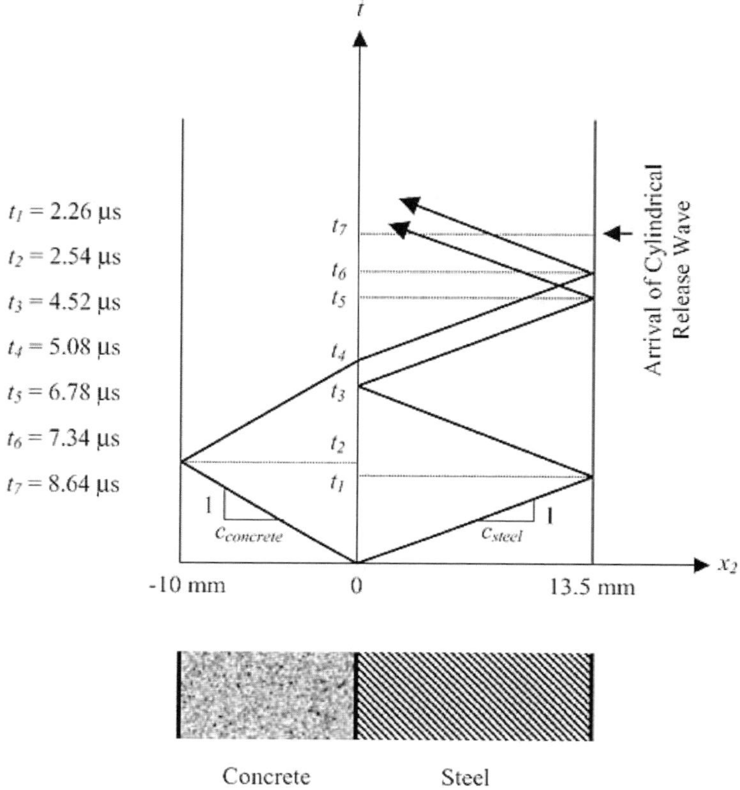

Figure 6.15. *Time-distance diagram for the plate impact experiment based on linear elastic material behavior [GRO 01]*

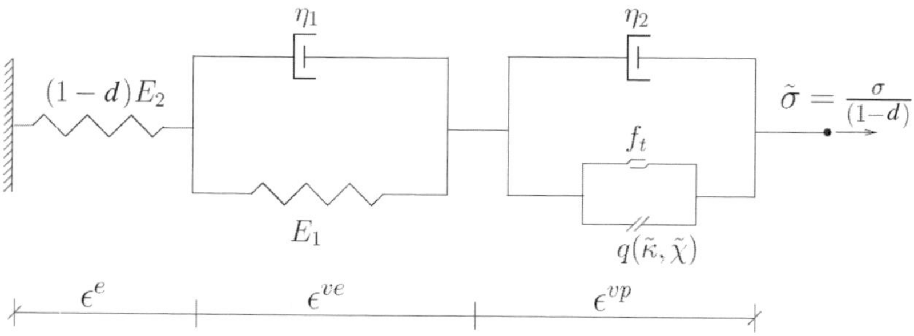

Figure 6.16. *Rheological model for the viscoelastic viscoplastic damage (VEVPD) model*

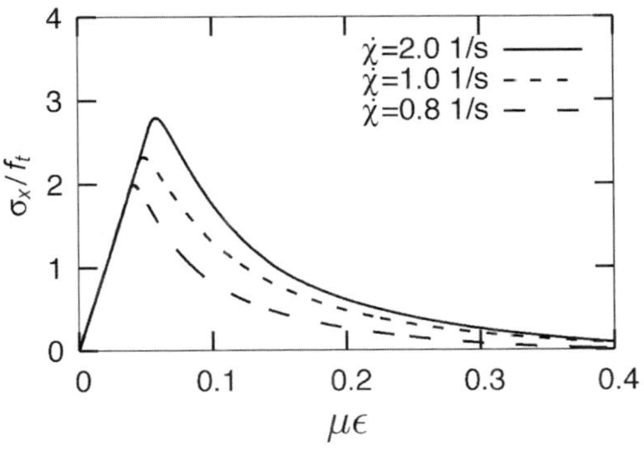

Figure 6.17. *Uniaxial response in tension for the VEVPD model proposed by Perdersen et al. [PED 08]*

6.2.3.2. *Quasi-edometric dynamic compression test [GAT 99]*

A very simple way to obtain lateral pressure consists of making quasi-edometric tests. Such a method has produced good results for sand testing [SEM 94; SEM 95] and has been extended to concrete in dynamic tests [GAT 99].

The specimen (Figure 6.12) is confined in a metallic ring and loaded by means of a hard-steel Hopkinson pressure bar (80 mm diameter, 6 m long) allowing for the testing of specimens large enough regarding the aggregate size. The constitutive law for the metal of the ring being known, transverse gages glued on its lateral surface allow for the measurement of the confining pressure. The hydrostatic and deviatoric responses of the specimen can then be computed. This method has been validated by several numerical simulations of tests involving a set of four different concrete-like behaviors and different friction coefficients between the cell and the specimen [FOR 08].

Figure 6.13 shows a comparison of results for the same concrete: dynamic and static quasi-edometric tests [GAT 99; FOR 08] and hydrostatic responses to pure hydrostatic loading [BUZ 98]. The concrete appears to be influenced by the loading rate and significantly depends on the loading path.

A high level of compaction strain (20%) can be produced with this testing device. It mostly operates in a one-dimensional strain situation. Stresses up to 1,000 MPa, confinement pressures up to 600 MPa and a strain-rate range between 100 and 500/s can be achieved to provide suitable data to evaluate the dynamic behavior of concrete and other rock-like materials under multiaxial dynamic loadings.

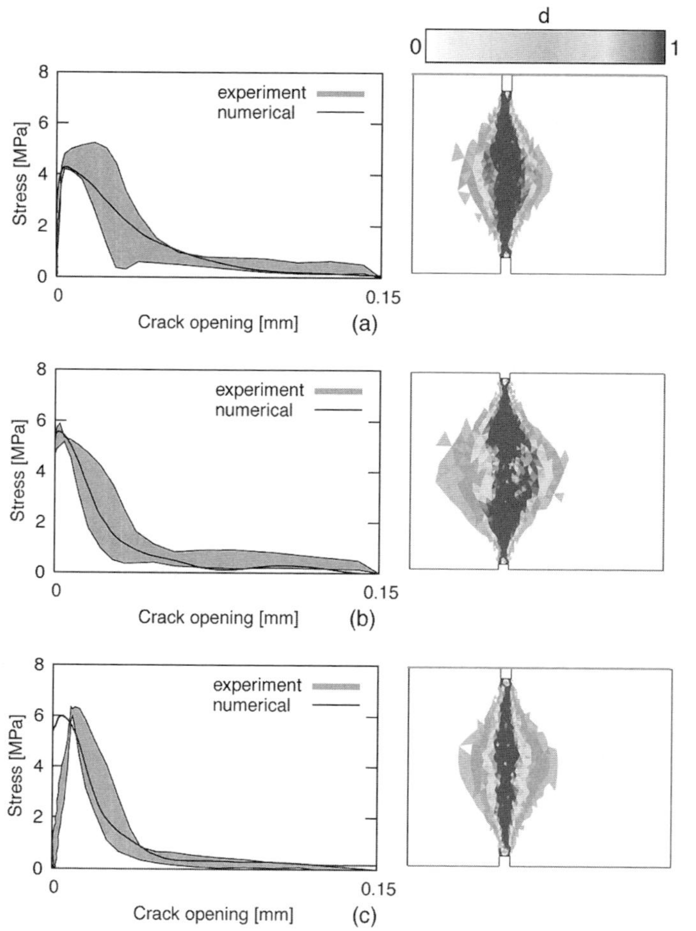

Figure 6.18. *Force-displacement curves (left) and corresponding fracture planes expressed in terms of the damage variable d (right) for the SHB apparatus for a) low, b) medium and c) high degree of saturation [PED 08]*

6.2.3.3. *Plate impact experiments [GRO 01]*

Plate impact provides a unique means to generate extremely high strain rates and high pressures under well-controlled conditions. In the experimental setup proposed by Grote *et al.* [GRO 01], normal impact experiments are conducted on circular disks of concrete to characterize their dynamic stress-carrying capacities. Average strain rates achieved are of the order of 10^4/s. Confining pressures are of the order of 1-1.5 GPa.

A schematic illustration of the impact configuration is shown in Figure 6.14. The specimens are 76.2 mm in diameter and 10 mm in thickness. The end surfaces are lapped flat and the actual thickness is within 0.0254 mm of 10 mm. The specimen is placed at the front of the projectile assembly. A gap between the disk specimen and the projectile tube is provided to allow a traction-free end condition for the back surface of the specimen during the impact process. The specimen impacts against an anvil plate made of hardened Hampden tool steel.

Upon impact, compressive stress waves are generated in both the specimen and the anvil plate. These waves propagate from the impact face toward the rear surface of the specimen and the rear surface of the anvil plate. Upon arriving at the rear free surfaces, these compressive waves are reflected as tensile waves. The reflected tensile waves then interfere destructively with the on-going incident compressive waves, reducing the compressive stresses in the specimen.

The time-distance diagram shown in Figure 6.15 indicates schematically how the waves propagate in the concrete specimen and the steel anvil plate during the impact process. The diagram is based on the one-dimensional wave propagation theory and the lines represent the longitudinal wave fronts at a given position and time. The construction of this diagram assumes that the materials involved are homogeneous and linearly elastic. This assumption does not consider the significant inelastic deformation and the material heterogeneity in the specimen. Therefore, this illustration should be viewed as an approximate representation for various wave fronts and a guide for experiment design. This diagram is also useful for the analysis and interpretation of experimental results. The experiment is designed such that the target steel plate remains elastic throughout the impact process. Although the stress and velocity at the impact face are not uniform due to the inhomogeneous specimen properties, elastic wave propagation in the target plate allows the stress and velocity to become more uniform as they approach the rear surface of the target plate. This wave propagation process can be used as a mechanism for obtaining an average measure for the stress history at the impact face.

Under the conditions of plate impact, significantly higher stresses are carried by concrete than what is observed for the materials under conditions of quasi-static, uniaxial compression. Based on the results of quasi-static uniaxial compression and plate impact experiments on mortar, approximately 42% of the strength enhancement is due to rate sensitivity and 58% is attributed to the effect of hydrostatic pressure.

Due to the material heterogeneity inherent in the concrete and mortar specimens, the stress and deformation are generally non-uniform. The plate impact configuration provides a means with which an average measure for the stress carried by the materials under dynamic conditions can be obtained experimentally.

6.3. Behavior modeling of concrete in dynamics

6.3.1. *Modeling of dynamic failure of concrete using a viscoelastic viscoplastic damage continuum model [PED 08]*

To reproduce the response of concrete structures exposed to extreme dynamic loading, reliable numerical material models, with rate effect accurately considered, are crucial. Inertia effects mainly cause rate-dependency in high-rate dynamics, whereas at moderate loading rates moisture in nano- and micropores influences the material properties. A constitutive model with viscous contributions to both the elastic and the inelastic parts, as proposed by [PED 08], is presented in this section. In dynamics, the viscoelastic part takes into account viscous phenomena due to moisture, such as dispersion, attenuation, and a frequency-dependent stiffness. Moreover, the viscoelastic contribution accounts for the Stefan effect, which is associated with the strength increase for intermediate loading rates [SER 98]. Conversely, on a macro level, the viscoplastic part represents irreversible thermodynamic changes, for example energy dissipation and change in stiffness, due to cracking in the microstructure. The coupled model shows objective results due to the viscous regularization of the plastic strains.

This model is an extension of the viscoelastic plastic model proposed by [SER 00] to account for viscoplasticity and damage, in which retardation of microcrack growth is taken into account. The corresponding rheological model is shown in Figure 6.16.

Strain decomposition is assumed:

$$\varepsilon = \varepsilon^{e} + \varepsilon^{ve} + \varepsilon^{vp}$$

where the strain tensor is split into elastic ε^{e}, viscoelastic ε^{ve} and viscoplastic ε^{vp} parts.

The viscoelastic strain contribution is expressed in a rate form according to

$$\dot{\varepsilon}^{ve} = \dot{\chi} b$$

where χ is an equivalent viscoelastic strain measure and b is defined as

$$b = \frac{s}{|s|}$$

where s the deviatoric stress tensor. The equivalent viscoelastic strain χ obeys the relation:

$$\eta_1^d \frac{\partial \chi}{\partial t} = |s| - E_1 \chi$$

where, with reference to the Kelvin element depicted in Figure 6.16, E_1 is the viscoelastic spring stiffness and η_1^d is the dynamic viscosity. The right-hand side of this equation can be interpreted as the macroscopic representation of the viscous effect due to moisture in nano- and micropores. It is similar to the hardening force in plasticity, which represents microcracking controlled by the equivalent plastic strain.

The viscoplastic strain contribution is expressed according to the formulation proposed by Perzyna. When the yield criterion is violated in the effective stress space, i.e. when f$\geq$0 with f the yield function, the viscoplastic strain rate is expressed in the associative form as:

$$\dot{\varepsilon}_{vp} = \frac{1}{\tau_2^d} \phi \frac{\partial f}{\partial \sigma}$$

The viscoplastic relaxation time is defined by $\tau_2^d = \eta_2^d / f_{tv}$ with η_2^d the viscosity of the damper element and f_{tv} is the tensile strength dependent on the viscoelastic strain rate $\dot{\chi}$. The over-stress function ϕ is defined as

$$\phi = \left(\frac{f}{f_{tv}} \right)^N$$

with f_{tv} the initial yield stress and N ($N \geq 1$) a real number.

An important component of this model is the hardening force, q, which defines the elastic domain of the material. This hardening force is made a function of both the equivalent viscoplastic strains and equivalent viscoelastic strains according to

$$q(\kappa, \chi, \dot{\chi}) = f_t g(\kappa) h(\chi, \dot{\chi})$$

where f_t is the static tensile strength. The function g, defined as an exponential softening curve and function of the equivalent viscoplastic strain κ, reads

$$g(\kappa) = \exp(-\beta_1 \kappa)$$

with β_1 a parameter governing the slope of the softening branch. The second function, h, reflects complex physical interaction between moisture and microstructural solid skeleton, one of the causes of the rate effect in concrete. The following viscous hardening force h proposed by [SER 00] has been adopted:

$$h(\chi,\dot{\chi}) = \left\{1 - \chi_a^d \log\left(\frac{\chi}{\chi_s}\right)\right\}$$

where, by using the log function, small values of the equivalent viscoelastic strain relative to χ_s result in an increase of the tensile strength.

As shown in the rheological model in Figure 6.16, the elastic stiffness E_2 is reduced by the damage variable d. The authors follow Ju [JU 89] in coupling plasticity and damage and by making use of the effective stress concept and the hypothesis of strain equivalence. In the context of a numerical procedure, the stress update relation at the end of a time step for the combined model reads

$$\sigma_{n+1} = (1 - d_{n+1})\tilde{\sigma}_{n+1}$$

where σ is the Cauchy stress tensor and d ($0 \leq d \leq 1$) a damage variable, which is updated through

$$d_{n+1} = 1 - \exp(-\beta_2 \kappa_{n+1})$$

β_2 regulates the slope of the damage evolution law, while κ_{n+1} is a measure of the deformation accumulated in the plastic regime.

The stress-strain responses related to the viscoelastic viscoplastic damage (VEVPD) model are shown in Figure 6.17 for three loading rates. As seen from the figure the model is able to represent the increase of the peak load when the loading rate increases.

The two viscous contributions aim to introduce rate effects due to moisture in micropores and micro-inertia. The viscoelastic contribution is related to the viscous hardening effect and is the main factor responsible for tensile strength increase. The viscoplastic part of the model regularizes the numerical problem due to the softening behavior.

In order to test their model, Pedersen *et al.* simulated the experimental tests obtained with the gravity-driven Split Hopkinson Bar (SHB) apparatus described in section 4.2.2; the results obtained are illustrated in Figure 6.18. The numerical force-displacement curves are in good agreement with the experimental results. However, the influence of the viscosity is more pronounced in dynamics than in statics as shown by the width of the FPZ; in the experiments the width of the FPZ in the SHB test is comparable to that obtained in static tests, but not in the numerical simulations.

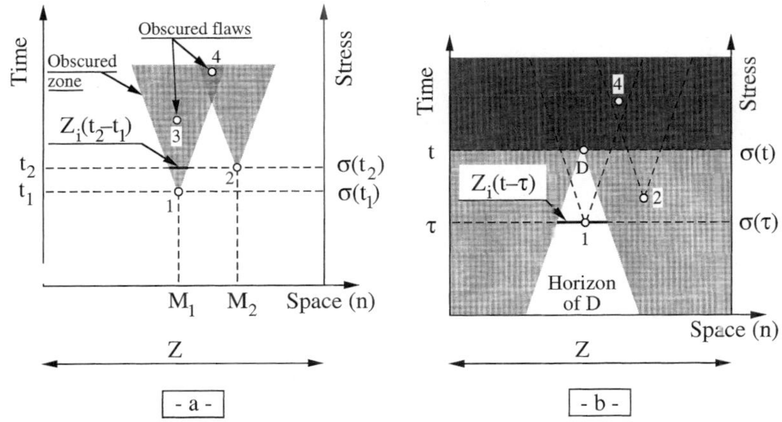

Figure 6.19. *a) Depiction of obscuration phenomena. b) Schematic of the obscuration/non-obscuration zones for a defect D*

6.3.2. *A damage model for the dynamic fragmentation of brittle solids [DEN 00]*

The aim of this section is to introduce all microscopic aspects of fracture to describe both static and dynamic loadings. The basis of the approach is an extended version of a Weibull model [WEI 39] applicable in both situations. Denoual and Hild [DEN 00] showed that the key distinction is between single and multiple fragmentation rather than static versus dynamic loadings. Their study was not designed for concrete-like materials, but the approach proposed has been successfully extended to them [FOR 03].

6.3.2.1. *Single fragmentation*

The failure of a ceramic in quasi-static tension is due to defects D_i of random location x defined by a failure stress $\sigma_i(x)$. When an equivalent stress $\sigma(x)$; e.g. maximum principal stress is greater than $\sigma_i(x)$; a crack emanating from the defect leads to the failure of the whole structure. Therefore, the ultimate strength of a ceramic specimen is not deterministic and a failure probability, P_F, must be defined and can be described by a Weibull law [WEI 39]:

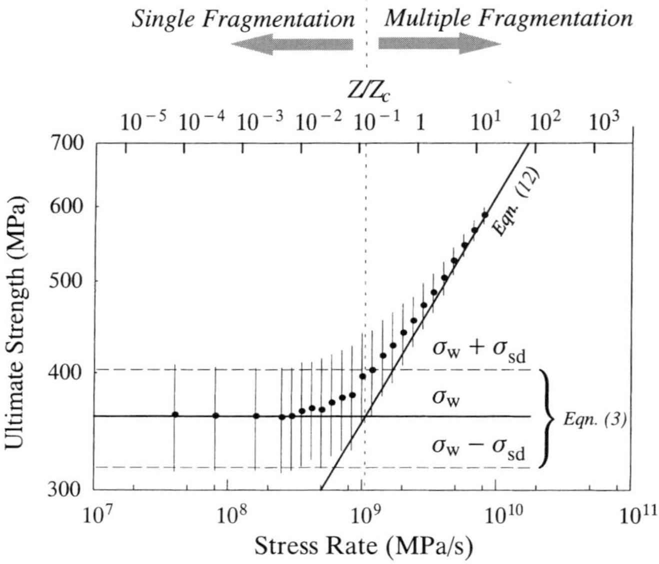

Figure 6.20. *Ultimate strength versus stress rate (SSiC)*

$$P_F = 1 - \exp\left[-\lambda_t(\sigma) Z_{eff}\right]$$

$$\lambda_t(\sigma) = \lambda_0 \left(\frac{\sigma}{S_0}\right)^m$$

where $\lambda_t(\sigma)$ is the defect density, m the Weibull modulus, S_0 the scale parameter relative to a reference density λ_0, and Z_{eff} the effective volume, surface, or length [DAV 73]. It can be noted that the previous formulation enters the framework of a Poisson point process of intensity λ_t.

6.3.2.2. *Multiple fragmentation*

In the bulk of an impacted ceramic, damage in tension is observed when the hoop stress induced by the radial motion is sufficiently large to generate fracture in mode I initiating on microdefects. When a fracture is initiated, the stress state around the propagating crack is a complex function of time, crack velocity and stress wave celerity. In order to simplify the following development, the shape of the affected (or interaction) zone (volume, surface or length) Z_i is supposed to be constant, i.e. all the interaction zones are self-similar and Z_i can be written as

$$Z_i(t-\tau) = S[kC(t-\tau)]^n$$

where kC is the velocity of a propagating crack, S a shape parameter, $C=\sqrt{E/\rho}$ a stress wave velocity (E is Young's modulus and ρ the mass density); therefore $kC(t-\tau)$ is a representative length of the relaxation zone at time, t, around a broken flaw at time τ. The power n=1, 2, 3 is the space dimension. The shape parameter S is defined in order to have $d\sigma/dt \leq 0$ in Z_i, i.e. no new nucleation can occur in Z_i. When the space dimension is n=3, it is assumed, as a first approximation, that the obscuration (or shielding) zones are spherical and $S=4\pi/3$. By studying the dynamic crack propagation, the value k=0.38 is usually assumed [KAN 85]. To understand why a crack nucleates, the interaction of a nucleated defect and other defects that would nucleate has to be modeled (more details are given in [DEN 00]).

To understand why a crack nucleates, the authors model the interaction of a nucleated defect and other defects that would nucleate. The space location of the defects is represented in a simple abscissa of an x-y graph where the y-axis represents time (or stress) to failure of a given defect. In this graph, a shaded cone represents the expansion of the interaction zone with time due to nucleation and propagation of a crack. A section Z (see Figure 6.19a) of a cone can be a volume, a surface, or a length, depending on the space dimension n. The defects outside the shaded cones can nucleate and produce their own increasing interaction zone (e.g. defects 1 and 2 of Figure 6.19a). Inside the cones, the defects that should have broken do not nucleate (e.g. defects 3 and 4 of Figure 6.19a) as they are shielded (or obscured). For a given flaw, D, a non-interaction zone can be defined so that, in this zone, a defect cannot obscure D (Figure 6.19b) and the horizon of D in which a defect will always obscure D. The total flaw density λ_t can, therefore, be split into two parts: λ_b (the broken flaws) and λ_{obsc} (the obscured flaws). Furthermore, they assume that a Poisson point process of intensity $\lambda_t(\sigma(t))$ models the distribution of total flaws in a zone Z. New cracks will initiate only if the defect exists in the considered zone and if no defects exist in its horizon:

$$\frac{d\lambda_b}{dt}(\sigma(t)) = \frac{d\lambda_t}{dt}(\sigma(t))[1-P_0(t)]$$

where P_0 is the probability of obscuration expressed as:

$$P_0(t) = 1-\exp(-\lambda_t(\sigma(t))\wp(t))$$

where $\wp(t)$ denotes the mean obscuration zone.

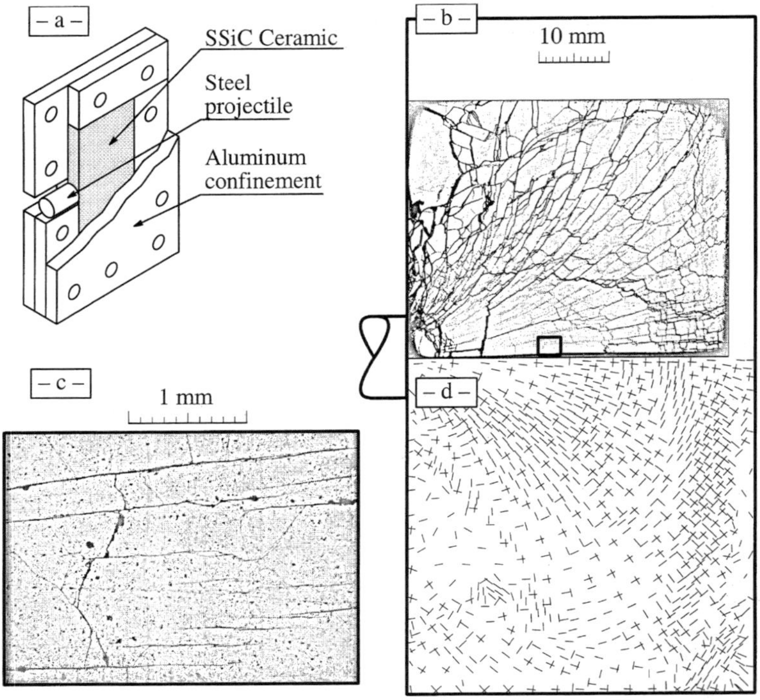

Figure 6.21. *a) Diagram of the ceramic confinement. b) Half top view of tile made of SSiC impacted at a velocity of 330 m/s. c) The magnification (of the boxed area of b)) shows many short radial cracks linked by short circumferential ones. d) Half bottom view of the corresponding numerical simulation. The lines depict the crack directions (perpendicular to the direction of the associated damage) [DEN 00]*

The variable P_0 can be used to define a damage variable in the framework of Continuum Damage Mechanics. By averaging over a representative zone, P_0 is equal to the damage variable D.

Figure 6.20 shows the evolution of the ultimate stress versus stress rate in a volume Z equal to Z_{eff} at 1.25 mm^3. The lines represent analytical solutions, while the dots and error bars represent Monte Carlo simulations of an underlying Poisson point process. For a stress rate within 0.5×10^8 MPa/s, the ultimate stress is not modified by the loading rate. During the single/multiple fragmentation transition, the difference between the solid lines and the Monte Carlo simulations does not exceed 10%. The standard deviation significantly decreases in the case of multiple fragmentations when the stress rate increases. Even if the ultimate strength has to be defined for both static and dynamic loadings by a mean value and a standard deviation, one can see that dynamic loadings lead to a more "deterministic" behavior in the multiple fragmentation regime. The probabilistic nature of quasi-static failure progressively vanishes with high stress rates and, therefore, allows a continuum description of fragmentation by using damage variables.

To test the model, Denoual and Hild [DEN 00] simulated an edge-on impact (EOI) with an aluminum confinement (called a sarcophagus) presented in Figure 6.21a. After impact, the tile is coated in an epoxy resin and polished for macroscopic and microscopic analyses. The post-mortem crack pattern is presented in Figure 6.21b and c. Different zones can be separated with respect to the crack density and orientation.

In front of the projectile, a small zone exhibits a randomly oriented crack pattern. In the bulk of the ceramic, long radial cracks with a second circumferential crack pattern superimposed on the first can be observed. In this last region, some radial cracks seem to kink in the circumferential direction, kink again, and propagate in the radial direction. The kinked fractures are made of two small (and hardly visible, see Figure 6.21c) cracks linked by a small circumferential crack. This complex crack pattern is then opened by the radial motion of the ceramic, widens, and becomes a long macroscopic "kinked" fracture. This phenomenon shows that the radial cracks appear prior to the circumferential ones, as the latter does not go through the former. Near the tile rear face, a third zone exhibits a high density of cracks. This thin strip remains at a constant distance from the rear surface, like a spalling zone. The numerical prediction is presented in Figure 6.21d. The cracking mechanism can be described as the superposition of a radially and a circumferentially oriented crack pattern, the former with a higher density than the latter. The radially oriented crack pattern is generated during the first microseconds of impact. It is followed by the circumferential ones due to the relaxation wave emanating from the projectile and the rear face of the ceramic tile. This result is in accordance with the experimental observations. The zone in front of the projectile, in which only few radial cracks seem to have nucleated, is well reproduced. Finally, the high-density crack zone resembling a spalling region is also well described, with a crack direction approximately in the radial and circumferential direction, as observed in the sarcophagus configuration.

6.3.3. *Anisotropic three-dimensional delay-damage model [GAT 08]*

High dynamic loadings lead to material degradation and structural failure. This is even more the case for concrete structures where the parts initially in compression break in tension due to wave propagation and reflection. The dissymmetry of the material behavior plays a major role in such cases, dissymmetry mainly due to damage-induced anisotropy. Loading-induced damage is most often anisotropic and one proposes here to take advantage of such a feature to build a damage model for concrete, dissymmetric in tension and in compression, 3D, suitable for dynamic computations. A single 2nd order tensorial damage variable D is considered with a damage law ensuring a damage rate proportional to the square of the positive part of the strain tensor.

6.3.3.1. *Viscous regularizations*

For many damage and plasticity models, the elasticity domain is defined through the introduction of a criterion function f such as the domain $f<0$ corresponds to elastic loading or unloading, and the consistency conditions $f = 0, \dot{f} = 0$ correspond to damage evolution and/or yielding. A regularization possibility is then to introduce a characteristic time that, altogether with the consideration of the laws of dynamics, indirectly defines a characteristic length. In the present case of elasticity coupled with damage this is simply done by introducing a viscosity law $\varepsilon_v = \varepsilon_v\left(\dot{D}\right)$ in the damage criterion. The damage evolution does not occur anymore at $f=0$, but at $f=\varepsilon_v>0$. A classical law for isotropic damage is the Norton-Perzyna power law, $\varepsilon_v = k\dot{D}^{1/n}$, with k and n the viscosity parameters (see also [DUB 94; GAT 02]). It leads to an unbounded damage rate often too high at high strain rates. It is possible to bound the damage rate, for instance by the maximum rate $\dot{D}_\infty = 1/\tau_c$ material dependent equal to the inverse of the characteristic time τ_c [LAD 98; ALL 97; SUF 04].

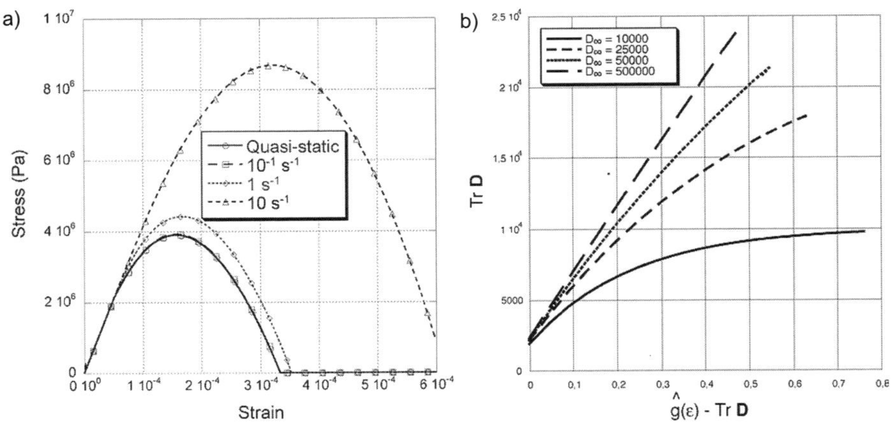

Figure 6.22. *a) Stress-strain curves for concrete at different loading rates. b) Saturation of the damage rate for different limit damage rate*

To gain this property, these authors rewrite the criterion surface as

$$f = \kappa^{-1}(\hat{\varepsilon}) - D$$

and define the viscosity law as

$$f = D_v > 0 \text{ with } D_v = -\frac{1}{b}\ln\left(\frac{\dot{D}_\infty - \dot{D}}{\dot{D}_\infty}\right)$$

from which derives the delay damage law, saturating at high strain rates:

$$\dot{D} = \dot{D}_\infty\left[1 - \exp\left(-b\left(\kappa^{-1}(\hat{\varepsilon}) - D\right)\right)\right]$$

The viscosity parameters, material dependent, are then $\dot{D}_\infty$ and b. This regularization is defined locally (i.e. at a structure Gauss point) and is well adapted for dynamics computations. The authors extend it to the case of induced anisotropic damage.

For concrete, the microcracks due to tension are mainly orthogonal to the loading direction, when the microcracks due to compression are mainly parallel to the loading direction. The damage state has then to be represented by a tensorial variable D. The use of a second-order damage tensor is more convenient for practical applications (as well as for the material parameters identification) and this is the choice made here. The damage anisotropy induced by either tension or compression is then simply modeled by the consideration of damage evolution laws ensuring a damage rate proportional to the positive part of the strain tensor, i.e. a damage governed by the principal extensions [MAZ 86; DRA 98].

The basis of the model presented in this section is the rate-independent anisotropic damage model proposed by [DES 07]. According to the thermodynamics framework, the single damage variable D is considered and a single set of material parameters is valid for tension and compression.

The full set of constitutive equations including anisotropic delay-damage reads:

Elasticity:

$$\varepsilon = \frac{1+\nu}{E}\tilde{\sigma} - \frac{\nu}{E}tr\tilde{\sigma}1 = \underline{E}^{-1} : \tilde{\sigma}$$

with E the Young's modulus, ν the Poisson's ration and $\underline{E}$ the Hooke's tensor.

Effective stress:

$$\tilde{\sigma} = \left[(1-D)^{-1/2}\sigma^D(1-D)^{-1/2}\right]^D + \frac{1}{3}\left[\frac{\langle tr\sigma\rangle_+}{1-trD} + \langle tr\sigma\rangle_-\right]1$$

where $(\bullet)^D$ denotes the deviatoric part, $\langle\bullet\rangle_+$ (resp. $\langle\bullet\rangle_-$) the positive (resp. negative) part of a scalar.

Mazars damage criterion:

$$f = \kappa^{-1}(\hat{\varepsilon}) - trD, \hat{\varepsilon} = \sqrt{tr\langle\varepsilon\rangle_+^2}$$

using the viscous regularization previously seen, one obtains

$$f = -\frac{1}{b}\ln\left(\frac{\dot{D}_\infty - tr\dot{D}}{\dot{D}_\infty}\right)$$

$\langle\varepsilon\rangle_+$ is the positive part of the strain tensor build from the positive Eigen strains. The material parameters $\dot{D}_\infty$ and b are the delay-damage parameters and the κ^{-1} function is set as

$$\kappa^{-1}(\hat{\varepsilon}) = aA\left[\arctan\left(\frac{\hat{\varepsilon}}{a}\right) - \arctan\left(\frac{\kappa_0}{a}\right)\right]$$

introducing κ_0 as damage threshold, A and a as damage parameters.

Induced damage anisotropy governed by the positive extensions:

$$\dot{D} = \dot{\lambda}\langle\varepsilon\rangle_+^2$$

The damage multiplier $\dot{\lambda}$ is determined from the damage criterion expression for *f>0*. The delay-damage law is recovered from previous equations and extended to induced anisotropy as:

$$tr\dot{D} = \dot{D}_\infty\left[1 - \exp\left(-b\left(\kappa^{-1}(\hat{\varepsilon}) - trD\right)\right)\right]$$

The use of a damage criterion function f written in terms of strains instead of stresses altogether with the logarithmic regularization allows for a simple implementation in a finite element computer code. Note that at the final stage of the numerical implementation the elasticity law needs to be inverted. This can be done in a closed form as:

$$\sigma = (1-D)^{1/2}\tilde{\sigma}(1-D)^{1/2} - \frac{(1-D):\tilde{\sigma}}{3-trD}(1-D)$$
$$+\frac{1}{3}\left[(1-trD)\langle tr\tilde{\sigma}\rangle_{++} + \langle tr\tilde{\sigma}\rangle_{-}\right]1$$

Figure 6.22a shows the monotonic stress-strain curves for concrete in tension. Quasi-static and dynamic responses (at different strain rates) are plotted. The material parameters adequately describing concrete quasi-static behavior are: E=42 GPa, ν=0.2, κ_0=5×10^{-5}, A=5×10^{3}, a=2.93×10^{-4}. The viscous regularization parameters are b=1 and $\dot{D}_\infty$ =50,000/s. The rate effect obtain with these parameters are closed to those experimentally observed.

Figure 6.22b shows different damage rate evolutions for $b\dot{D}_\infty = 50{,}000$ constant.

The numerical scheme of the model is fully implicit, and therefore, robust, but it has the main advantage of the explicit schemes: a local iterative process is not required as the exact solution of the discretized constitutive equations is explicit. Moreover when damage reaches large values, one must be careful to ensure the damaged elastic tensor remains positively defined. This is done for induced anisotropic damage by using a specific procedure for the numerical control of rupture [DES 07].

The anisotropic delay-damage model has been implemented in the explicit finite element code LS-Dyna. In order to evaluate the ability of the anisotropic damage model to describe concrete behavior in a rather complex case, but representative of an industrial application, a test in which a projectile impacts a concrete slab has been carried out [GAT 08]. The projectile is a cylinder representative of a Cessna engine (mass=200 kg, velocity=83.3 m/s, cross-section=1 m^2) with an elastic behavior. Figure 6.23 shows the finite element mesh used for the simulations on a 4 m width and 0.5 m thick slab. The slab is meshed with 24,000 three-dimensional underintegrated elements and the reinforcements are represented by 2,300 truss-bars. The impacted area has a refined mesh whereas the remaining part of the slab has a coarse one.

Figure 6.24 illustrates the damages D11, D22, and D33 on the slab. It can be noted that due to the symmetry condition, damages D11 and D22 have a similar pattern. Damage D33 represents the cracks in the slab thickness and is representative of the scabbing phenomenon. In the simulation, damages D11 and D22 are quite large exhibiting a shear rupture of the concrete slab with the apparition of a punch cone as experimentally observed in cases of thin slabs. At the same time, damage D33 remains small and does not exhibit scabbing. Theses results are in a good agreement with the expected results.

6.3.4. *Damage coupled to plasticity for high strain rate dynamics [GAT 02]*

Existing constitutive relations for alloys and ceramics inspire most approaches, where a split between the spherical and deviatoric parts of the stress-strain relations is assumed. The deviatoric part is often modeled with a plasticity-based, or viscoplastic model, whereas the spherical part, also called "equation of state", is fitted with a cap model [LUB 96]. A constitutive relation for concrete in dynamics based on viscoplasticity combined with rate-dependent continuum damage is presented in this section. This relation is restricted to cases with moderate strain rates in concrete, corresponding typically to explosions and impacts of projectiles at a velocity less than 350 m/s, yielding a hydrostatic pressure in the material that is less than 1 GPa.

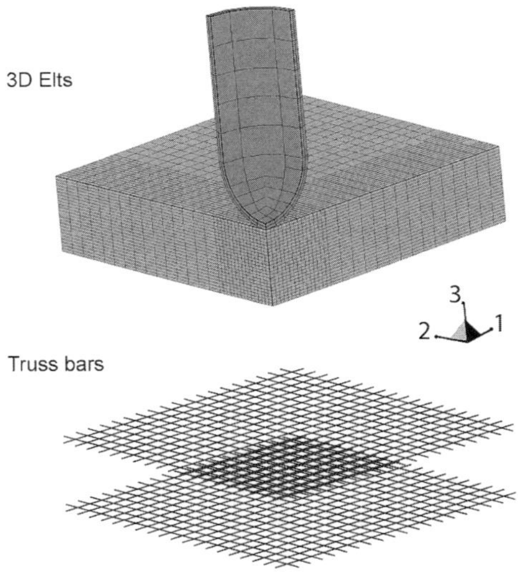

Figure 6.23. *Finite element mesh*

6.3.4.1. *Constitutive equations*

Let us consider the case of an impact on a concrete slab and look at the various modes of deformation and failure in concrete. Near the striker, inertial forces in concrete are such that the material is confined. Concrete experiences a very high state of triaxial compression. The equation of state (e.g. the curve relating the volumetric deformation to the hydrostatic stress) exhibits an increase of the tangent and decreasing volumetric moduli with residual plastic strains upon unloading. The hydrostatic stress at a given volumetric strain increases with the strain rate. It is generally accepted that the observed hardening phenomenon is due to irreversible material compaction at a given strain rate.

Distant from the impactor, concrete is subjected to compression with a moderate amount of confinement. Its response is essentially controlled by microcracking in compression and internal friction. Constitutive relations based on plasticity or plasticity coupled to damage can capture this state of stress.

The third failure mechanism is scabbing and occurs following the impacted structure. This failure mechanism is almost rate dependent.

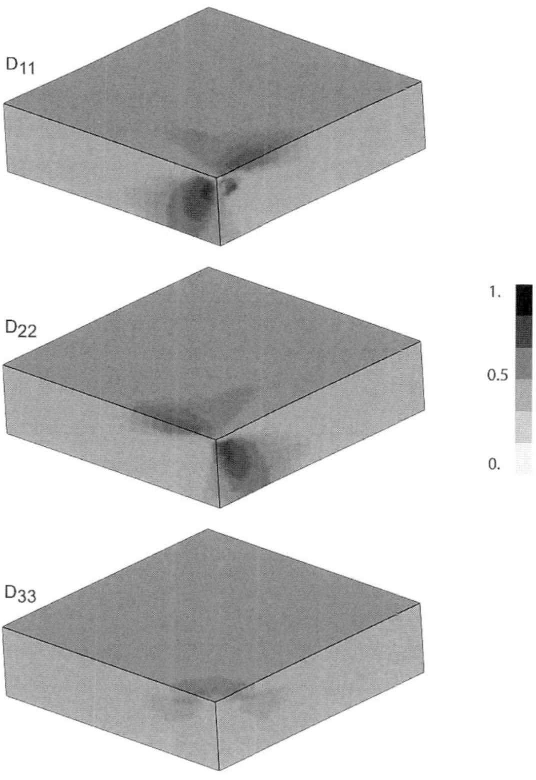

Figure 6.24. *Damage map of D11, D22, and D33 into the slab*

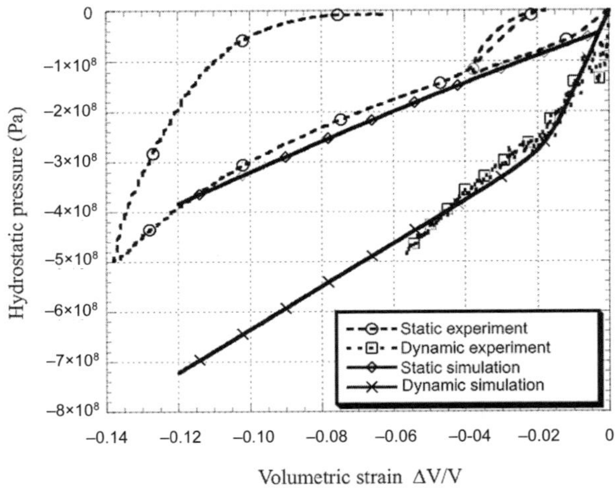

Figure 6.25. *Calibration of the viscosity parameter on the hydrostatic response of the model*

An accurate description of the response of concrete subjected to impacts and explosions requires the combination of at least the three of the abovementioned features. As a consequence, the constitutive relations presented herein are based on three main characteristics:

The variation in porosity of concrete is taken into account with the help of a homogenization technique. Thus, the elastic properties of the material are functions of the variation of porosity, especially during hydrostatic loadings.

Microcracking is captured with a rate-dependent isotropic damage model [DUB 96]. Rate effects are necessary to reproduce the results of dynamic experiments (mostly dynamic tensile tests). In addition, rate dependency preserves well posedness of the equations of motion when strain softening occurs [SLU 93].

In order to describe the material response in hydrostatic compression, a viscoplastic model is implemented, based on Perzyna's approach associated with a modified Gurson's yield function [NEE 84]. The modification of the plasticity model by the rate effect is needed to capture the increase of stress with strain rate in hydrostatic compression experimentally observed in dynamics [GAT 99].

These mechanical effects are combined in the relationships that relate the stresses to the elastic strains:

$$\sigma_{ij} = (1-D)\left[K\varepsilon_{kk}\delta_{ij} + 2G\left(\varepsilon_{ij} - \frac{1}{3}\varepsilon_{kk}\delta_{ij}\right)\right]$$

where the shear *G* and bulk moduli *K* are defined by Mori-Tanaka's expressions:

$$K = \frac{4K_M G_M (1-f^*)}{4G_M + 3K_M f^*}$$

$$G = \frac{G_M (1-f^*)}{1 + \frac{6K_M + 12G_M}{9K_M + 8G_M} f^*}$$

where K_M and G_M are the bulk and shear moduli of the material without pores, respectively.

In this model the damage is isotropic and represented by a scalar variable *D*. The damage growth is defined by:

$$D = \alpha_c D_c + \alpha_t D_t$$

where D_c and D_t represent the compressive and tensile damage respectively, while α_c and α_t are parameters defined as:

$$\begin{cases} \textit{uniaxial tension} & \alpha_t = 1, \alpha_c = 0 \\ \textit{uniaxial compression} & \alpha_t = 0, \alpha_c = 1 \\ \textit{general loading} & \alpha_t + \alpha_c = 1 \end{cases}$$

The definition of these parameters is given in the original work of Mazars. The growth of the two damage variables is governed by the elastic equivalent strain [MAZ 86]:

$$\tilde{\varepsilon}^e = \sqrt{\sum_i \left\langle \varepsilon_i^e \right\rangle_+^2}$$

where ε_i^e is the *i*th component of the tensor of the principal strains and $\left\langle x \right\rangle_+$ is the positive part of *x*.

The growth of D_c and D_t is defined by the following equations, which are similar to the equations used by Dube [DUB 96]:

$$\dot{D}_{c,t} = \left(\frac{\left\langle \tilde{\varepsilon}^e - \varepsilon_{D0} - \frac{1}{a_{c,t}} \left(\frac{D_{c,t}}{1 - D_{c,t}} \right)^{1/b_{c,t}} \right\rangle}{m_{D_{c,t}}} \right)^{n_{D_{c,t}}}$$

where m_{Dt}, n_{Dt}, m_{Dc}, and n_{Dc} are material parameters that control the rate effect. a_c, a_t, b_c, b_t are material parameters that govern the growth of damage in quasi-static tension and compression and ε_{D0} is the initial threshold of damage.

The experiments showed that there was a dependence of the loading rate on the curve relating the volumetric strain to the hydrostatic stress [GAT 99], in addition to permanent plastic strains and to the material compaction which induces an increase of the bulk and shear moduli of the material. It is thus pertinent to use a viscoplastic model. Within the classical framework of (small strain) elastoplasticity, we use the basic assumption of additive strain decomposition:

$$\dot{\varepsilon}_{ij} = \dot{\varepsilon}_{ij}^{e} + \dot{\varepsilon}_{ij}^{vp}$$

where $\dot{\varepsilon}_{ij}$ is the total strain rate, $\dot{\varepsilon}_{ij}^{e}$ the elastic one and $\dot{\varepsilon}_{ij}^{vp}$ the viscoplastic one.

The viscoplastic strains are obtained following Perzyna's approach:

$$\dot{\varepsilon}_{ij}^{vp} = \lambda \frac{\partial F_{NT}}{\partial \sigma_{ij}}$$

F_{NT} is the modified Gurson's yield function proposed by Needleman and Tvergaard [NEE 84]:

$$F_{NT}\left(\sigma_{ij}, \sigma_M, f^*\right) = \frac{3J_2}{\sigma_M^2} + 2q_1 f^* \cosh\left(q_2 \frac{I_1}{2\sigma_M} \right) - \left(1 + (q_3 f)^2\right) = 0$$

where σ_M is the stress in concrete without voids and q_1, q_2, q_3 are scalars parameters.

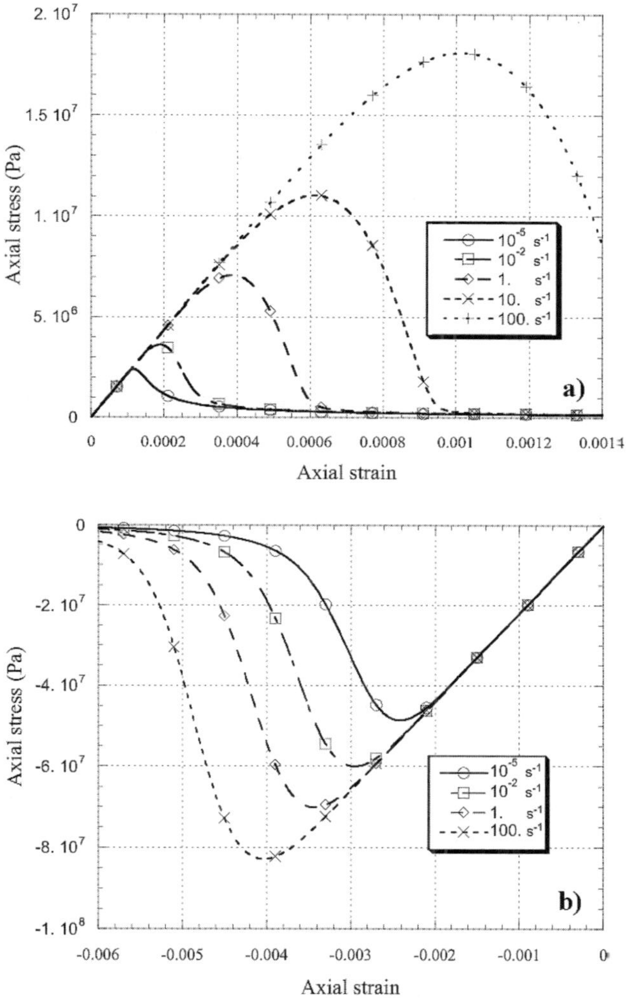

Figure 6.26. *Model response for several strain rates: a) uniaxial tension test; b) uniaxial compression test*

Colantonio and Stainier [COL 96] proposed a similar model in which the definition of the plastic multiplier accounts for the variation of porosity of the material. The same approach has been followed and the viscoplastic multiplier is noted as:

$$\dot{\lambda} = \frac{f^*}{1-f^*}\left\langle \frac{F_{NT}}{m_{vp}} \right\rangle^{n_{vp}}$$

where m_{vp} and n_{vp} are material parameters. As in the model by [BUR 00], the evolution of the porosity is controlled by the irreversible volumetric strain:

$$\dot{f}^* = k(1-f^*)f^*\dot{\varepsilon}_{kk}^{vp}$$

where a parameter k is introduced enable calibration of the velocity with which porosity is closed. Figure 6.25 shows a comparison of the hydrostatic responses between static and dynamic simulations and the corresponding experiments.

Figure 6.26a shows the response of the model in uniaxial tension tests carried out at various strain rates. The response is strongly dependent on the rate of strain, which is in agreement with experiments. Figure 6.26b shows the response of the model in uniaxial compression. It can be noted that the compression behavior is only slightly dependent on the strain rate. This is quite consistent with experimental data. It should be noted that the model response is modified in compression, due to the coupling with the viscoplastic part of the model.

Figure 6.27a shows the response of the model for hydrostatic compression followed by hydrostatic tension. A hardening behavior can be seen when the porosity of the material decreases. During unloading, the modulus of elasticity of material is higher than at the beginning of the loading (this is due to the decrease of porosity). Finally, when tension is reached, the rate-dependent damage model is recovered, with the elastic constants K and G, which have increased due to compaction. A static loading and a dynamic loading were carried out in order to show the influence the strain rate on the model response.

With regard to the material behavior in uniaxial compression, the values of the model parameters (particularly m_{Dt} and σ_{M_0}) provide a behavior that is either coupled or not, in the sense that the viscoplastic criterion and the damage criterion can be activated at the same time or not. Figure 6.27b shows, in the case of a dynamic loading, a curve where the model parameters are such that, in one case, the damage criterion is activated only (no irreversible strains), and in the second case, the two criteria are activated simultaneously. In the second case, a decrease of stiffness occurs at the same time that incremental viscoplastic strains are non-zero. Obviously, the first case is much easier to handle as far as calibration is concerned because viscoplasticity is restricted to cases where the material is subjected to compression with a significant amount of confinement.

In order to evaluate the model on a case representative of a real application, a test where an explosive is placed in contact with a concrete slab was carried out [GAT 02]. For the sake of simplicity, the explosive, as well as the concrete slab, were axisymmetric. The circular concrete slab had a diameter of 1 m and a thickness of 20 cm (Figure 6.28).

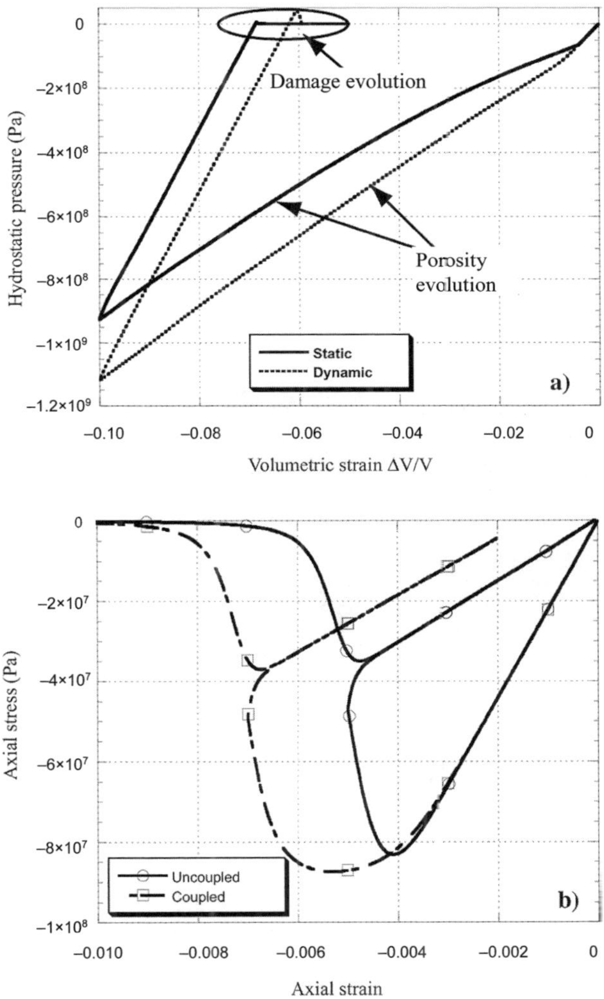

Figure 6.27. *Model response for several strain rates: a) uniaxial tension test, b) uniaxial compression test*

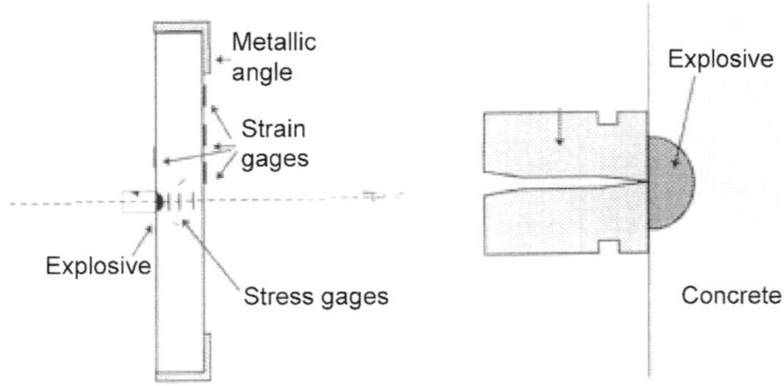

Figure 6.28. *Explosion at the surface of a concrete slab: description of the experimental device*

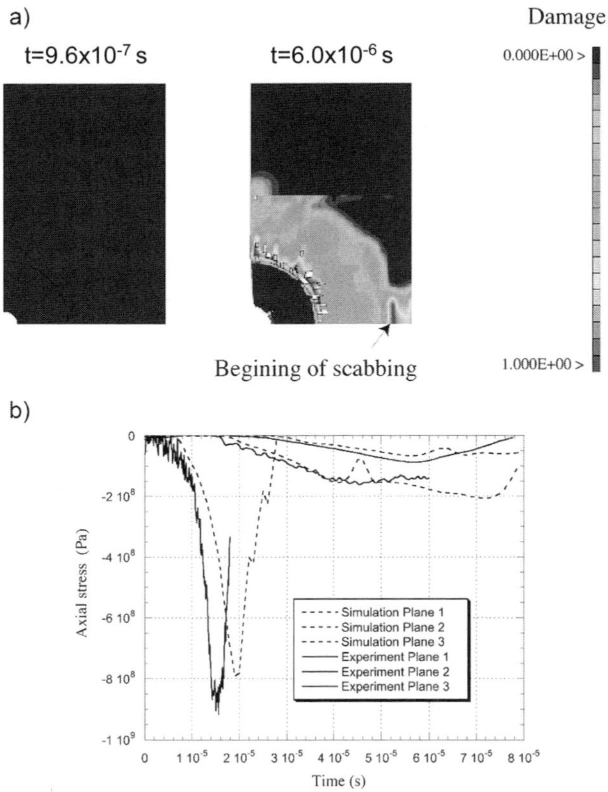

Figure 6.29. *a) Damage map. b) Comparison between the simulation and the experiment: axial stress on the three planes versus time*

Figure 6.29a shows the evolution of the damage during the simulation. Although a significant mesh effect was observed at the interfaces between the regions of the finite element model where the element density is changed abruptly, the computed size of the scab is: diameter=26 cm, depth=4 cm. This is quite similar to the experimental values measured on the slabs equipped with the embedded strain gages. Figure 6.29b shows the evolution of axial stress versus time for three planes in the thickness of the concrete slab. One can see that the model is also able to adequately capture the velocity and amplitude of the compression wave within the thickness of the slab.

6.4. Bibliography

[ALL 97] ALLIX O., DEÜ J.F., "Delay-damage modelling for fracture prediction of laminated composites under dynamic loading", *Eng. Trans.*, vol. 45, pp. 29–46, 1997.

[BAI 94] Bailly P., Une modélisation d'un matériau fragile avec prise en compte d'effets dynamiques, *Compte Rendu Académie des Sciences, Série II*, pp. 727-732, 1994.

[BIS 91] BISCHOFF P.H., PERRY S.H., "Compressive behaviour of concrete at high strain rates", *Mater. Struct.,* vol. 24, pp. 425-450, 1991.

[BRA 01] BRARA A., CAMBORDE F., KLEPACZKO J.R., MARIOTTI C., "Experimental and numerical study of concrete at high strain rates in tension", *Mech. Mater.* vol. 33, pp. 33-45, 2001.

[BRA 06] BRARA A., KLEPACZKO J.R., "Experimental characterization of concrete in dynamic tension", *Mech. Mater.*, vol. 38, pp. 253-267, 2006.

[BUR 00] BURLION N., GATUINGT F., PIJAUDIER-CABOT G., DAUDEVILLE L., "Compaction and tensile damage in concrete: constitutive modelling and application to dynamics", *Comput. Meth. Appl. Mech. Eng.*, vol. 183, pp. 291-308, 2000.

[BUZ 98] BUZAUD E., *Performances Mécaniques et Balistiques du Microbéton MB50*, DGA/Centre d'Etudes de Gramat report, 1998.

[CAD 01] CADONI E., LABIBES K., ALBERTINI C., BERRA M., GIANGRASSO M., "Strain rate effect on the tensile behaviour of concrete at different relative humidity levels", *Mater. Struct.*, vol.34, pp. 21-26, 2001.

[CAM 00] CAMBORDE F., MARIOTTI C., DONZÉ F.V., "Numerical study of rock and concrete behaviour by discrete element modeling", *Comp. Geotech.*, vol. 27, pp. 225-2474, 2000.

[COL 96] COLANTONIO L, STAINIER L., "Numerical integration of visoplastic constitutive equations for porous materials", in DESIDERI J-A, LE TALLEC P, ONATE E, PERIAUX J, STEIN E, *Numerical Methods in Engineering 96*, Wiley, pp. 28–34, 1996.

[DAV 73] DAVIES D.G.S., "The statistical approach to engineering design in ceramics", *Proc. Brit. Ceram. Soc.*, vol. 22, pp. 429-452, 1973.

[DEN 00] DENOUAL C., HILD F., "A damage model for the dynamic fragmentation of brittle solids", *Comput. Meth. Appl. Mech. Eng.,* vol.183, pp. 247-258, 2000.

[DES 07] DESMORAT R., GATUINGT F., RAGUENEAU F., "Local and nonlocal anisotropic damage model for quasi-brittle materials", *Eng. Fracture Mech.*, vol.74, pp. 1539-1560, 2007.

[DRA 98] DRAGON A., HALM D., "An anisotropic model of damage and frictional sliding for brittle materials", *Eur. J. Mech., A/Solids*, vol. 17, pp. 439–460, 1998.

[DUB 96] DUBÉ J.F., PIJAUDIER-CABOT G, LA BORDERIE C. "A rate dependent damage model for concrete in dynamics". *J. Eng. Mech.* ASCE, vol. 122, pp. 939-947, 1996.

[DUB 97] DUBÉ J.F., KANJI NANJI A., WIELGOSZ, "Comportement du béton en dynamique rapide", J. Phys. IV, (Suppl. to *J. Phys.* III 1997), 1997.

[FOR 03] FORQUIN P., Endommagement et fissuration de matériaux fragiles sous impact balistique, role de la microstructure, PhD thesis, ENS Cachan, 2003.

[FOR 08] FORQUIN P., GARY G., GATUINGT F. "A testing technique for concrete under confinement at high rates of strain", *Int. J. Impact Eng.*, vol. 35, pp. 425-446, 2008.

[GAR 98] GARY G., BAILLY P., "Behaviour of quasi-brittle material at high strain rate. Experiment and modeling", *Eur. J. Mech., A/Solids*, vol. 17, pp. 403-420, 1998.

[GAT 99] GATUINGT F., Prévision de la rupture des ouvrage en béton sollicités en dynamique rapide, PhD thesis, ENS Cachan,1999.

[GAT 08] GATUINGT F., DESMORAT R., CHAMBART M., COMBESCURE D., GUILBAUD D., "Anisotropic 3D delay-damage model to simulate concrete structures", *Rev. Eur. Méc. Numér.*, vol. 17, pp. 749-760, 2008

[GAT 02] GATUINGT F., PIJAUDIER-CABOT G., "Coupled damage and plasticity modelling in transient dynamic analysis of concrete", *Int. J. Numer. Anal. Meth. Geomech.*, vol.26, pp. 1-24, 2002.

[GRO 01] GROTE D.L., PARK S.W., ZHOU M., "Dynamic behavior of concrete at high strain rates and pressures: I. experimental characterization", *Int. J. Impact Eng.*, vol. 25, pp. 869–886, 2001.

[JU 89] JU J.W., "On energy-based coupled elastoplastic damage theories: constitutive modeling and computational aspects", *Int. J. Solids Struct.*, vol. 25, pp. 803–833, 1989.

[KAN 85] KANNINEN M.F., POPELAR C.H., *Advanced Fracture Mechanics*, Oxford University Press, 1985.

[KLE 01] KLEPACZKO J.R., BRARA A., "An experimental method for dynamic tensile testing of concrete by spalling", *Int. J. Impact Eng.*, vol. 25, pp. 387-409, 2001.

[KOT 83] KOTSOVOS M.D., "Effect on testing techniques on the post-ultimate behaviour of concrete in compression", *Mater. Constr.*, vol.16, 1983.

[LAD 98] LADEVÈZE P., ALLIX O., GORNET L., LEVEQUE D., PERRET L., "Computational damage mechanics approach for laminates: identification and comparison with experimental results", in VOYIADJIS G., *Damage Mechanics in Engineering Materials. Section A*, Elsevier, 1998.

[LeN 96] LE NARD H., BAILLY P., "Simulation de la compression dynamique de cylindre en béton", in *Communication Congrès GEO*, Aussois, France, 1996.

[LUB 96] LUBARDA VA, MASTILOVIC S, KNAP J., "Brittle-ductile transition in porous rock by Cap model", *J. Eng. Mech.* ASCE, vol. 122, 1996.

[MAL 91] MALVERN L. E., JENKINDS D. A., TANG T., MCLURE S., "Dynamic testing of laterally confined concrete", in S. P. SHAH, S. E. SWARTZ, M. L. WANG, *Micromechanics of Failure of Quasi Brittle Materials*, Elsevier Applied Science, pp. 343-352, 1991.

[MAZ 86] MAZARS J., "A description of micro and macroscale damage of concrete structures", *J. Eng. Fracture Mech.*, vol. 25, pp. 729-737, 1986.

[MAZ 90] MAZARS J., BERTHAUD Y., RAMTANI S., "The unilateral behaviour of damaged concrete", *Eng. Fract. Mech.*, vol. 35, pp. 629–635, 1990.

[NEE 84] NEEDLEMAN A, TVERGAARD V., "An analysis of ductile rupture in notched bars", *J. Phys. Mech. Solids*, vol. 32, pp. 461-490, 1984.

[PED 08] PEDERSEN R.R., SIMONE A., SLUYS L.J., "An analysis of dynamic fracture in concrete with a continuum viscoelastic visco-plastic damage model", *Eng. Fracture Mech.*, vol. 75, pp. 3782-3805, 2008.

[ROS 96] ROSS A.C., JEROME D.M., TEDESCO J.W., HUGHES, M.L., "Moisture and strain rate effects on concrete strength", *ACI Mater. J.*, vol. 93, pp. 293-300, 1996.

[ROS 91] ROSSI P., "Influence of cracking in the presence of free water on the mechanical behaviour of concrete", *Mag. Concrete Res.*, vol. 43, pp. 53-57, 1991.

[ROS 92] ROSSI P., VAN MIER J.G.M., BOULAY C., LE MAOU F., "The dynamic behaviour of concrete: influence of free water", *Mater. Struct.*, vol. 25, pp. 509-514, 1992.

[ROS 97] ROSSI P., "Strain rate effects in concrete structures: the LCPC experience", *Mater. Struct.*, (Suppl.), pp. 54-62, 1997.

[SEM 94] SEMBLAT J. F., Sols sous sollicitations dynamiques et transitoires : réponse dynamique aux barres de Hopkinson, propagation d'onde en milieu centrifuge, PhD thesis, l'Ecole Polytechnique, LCPC, 1994.

[SEM 95] SEMBLAT J. F., GARY G., LUONG M. P., "Dynamic response of sand using 3D Hopkinson bar", in *Proceedings of IS-TOKYO '95 First International Conference on Earthquake Geotechnical Engineering*, Tokyo, 1995.

[SER 98] SERCOME J., ULM FJ, TOUTLEMONDE F., "Viscous hardening plasticity for concrete in high rate dynamic", *J. Eng. Mech.*,vol.124, pp. 1050-1057, 1998.

[SER 00] SERCOMBE J, ULM FJ, MANG HA., "Consistent return mapping algorithm for chemoplastic constitutive laws with internal couplings", *Int. J. Num. Meth. Eng.*, vol. 47, pp. 75–100, 2000.

[SLU 92] SLUYS LJ., Wave propagation, localisation and dispersion in softening solids, PhD thesis, Delft University of Technology, 1992.

[SUF 04] SUFFIS A., Développement d'un modèle d'endommagement à taux de croissance contrôlé pour la simulation robuste de structures sous impact, PhD thesis, INSA Lyon, 2004.

[TOU 94] TOUTLEMONDE F., Résistance au choc des structures en béton, PhD thesis, École National des Ponts et Chaussées, 1994.

[WEI 39] WEIBULL W., "A statistical theory of the strength of materials", *Roy. Swed. Inst. Eng. Res.*, vol. 151, 1939.

[WEE 01] WEERHEIJM J., DOORMAAL J.C.A.M., VAN DE KASTEELE R.M., *Development of a New Test Set-Up for Dynamic Tensile Tests on Concrete Under High Loading Rates. Part 1: Results of Test Series A*, TNO, 2001.

[WEE 03] WEERHEIJM J., DOORMAAL J.C.A.M., VAN DE KASTEELE R.M., *Development of a New Test Set-Up for Dynamic Tensile Tests on Concrete Under High Loading Rates. Part 2: Results and Evaluation of Test Series B and C on Concrete*, TNO, 2003.

PART 3

Time-Dependent Response of Concrete

Chapter 7

Concrete at an Early Age: the Major Parameters

7.1. Introduction

Concrete undergoes many phenomena during the first dozen hours following batching and set up by pouring or projecting (shooting). Concrete progressively hardens due to the development of hydrate connections and is subjected to different thermal (thermal curing, ambient temperature), hydric (drying, wet curing), or mechanical (unmolding, transport, pre-strain, loading) stresses. As a result, its behavior strongly depends on its composition and the surrounding conditions, called "boundary conditions", during its early age (a few days after production). The aim of this chapter is to describe these two groups of influential parameters.

7.2. Influence of the composition of concrete

For given conservation conditions (thermal, hydric, etc.), characteristics of a concrete formulation depend on its composition [LAR 00]. The characteristics at early age influence the mechanical behavior of concrete, either directly as material parameters (strength, elasticity modulus, thermal dilatation coefficient, etc.), or indirectly *via* the state variables of the material (heat production of concrete, which more or less influences the temperature of the material depending on the conservation conditions).

Chapter written by Vincent WALLER and Buqan MIAO.

These variables can be expressed as equations or models between the composition of concrete (characteristics of the constituents and proportions in the mix) and the mix characteristics. They significantly influence the mix-design process, which determines the composition of concrete to meet the required specifications, including technical (workability, strength, etc.), normative (durability, etc.), and economical (minimum cost, etc.).

The following sections present models estimating different characteristics influencing the temperature of concrete (heat release, heat capacity) or its mechanical behavior (elastic modulus, expansion coefficient, strength, etc.) at early ages.

7.2.1. *Characteristics related to the temperature of concrete*

Temperature variation of early age concrete depends on the heat produced by the reactions occurring within concrete (cement hydration, hydraulic or pozzolanic reactions of the mineral additions), on the thermal dispersion of concrete towards the surroundings and, finally, on the heat capacity of the material.

These variables are connected by the following equation:

$$\Delta T_{th}(t)=\int_0^t \frac{\dot{Q}(\tau)+\dot{Q}^0(\tau)}{C^{th}(\tau)}d\tau \qquad [7.1]$$

where:

– time t corresponds to the age of concrete (we take the batching time of concrete as the origin of time), in s;

–$\Delta T_{th}(t)$ is the temperature rise of the considered concrete element at time t, in kelvin;

–$C^{th}(\tau)$ is the heat capacity of the considered element at time τ, in joules per kelvin;

–$\dot{Q}(\tau)$ is the produced heat flow by both hydration and pozzolanic reactions at time τ, in joules per second;

– $\dot{Q}^0(\tau)$ is the heat flow exchanged with the surroundings at time τ, in joules per second.

It is possible to consider an average value of the heat capacity of concrete, constant during the whole hydration reaction. The temperature rise of concrete can then be simplified to:

$$\Delta T_{th}(t) = \frac{Q(t)+Q^0(t)}{C^{th}} \qquad [7.2]$$

where:

– C^{th} is the average heat capacity of the element during the whole hydration process, in joules per kelvin;

– $Q(t)$ is the amount of heat produced by the hydration and pozzolanic reactions within concrete between batching and time t, in joules;

– $Q^0(t)$ is the amount of heat exchanged with the surroundings between batching and time t, in joules.

This last term, $Q^0(t)$, depends on the boundary conditions of the considered element. This aspect is presented in section 7.3.

7.2.1.1. *Heat production*

Heat production that occurs within a setting concrete is the consequence of different exothermic chemical reactions: the hydration reaction of cement and hydraulic or pozzolanic reactions of possibly present mineral additions[1]. We then have:

$$Q(t) = Q_c(t) + Q_z(t) \qquad [7.3]$$

where:

– $Q_c(t)$ is the cumulated heat produced by cement hydration at time t, in joules;

– $Q_z(t)$ is the cumulated heat produced by pozzolanic reactions at time t, in joules.

1 We only consider the case of pozzolans (silico-aluminous “Class F” fly ash or silica fume) as long as cement is of CEM I type (non-composed). In the case of slags, see [SCH 95a]. Limestone or siliceous fillers have to be considered as aggregates.

The amount of heat $Q_c(t)$ can also be written as follows:

$$Q_c(t) = \underbrace{\underbrace{r(t) \cdot \xi(\infty)}_{\xi(t)} \cdot c}_{c_h(t)} \cdot q_c \qquad [7.4]$$

where:

– $r(t)$ is the extent of the cement's hydration reactions at time t; by construction: $r(0)=0$ and $r(\infty)=1$;

– $\xi(\infty)$ is the cement's final degree of hydration;

– $\xi(t)$ is the cement's degree of hydration at time t;

– c is the amount of cement within the considered element, in kilograms;

– $c_h(t)$ is the amount of hydrated cement at time t, in kilograms;

– q_c is the cement's heat of hydration, in joules per kilogram.

As well, regarding the amount of heat $Q_z(t)$:

$$Q_z(t) = \underbrace{\underbrace{r_z(t) \cdot \xi_z(\infty)}_{\xi_z(t)} \cdot z}_{z_r(t)} \cdot q_z \qquad [7.5]$$

where:

– $r_z(t)$ is the extent of the pozzolanic reaction at time t; by construction: $r_z(0) = 0$ and $r_z(\infty) = 1$;

– $\xi_z(\infty)$ is the final degree of pozzolanic activity;

– $\xi_z(t)$ is the degree of pozzolanic activity at time t;

– z is the amount of pozzolan within the considered element in kilograms;

– $z_r(t)$ is the amount of pozzolan which has reacted at time t, in kilograms;

– q_z is the pozzolan's heat of reaction, in joules per kilogram.

This leads to the following equation of the total amount of cumulated heat:

$$Q(t) = r(t)\,\xi(\infty)\,c\,q_c + r_z(t)\,\xi_z(\infty)\,z\,q_z \qquad [7.6]$$

The final amplitude $Q(\infty)$ of the total heat produced can be obtained by considering the final hydration state of concrete ($r(\infty)=r_z(\infty)=1$):

$$Q(\infty) = \xi(\infty)\, c\, q_c + \xi_z(\infty)\, z\, q_z \qquad [7.7]$$

7.2.1.1.1. Degree of hydration of the binders

Without any pozzolan, the final degree of hydration of the cement (that is to say after an infinite theoretical hydration time) depends on the *water to cement (w/c)* weight ratio[2]. Hydration needs about 0.43 g of water per gram of hydrated cement (0.22 g of chemically bonded water + 0.21 g of adsorbed water unavailable for the rest of hydration) [POW 46; POW 47]. We should then get the final degree of hydration $\xi(\infty)$ of the cement equal to 1 for *w/c* above 0.43 and a $\xi(\infty)$ proportional to *w/c* below.

For practical purposes, the final degree hydration of cement never reaches the value of 1 even with *w/c* ratios higher than 0.43 because of progressive densification of the layer of the hydrates around the grains of the anhydrous cement, along the hydration progress. The following empirical model, built on the basis of many experimental results, allows us to adequately estimate this final value [WAL 00]:

$$\xi(\infty) = 1 - \exp(-3.3\, w/c) \qquad [7.8]$$

where *w/c* is the weight ratio between the amounts of water and cement within concrete.

For instance, when *w/c* =0.4, we have $\xi(\infty)$=0.73 and when *w/c*=0.6 we have $\xi(\infty)$=0.86.

When a pozzolan is present, it reacts with the lime released by the hydration of the cement (0.3 g of produced lime per gram of hydrated cement). A gram of fly ash reacts with the same weight of lime, whereas 1.3 g of lime is used per gram of silica fume that has reacted. The final degree of pozzolanic activity depends then on the ratio between the available amount of lime, depending itself on the cement's final degree of hydration, and the amount of pozzolan [WAL 00]. For a silica fume:

$$\xi_z(\infty) = \min\left(1\,;\, \frac{A\, \xi(\infty)}{z/c}\right) \qquad [7.9]$$

2 In this case, we suppose that there is no external water addition; otherwise, the remaining anhydrous cement could potentially become hydrated.

where:

– z/c is the *pozzolan to cement* weight ratio within concrete;

– $A = 0.23$ for a silica fume, and 0.3 for a fly ash.

When pozzolan is present, the final cement's degree of hydration $\xi(\infty)$ is lower because of water adsorption during the pozzolanic reactions [JUS 92; JUS 94]. The reduced value can be calculated as follows [WAL 00]:

$$\xi(\infty) = 1 - \exp[-3.3(w/c - \delta)] \qquad [7.10]$$

with:

$$\delta = D\exp(1.6\, w/c)\xi_z(\infty) z/c \qquad [7.11]$$

where D=0.6 for a silica fume, and 0.4 for a fly ash.

In that case as the final cement's degree of hydration also influences δ through $\xi_z(\infty)$, it is thus defined by an implicit function "x = f(x)", which can be easily resolved iteratively.

7.2.1.1.2. Heats of hydration

Cement's heat of hydration can be estimated based on the composition of its phases:

$$q_c = \sum_{i=phases} (\varphi_i \; q_i) \qquad [7.12]$$

where:

– q_c is the cement's heat of hydration, in J/kg;

– q_i is the heat of hydration of the cement's phase i, in J/kg;

– φ_i is the weight ratio of phase i within cement.

The heats of hydration of the phases of cement have been known for a long time [BRI 54; LEA 70; LER 34; THO 30; VER 50; VER 60]. They are presented in Table 7.1.

As an example, for a cement containing 58% of C_3S, 15% of C_2S, 7.8% of C_3A and 8.4% of C_4AF, it gives a heat of hydration q_c of 456 kJ/kg.

Phases of cement	Heat of hydration q_i (kJ/kg)
C_3S	510
C_2S	260
C_3A	1 100
C_4AF	410

Table 7.1. *Heats of hydration of the different phases of cement*

The heats of reaction of pozzolans are also dependent on their phase composition and vitreous material content, which are actually the only reactive components. However, the measures performed on three silica fumes and five Class F fly ashes have shown that heat of reaction is relatively constant for a given type of pozzolan: from 810 to 910 kJ/kg for silica fumes, and from 500 to 630 kJ/kg for fly ashes [WAL 00]. We can find similar values in the literature [BRI 54; MEJ 90; BRU 92]. The heat produced by pozzolan reactions represents a limited proportion of the total heat produced during the hydration process; therefore, it is reasonable to use a specific value for a given pozzolan type. The proposed values are given in Table 7.2.

7.2.1.1.3. Kinetics

Studying kinetics of heat production during the setting of concrete involves studying the functions $r(t)$ and $r_z(t)$. Hydration reactions are thermo-activated, that is to say accelerated when temperature increases. Arrhenius law is known to correctly express this dependence [FRE 77; REG 80; CAR 82; CAR 84; CHA 94; TOR 95; TOR 96; ROU 96; ALO 98a]. It involves a parameter that characterizes the thermo-activation, the activation energy, E_a, described as "apparent" in the case of hydration reactions. In practical use, in order to entirely characterize functions $r(t)$ and $r_z(t)$, the evolution of these functions under isothermal conditions at a reference temperature, usually 20°C (293.16K), as well as the associated activation energies have to be known. Regarding cement, we have:

$$r(t) = r\left[t_{eq20}(t)\right] \qquad [7.13]$$

with:

$$t_{eq20}(t) = \int_0^t \exp\left\{\frac{E_a}{R}\left[\frac{1}{293.16} - \frac{1}{T_{th}(\tau)}\right]\right\} d\tau \qquad [7.14]$$

where:

– $r(t_{eq20})$ is the function giving the evolution of the extent of the hydration of the cement as a function of 20°C equivalent age;

– $t_{eq20}(t)$ is the 20°C equivalent age of concrete at time t, i.e. the time (in seconds) necessary to reach the same extent of hydration at 20°C as that reached at time t under real curing conditions;

– E_a is the apparent activation energy of the hydration reactions in joules per mole;

– R is the ideal gas constant, which is equal to 8.31 J/mol/K;

– $T_{th}(\tau)$ is the temperature of the element considered at time τ, in kelvin.

Pozzolan type	Heat of reaction q_z (kJ/kg)
Silica fume	860
Class F fly ash	560

Table 7.2. *Heat of reaction of pozzolans*

We can deduce function $r(t_{eq20})$ from a heat production curve, measured on a concrete or a mortar containing the considered cement, and from the apparent activation energy E_a. This energy is determined using two heat or strength curves measured on the same concrete or mortar containing again the considered cement, but for two different thermal histories [ALO 98b; KAD 00; BRO 02; WIR 02; ALO 03]. When we have no available data, a standard mortar heat curve obtained from the cement manufacturer may be used, with a fixed activation energy of 33,240 J/mol (E_a/R = 4,000K).

The kinetics of the extent of the hydration reactions depends on the w/c ratio of concrete: the reaction rate $\dot{r}$ is all the higher when the w/c ratio is lower. Thus a correction has to be undertaken on the equivalent age if the function $r(t_{eq20})$ is obtained on a mortar or a concrete with a w/c ratio different from that of the concrete to be characterized:

$$t_{eq20}(t) = t_{eq20}\left[t,(w/c)_0\right]\left[\frac{(w/c)_0}{w/c}\right]^{0.62} \quad [7.15]$$

where $t_{eq20}[t,(w/c)_0]$ and $(w/c)_0$, respectively, are the 20°C equivalent age, in seconds, and the w/c ratio of the concrete in which the function $r(t_{eq20})$ and the cement's apparent activation energy E_a have been determined.

When a pozzolan is used, we can again consider a unique function for a given pozzolan type [WAL 00]:

$$r_z\left(t_{eq20}\right)=\frac{1}{1+\left(\dfrac{\tau_z}{t_{eq20}}\right)^n} \quad [7.16]$$

The values of the parameters τ_z and n, and of the apparent activation energies to be used for the calculation of the 20°C equivalent age are given in Table 7.3.

Pozzolan type	**Apparent activation energy Ea_z/R (K)**	**τ_z (h)**	**N**
Silica fume	11,400	250	0.9
Class F fly ash	12,000	770	1

Table 7.3. *Apparent activation energies and kinetics parameters of pozzolanic reactions*

In the presence of mineral addition, especially pozzolan additions, the kinetics of the hydration reactions are also affected: the reaction rate $\dot{r}$ is higher when the amount of addition is higher. Thus a correction to the equivalent age has to be performed if the function $r(t_{eq20})$ was determined on a mortar or a concrete with a z/c ratio different to that of the concrete to be characterized:

$$t_{eq20}(t)=t_{eq20}\left[t,(z/c)_0\right]\left[\frac{1+\beta_a\ z/c}{1+\beta_a\ (z/c)_0}\right] \quad [7.17]$$

where:

– $t_{eq20}[t,(z/c)_0]$ and $(z/c)_0$, respectively, are the 20°C equivalent age, in seconds, and the pozzolan to c/w ratio of the concrete in which the function $r(t_{eq20})$ and the apparent activation energy E_a of the cement have been determined;

– β_a is the factor expressing the acceleration due to the presence of addition (given in table 7.4).

NB: This acceleration also occurs when the addition does not involve any pozzolanic element (e.g. limestone filler). It is then easy to switch the pozzolan quantity z to the addition amount a in the previous equation. Some chemical admixtures also influence, intentionally or unintentionally, the kinetics (acceleration, delay). As we do not precisely know this influence *a priori*, the function $r(t_{eq20})$ should be determined in the presence of this admixture, directly in the concrete or at least in a concrete-equivalent mortar [SCH 00].

7.2.1.2. *Heat capacity*

The heat capacity of concrete can have a wide range of values (from 800 to 1,400 J/K/kg [NEV 00]), due to mix-design and to the nature of the different materials used, but there is not a valid average value for all concrete.

Actually, the discrepancy between the predicted heat capacity of a concrete and the reality (up to 30 %) would not match the precision obtained using a model of the concrete. Therefore, it is necessary to evaluate the heat capacity of every concrete.

The equation for the heat capacity of a concrete can be written as:

$$C^{th} = \sum_{i=constituents} m_i c_i^{th} \quad [7.18]$$

where:

– C^{th} is the heat capacity of the considered concrete element, in joules per kelvin;

– m_i is the amount of constituent i within the considered element, in kilograms;

– c_i^{th} is the specific heat capacity of constituent i, in joules per kelvin per kilogram.

Addition type	β_a
Silica fume	4.0
Class F fly ash	1.5
Limestone filler	0.5

Table 7.4. *β_a factor showing the acceleration of the hydration reactions of cement due to the presence of additions*

The literature suggests several methods to estimate the heat capacity of concrete during hydration. Most studies consider the decrease in heat capacity observed during hydration [FRE 82; BAS 95; SCH 95b; SCH 03a; MOU 03] is related to the binding of water to the hydrates [WHI 78, MOR 89, HAM 93]. However, there is no consensus on the extent of the reduction [TRI 82; JOL 94; SCH 95a]. In addition, none of the studies considered the specific heat capacities of the constituents as functions of the temperature, but the specific heat capacity can vary significantly during the hydration of concrete, particularly under adiabatic conditions.

Based on the data provided by Babushkin *et al.* [BAB 85], which list the specific heat capacities of a wide range of simple and complex chemical elements as functions of the temperature, it is possible to consider both the reaction rate and the temperature in the calculation of the heat capacity of concrete [WAL 00]. The hydration process causes a decrease in the heat capacity of concrete by about 5 to 10%, which is greater or less significant depending on whether the binding agent contains fly ash or silica fumes or is made only of cement. Conversely, the temperature rise of concrete causes an increase in its heat capacity up to 10% under adiabatic conditions.

Under these adiabatic conditions, this compensation makes it possible to assimilate the average value of the heat capacity during the whole hydration reaction to the heat capacity of fresh concrete (see equation [7.18]). Under semi-adiabatic conditions (thin structures), this approach can lead to an overestimation of the average heat capacity of concrete, giving an underestimation of the temperature rise of concrete.

To limit these errors , we can consider a reduced specific heat capacity for water [AFN 88] (see Table 7.5). Table 7.5 gives the values at a temperature of 20°C of the constituents of a concrete. For instance, a cubic meter of concrete containing 1,800 kg of siliceous aggregates, 350 kg of cement, and 180 l of water, has a heat capacity C^{th} of 2,333 J/K.

Constituent	**Specific heat capacity (J/K/kg)**
Siliceous aggregates, fine siliceous	730
Limestone aggregates, fine limestone	840
Dolomitic aggregates	890
Anhydrous cement	760
Silica fume	730
Fly ash	730
Water	4,186
Water [AFN 88]	3,760

Table 7.5. *Specific heat capacities of the constituents of a concrete at 20°C*

7.2.1.3. *Coefficient of thermal expansion*

The coefficient of thermal expansion of traditional or high-performances (HPC) concretes rapidly becomes stable once hydration has commenced [LAP 93; KAD 02; SCH 03a]. Therefore a constant value is assumed during the whole hydration reaction, which is equal to the final value. Table 7.6 presents common values depending on the type of aggregates used [NEV 00].

7.2.2. *Mechanical characteristics*

7.2.2.1. *Compressive strength at early age*

Compressive strength, at anytime, directly depends on the composition of concrete or mortar: strength class of the cement, w/c ratio, nature of the aggregates, volume of the paste, presence of additions, etc. [LAR 00]. Nevertheless, some constituents (typically chemical admixture) may induce a strong delaying effect, which has a major influence on the strength at the very early ages. Classical strength models are often ineffective in these cases.

To determine the evolution of strength at this point compression tests are performed, or the strength is estimated by using a linear equation between the hydration reactions rate and the strength [BYF 80; TOR 92; SCH 96; EMB 98; SCH 03b].

The equation is valid once a certain reaction rate r_0 is reached; below this, the strength value is zero:

$$\begin{cases} When\ r_{tot} < r_0 : f_c = 0 \\ When\ r_{tot} \geq r_0 : f_c = f_c(\infty)\dfrac{r_{tot} - r_0}{1 - r_0} \end{cases} \quad [7.19]$$

where:

– r_{tot} is the global extent of hydration of concrete;

–r_0 is a percolation threshold below which the strength is equal to zero;

– f_c is the compressive strength of concrete (MPa);

– $f_c(\infty)$ is the final compressive strength of concrete (MPa).

Aggregates of concrete	**Coefficient α of concrete (K^{-1})**
Limestone	6×10^{-6}
Granite	9×10^{-6}
Sandstone	10×10^{-6}
Quartzite	12×10^{-6}

Table 7.6. *Average value of the final coefficient of thermal expansion of concrete*

When there is no pozzolanic addition, we can consider:

$$r_{tot}(t_{eq20}) = r(t_{eq20}) \quad [7.20]$$

Otherwise, we will consider:

$$r_{tot}(t_{eq20}) = \frac{Q(t_{eq20})}{Q(\infty)} \quad [7.21]$$

We then just have to know two early age strength values and the corresponding 20°C equivalent ages in order to determine the parameters of equation [7.19], and thus the evolution of the early age compressive strength.

7.2.2.2. *Tensile strength at early age*

Tensile strength can be estimated from the compressive strength using the following equation [OLU 91a; OLU 91b]:

$$f_t = k_t \, (f_c)^n \quad [7.22]$$

where:

– f_t is the tensile strength of concrete (MPa);

– k_t and n are two coefficients.

Recent results [LAR 00] suggest the value $n = 0.57$. Coefficient k_t depends only on the nature of the aggregate and has to be calibrated using known results. It varies from about 0.35 for limestone aggregates to 0.45 for flints or basalts, on sealed specimens. For some specimens cured in water, coefficient k_t slightly increases.

7.2.2.3. *Elastic modulus and Poisson's ratio*

As is the case with tensile strength, power laws relate the elastic modulus of concrete and its compressive strength [SCH 96; NEV 00]. Once a sufficient estimation of the compressive strength has been made, the "tri-spheres" model [LAR 92, LER 96] can be used to obtain a more precise value for the elastic modulus [LAR 00]:

$$E = E_m \left(1 + 2\,g \frac{E_g^{\,2} - E_m^{\,2}}{(g^* - g)\,E_g^{\,2} + 2\,(2 - g^*)\,E_g E_m + (g^* + g)\,E_m^{\,2}} \right) \quad [7.23]$$

where:

– E is the elastic modulus of concrete; in MPa;

–E_m is the elastic modulus of the concrete matrix determined from the compressive strength f_c of concrete (MPa):

$$E_m = 226\, f_c \qquad [7.24]$$

– E_g is the elastic modulus of the aggregates (MPa);

– g is the volume fraction of the aggregates in concrete;

– g^* is the maximal packing density of the aggregates combination.

Poisson's ratio does not vary significantly so an average value, v =0.2, represents most concretes [NEV 00].

7.2.2.4. *Shrinkage*

Before setting, shrinkage of a cement paste results from Le Chatelier contraction; after setting, it results from the increase in the tensile strain in water, due to chemical water consumption (hydration) leading to *endogenous* or *autogenous* shrinkage or to physical water consumption (drying) producing *drying* shrinkage. Within concrete, the paste shrinks around inclusions, i.e. around the aggregates. Shrinkage is then closely related to the composition of the paste and to the elastic characteristics of the aggregates. As aggregates represent an obstacle, the shrinkage of concrete for a given quality of cement paste is lower the more aggregates there are. This influence can be expressed as [NEV 00]:

$$\varepsilon_{b,tot} = \varepsilon_{p,tot}\left(1-g\right)^n \qquad [7.25]$$

where:

– $\varepsilon_{b,tot}$ is the total shrinkage of concrete;

– $\varepsilon_{p,tot}$ is the total shrinkage of the paste;

– n is a constant, which ranges from 1.2 to 1.7.

The elastic modulus of the aggregates also influences the shrinkage of concrete. This shrinkage is lower when the modulus is higher for a given paste [NEV 00]. The tri-spheres model, presented above, can be used to summarize these two characteristics [LAR 00].

Drying shrinkage depends on the relative humidity present at the boundaries of the considered concrete element. This shrinkage is stronger when the amount of free water within the paste is significant. Therefore, regular concretes have a more significant drying shrinkage than high-performance concretes.

Autogenous shrinkage comes from the paste composition. The autogenous shrinkage is more important in high-performance concretes than in traditional concretes because the pores are of a smaller average size and the w/c ratio is lower [TAZ 95]. With the same reasoning, the presence of silica fumes increases this type of shrinkage [BEN 00]. The characteristics of cement naturally play a major role: autogenous shrinkage is increased by increases in the Blaine surface or decreases of free lime [BAR 01]. The influence of the C_3A level has also been proven [TAZ 95]. With a high SO_3/K_2O, ratio, the increase of C_3A increases shrinkage, with a low ratio, the reverse effect is obtained [BAR 00]. The addition of expansive elements allows us to describe shrinkage [SCH 02].

The kinetics of autogenous shrinkage at an early age is sometimes expressed as a function of the hydration reaction rate [TOR 96; MOU 03] (see also Chapter 8), of the strength [LAR 00], or of the 20°C equivalent age through a law taking into account the other formulation parameters [TAZ 00] :

$$\varepsilon_{b,end}\left(t_{eq20}\right) = B\,\beta\left(t_{eq20}\right) \qquad [7.26]$$

where:

– $\varepsilon_{b,end}(t_{eq20})$ is the autogenous shrinkage at 20°C equivalent age t_{eq20};

– B is the final autogenous shrinkage of concrete:

$$\begin{cases} When\ 0.2 \le w/c \le 0.5: & B = 3{,}070\times10^{-6}\exp(-7.3\,w/c) \\ When\ w/c > 0.5: & B = 80\times10^{-6} \end{cases}$$

$$\begin{cases} When\ t_{eq20} \le t_0: & \beta\left(t_{eq20}\right) = 0 \\ When\ t_{eq20} > t_0: & \beta\left(t_{eq20}\right) = 1-\exp\left[-F\left(t_{eq20}-t_0\right)^n\right] \end{cases}$$

where:

– t_0 is the setting time of concrete, in seconds;

– F and n are two parameters.

7.2.2.5. *Basic creep*

Under loading, concrete is instantaneously deformed. Basic creep of concrete consists of an increase, after loading, of the deformations of concrete when left under a constant load and without any exchange of water with the surroundings. For mid-term deformations, this phenomenon could be caused by redistribution induced by evaporation of water from the capillary pores. Long-term deformations could be due to a process of dislocations at the level of the solid surfaces of the hydrates [GUE 97].

As deformations occur within the paste, creep, like shrinkage (see above), is low when there is a large amount of aggregates in the concrete. In addition, for a given age of loading, experiments show that the amplitude of the deformations is proportional to the ratio between applied stress and compressive strength, except in the case of loadings at early ages [NEV 00].

Creep modeling is carried out using laws, or kernels, which give the response of concrete for a unit strain at different loading ages [TOR 96]. The loading/unloading history is then taken into account, for instance, using the equivalent time method which gives, after every loading increment, the age when concrete should have been loaded under the final strain to reach its present deformation state [ACK 89; ACK 92; GUE 96; SCH 00b].

7.3. Consequences of boundary conditions

As is the case in many other chemical reactions, hydration of cement depends on the surrounding conditions, in particular, on temperature and humidity. Two significant phenomena occur in the hydration reaction: hardening and dimensional variations. As concrete strains are very often restrained, cracking can occur as a function of the stiffness of the concrete pieces.

In this section, we will focus on the effects of temperature, humidity, and restrained deformations on the behavior of concrete at an early age.

7.3.1. *Consequences of thermal boundary conditions*

The external temperature (room temperature or imposed temperature) influences the behavior of concrete at early age at three different levels. First, as hydration of cement is a thermo-activated reaction, a variation in the external temperature has direct consequences on the properties of the material. The temperature field due to the hydration of cement or to a variation of the external temperature is often non-uniform within the concrete pieces. This non-uniform temperature field has two consequences on concrete from a mechanical point of view:

– formation of self-balanced strain field, usually with strong tensile stresses towards the surface in the case of massive pieces, involving skin cracking;

– global deformation of the piece.

Usually, as global deformation of a piece of a structure is hindered by the other parts, additional strains are formed at the structural level within concrete as a function of the rigidity of these parts.

We will first focus on the consequences of the temperatures at the level of the piece. The effect at the structural level will be treated in section 7.3.3.

7.3.1.1. *At the material level*

Temperature directly influences the behavior of concrete at early age and hardened concrete. It is well known that high temperature on a young concrete leads to lower strengths and to a reduction in the setting time [SPR 78].

Although high temperatures during the setting of concrete increase the strength at early age, it can have the reverse effect on the strength after 7 days. An experimental study carried out by Price [PRI 51] has shown that the difference in the compressive strength at 28 days can reach 10% between a concrete subjected to 46°C for 2 hours once set up.

High temperature during the first days can also have harmful consequences on the long-term compressive strength. Richardson [RIC 91] reported a strength loss at 28 days of 10% after a maturing process that lasted 1 day at 38°C, and of 22% after a similar maturing process that lasted 3 days.

Other researchers [KLE 58; VER 68; ALE 68; ALE 72; BYF 80; REG 80; MAM 70; ALO 98a] observed the same phenomena. Interpretation of these phenomena, in particular, the weakest long-term strength, was the subject of numerous research works in the 1970s and 1980s. Laplante summarized the main results of these studies [LAP 93]. These studies elucidated the effect of the maturing temperature on concrete at an early age. The results of Byfors [BYF 80] should be highlighted, which clearly showed that the increase of the compressive strength of concrete at an early age can be explained by the maturing process (equivalent age), at least regarding moderate maturing temperatures, between 10 and 40°C (this part will be developed in Chapter 8).

Temperature also influences the delayed behavior of concrete at early age. Little data has been published up to now regarding the effect of maturing temperature on the autogenous shrinkage of concrete. Nevertheless, it appears that a correlation can be found between autogenous shrinkage and maturity (equivalent age) of concrete for different maturing temperatures [LAP 93] (see Chapter 8).

Regarding desiccation shrinkage, high temperatures speed up the drying process, thus implying more significant shrinkage for the same duration of drying.

Basic creep increases with temperature for a concrete of advanced maturity [BYF 80]. Regarding concrete at early age, the temperature effect is much less obvious as high temperature causes a twin effect:

– acceleration of the hydration of cement, thus an increase of the strength of concrete and a decrease of basic creep;

– increase of basic creep as temperature increases the kinetics of creep as is the case with a concrete of advanced maturity.

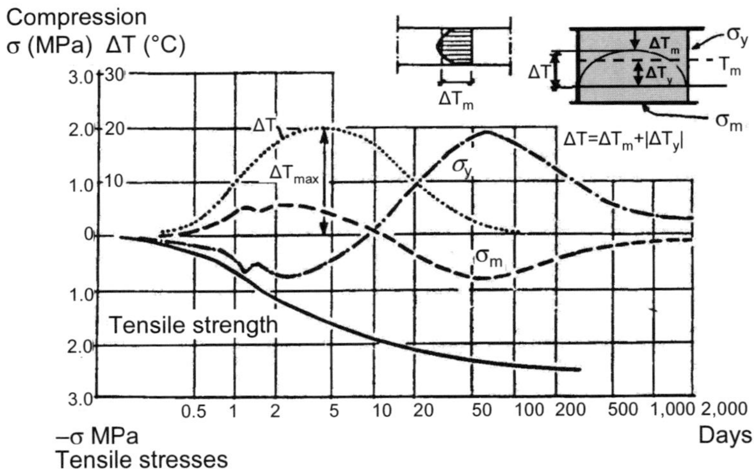

Figure 7.1. *Calculated thermal stresses within a concrete element of 2 m thickness [BER 82]*

7.3.1.2. *At the structural element level*

Hydration of concrete leads to an increase in temperature within concrete during the first days. This temperature field, rapidly balanced at the boundary, is not uniform within the piece. The maximal temperature can be higher than 65°C in massive pieces, depending on the formulation of concrete and the temperature of the room. This implies a risk of delayed ettringite formation.

Presently, we can measure, with a precision of 0.1°C, the evolution of the temperatures at different points of a concrete element. Numerical simulation software also enable us to predict the increase in temperatures within a concrete element due to the thermal treatment or to the hydration of cement. Starting from the evolution of the temperature field, different formulations [BYF 80; LAP 93] based on the concept of maturity, allow us to estimate the mechanical properties of concrete at any time and at any point of the element (see also Chapter 8). With the finite elements method, the stress and deformation fields can be determined within concrete. Figure 7.1 [BER 82] illustrates the evolution of the stresses within a 2 m thick concrete slab, with an imposed maximal temperature gap of 20°C between the surface and the center of the slab. The stresses thus obtained are self-balanced with a strong tension at the surface (close to 1 MPa at the age of 3 days). In reality, temperature gaps higher than 40°C have been observed [COO 92].

Tensile stress at the surface would be much stronger and probably could be higher than the tensile strength of the material. Risk of skin cracking is therefore very high.

7.3.1.3. *At the global structural level*

Within a concrete structure, the different parts are connected and become deformed together.

Thus, deformation of a part under the effect of the variation of the room temperature depends on the rigidity of the parts connected with it, in other words, on the mechanical boundary conditions. This issue is presented in detail in section 7.3.3.

7.3.2. *Consequences of hygrometric boundary conditions*

Like temperature, humidity affects the behavior of a concrete at the material and structure levels for the following reasons:

– hydration of cement depends on both humidity and temperature;

– for a concrete exposed to a relatively dry medium, drying occurs from the surface towards the core of the concrete, thus creating a non-uniform humidity field, as occurs when room temperature drops;

– concrete contracts because of humidity loss and dilates because of humidification, as occurs for a change of temperature.

Compared with heat transfer, drying is a slow process and thus more sensitive to the dimensions of the pieces. It is almost impossible to obtain a concrete sample with a uniform humidity field within an unsaturated medium. A sample with a non-uniform humidity field cannot be considered as representative of the concrete material. As a consequence, we cannot differentiate the humidity effect at the material level and at the piece level. In what follows, we focus on the surrounding humidity at the material level and at the piece level of concrete pieces that can freely deform. The effect at the structure level will be treated in section 7.3.3.

It is also worth noting that drying is an important process before the setting of concrete, which can lead to cracking due to plastic shrinkage (a phenomenon similar to the drying of clay). This phenomenon is especially sensitive for high-performance or very high-performance concretes [BAR 82], but also self-compacting concrete (SCC) [TUR 04]. Curing of concrete usually allows us to avoid this phenomenon.

7.3.2.1. *Drying and humidity field*

When a concrete, initially saturated in water, is exposed to dry conditions (with regard to concrete), evaporation (drying) occurs from the surface towards the core, thus forming a non-uniform humidity field. Currently, several experimental methods are available to determine the humidity as a function of the depth within a piece of concrete, subjected to drying. Some numerical tools can also be found. They allow us to simulate the evolution in time of the humidity fields within concrete.

As it is the case of temperatures presented in the previous section, a non-uniform humidity field has two consequences on concrete from a mechanical viewpoint:

– self-balanced stress field as concrete gets deformed under variation of the humidity;

– apparent shrinkage (or swelling).

Stress fields due to a non-uniform humidity field within a concrete sample submitted to drying can also be determined using the finite element methods. The stress field is characterized by compressive stress in the core, balanced by usually very high tensile stresses within the superficial zone. If we assume that concrete is perfectly elastic, tensile stresses thus calculated can sometimes reach 20 MPa, which is largely higher than the traction strength of concrete, and inevitably triggers skin cracking.

The stresses, previously calculated, depend on the hydric contraction coefficient of concrete. Nevertheless, we do not have enough specific data on this coefficient as it is very hard to determine experimentally: the humidity within the concrete piece cannot be uniformly varied (see [TOR 99] for an estimation of this coefficient). In addition, as the zones where the calculated stress is higher than the tensile strength of concrete get cracked, it is more difficult to evaluate the stresses due to drying. As a consequence, determination of stresses due to drying using experimental means becomes important to confirm the results of the numerical simulation and also to estimate the hydric stress coefficient. Figure 7.2 [MIA 88] shows the distribution of the stresses due to drying within the slabs of different thicknesses, distribution obtained using the method of the stress liberation, which was developed at the LCPC (Laboratoire Central des Ponts et Chaussées) [ABD 85].

Regarding a green concrete, drying has an additional effect: non-homogeneity of the mechanical properties. It has been shown [POW 47] that hydration of cement is significantly slowed down when the relative humidity in the pores of the paste is lower than 80%. As humidity is reduced by drying, the surface concrete becomes less hydrated than the core one. Coupling with the hydration reaction, which is slowed down or stopped by drying, can also be found [OH 03].

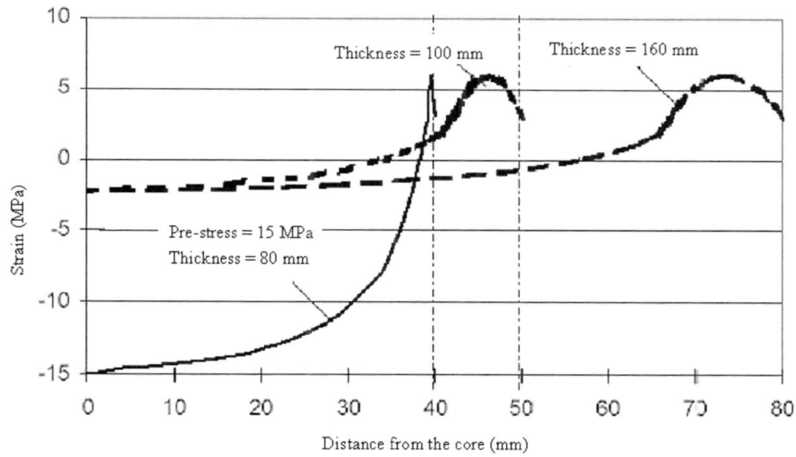

Figure 7.2. *Stresses measured using the stress liberation method within slabs of different thicknesses, matured for 56 days at 20°C and 50% relative humidity [MIA 88]*

In reality, the properties that we measure on samples are influenced by these surrounding humidity effects (strains, cracking and non-homogeneity) and results of these measurements mainly depend on the size of these sample.

7.3.2.2. *Effects on the mechanical properties*

According to a survey carried out by the Portland Cement Association [PCA 95], the compressive strength at 180 days of a concrete, matured in water for 7 days and then in ambient air until the test was performed was 20% higher than the strength of a similar concrete matured in water for only 3 days. A concrete continually matured in water has a strength two times higher than the same concrete matured in air. The effect of hygrometric conditions at an early age is then extremely important regarding the compressive strength of concrete. Obviously, the intrinsic effect of the humidity on the strength of concrete and the structure effect due to the humidity gradients lead to a superficial cracking of the samples. More details can be found in a survey carried out by Torrenti [TOR 96].

7.3.2.3. *Effects on durability (permeability)*

Permeability (and thus durability) of concrete is also very sensitive to the hygrometric conditions at an early age. Premature exposition of concrete to drying has huge consequences on its permeability. During an experiment, Whiting [WHI 88] has shown that the water permeability of a concrete that was matured for 1 day in water was three times higher than the permeability of the same concrete matured for 7 days in water. More information can be found in the following references [BAR 92; YSS 95].

On a structural level, boundary hygrometric conditions have the same effects as the thermal conditions and will be presented in the next section.

7.3.3. *Consequences of mechanical boundary conditions*

Deformations of thermal or hygrometric origin within a concrete piece can also induce strains within itself if it cannot deform freely. The stresses thus generated depend on the displacement conditions set at the boundaries [ROS 98]. For instance, the stress variation, $\Delta\sigma$, due to a temperature variation, ΔT, within a one-dimensional piece with a fixed end can be evaluated using a simplistic method, which neglects shrinkage, creep, and the variation of the properties (in particular, the evolution of the modulus):

$$\Delta\sigma = \alpha \, E \, \Delta T$$

where α and E are the thermal dilatation coefficient and the elasticity modulus of the material, respectively.

For instance, the stress due to a temperature decrease of 10°C within a concrete piece with an elasticity modulus of 10 GPa and a tensile thermal dilatation coefficient of 12×10^{-6}/°C has a value of 1.2 MPa, which is usually higher than the tensile strength of a concrete at an early age. More details can be found in reference [MIA 97].

Drying shrinkage of concrete at an early age can yield similar effects and increases the cracking risk.

Within a piece belonging to a concrete hyperstatic structure, deformation of this piece is partially prevented because of the deformation of the other pieces connected to its ends. The stress due to a change of temperature (or due to shrinkage) within the piece depends on the relative stiffness of the other pieces.

Figure 7.3 shows the iso-values of the stresses within a concrete slab due to a shrinkage of 100×10^{-6} with an elasticity modulus of 10 GPa (representing a green concrete at about 12 hours) under different boundary conditions at the ends [MIA 01].

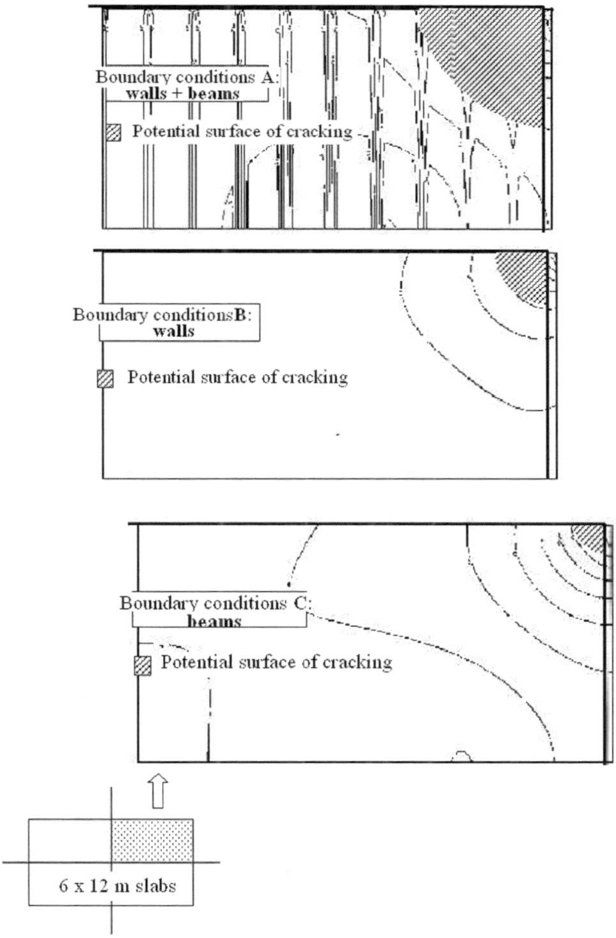

Figure 7.3. *Iso-values of the principal stress within fresh concrete slobs under different boundary conditions at the ends of the potential cracking zone[MIA 01]*

At this age, the tensile strength of concrete is of about 1 MPa, and there is then a real risk of cracking at the corners of the slabs, with a cracking potential surface (surface of the zone where traction strains is higher than 1 MPa) 16 times higher in case C than in case A.

In reality, once concrete is set up, it stays for some time within the form to reach a certain maturity before form removal.

At this stage, two effects occur at the same time: the effect of temperature rise due to the hydration of the cement, and the drying effect following form removal.

7.4. Conclusion

Regarding concrete at early age, thermal and hygrometric boundary conditions affect the hydration process and, subsequently, have direct consequences on the properties of the material. These boundary conditions also create non-uniform temperature and humidity fields within concrete pieces with deformation and stress fields, which can cause skin cracking. Depending on the mechanical boundary conditions, these effects generate additional strains, which can cause cracking on a structural level. These deformations, stresses, and cracks are the main causes of weakening of the mechanical properties and durability of concrete.

7.5. Bibliography

[ABD 85] C. ABDUNUR, "Mesure de contraintes sur *Ouvrages d'Art* par une méthode de libération miniaturisée", *Bull. Liaison LPC,* vol. 138, pp. 5-14, 1985.

[ACK 86] P. ACKER, C. FOUCRIER, Y. MALIER, "Temperature-related mechanical effects in concrete elements and optimization of the manufacturing process, concrete at early age", *ACI SP*, vol. 95, pp. 33-47, 1986.

[ACK 89] P. ACKER, M. Y. LAU, F. COLLET, "Comportement différé du béton: validation expérimentale de la méthode du temps équivalent", *Bull. Liaison LPC*, vol. 163, pp. 34-39, 1989.

[ACK 92] P. ACKER, R. EYMARD, "Fluage du béton : un modèle plus performant et plus simple à introduire dans les calculs numériques", *Ann. l'ITBTP*, vol. 507, pp. 57-75, 1992.

[AFN 88] AFNOR, "Mesure de la chaleur d'hydratation des ciments par calorimétrie semi-adiabatique dite méthode du calorimètre de Langavant", 15-436, NF, 1988.

[ALE 72] J. ALEXENDERSON, *Strength Losses in Heat Cured Concrete*, Swedish Cement and Concrete Research Institute, no. 43, 1972.

[ALO 03] L. D'ALOÏA *et al.*, "Résistance du béton dans l'ouvrage : la Maturométrie", in *Techniques et Méthodes des Laboratoires des Ponts et Chaussées*, LCPC – CALIBE National Project Civil Engineering Plan – IREX editions, 2003.

[ALO 98a] L. D'ALOÏA, Détermination de l'énergie d'activation apparente du béton dans le cadre de l'application de la méthode du temps équivalent à la prévision de la résistance en compression au jeune âge: approches mécanique et calorimétrique, simulations numériques, PhD Thesis, INSA of Lyon, 1998.

[ALO 98b] L. D'ALOÏA, G. CHANVILLARD, "Determination of the apparent activation energy of concrete: E_a. Semi-adiabatic tests of heat development", in *4th CANMET/ACI/JCI International Symposium on Recent Advances in Concrete Technology*, pp. 561-586, V. M. Malhotra editions, 1998.

[BAB 85] V. I. BABUSHKIN, G. M. MATVEYEV, P. MCHEDLOV-PETROSSYAN, *Thermodynamics of Silicates*, Springer-Verlag, 1985.

[BAR 01] L. BARCELO, Influence des caractéristiques des ciments sur la structuration et le comportement dimensionnel des matériaux cimentaires au jeune âge, PhD Thesis, ENS Cachan, 2001.

[BAR 82] J. BARON, "Les retraits de la pâte de ciment", in *Le Béton Hydraulique*, ENPC editions, 1982.

[BAR 92] J. BARON, J. P. OLLIVIER, *La Durabilité des Bétons*, ENPC editions, 1992.

[BAS 95] G. BASTIAN, A. KHELIDJ, "Propriétés thermophysiques d'un beton fraîchement coulé", *Bull. Liaison LPC*, vol. 200, pp. 25-35, 1995.

[BEN 00] A. BENTUR, "Early age shrinkage and cracking in cementitious systems", in V. BAROGHEL-BOUNY & P.C. AÏTCIN, *Shrinkage of Concrete – Shrinkage 2000*, pp. 1-20, RILEM, Paris, 2000.

[BRI 54] C. BRISI, "Tonalità termica selle reazioni di idratazione dei costituenti del clinker di Portland", *La Ricerca Scientifica*, vol. 24, pp. 1436-1444, 1954.

[BRO 02] M. BRODA, E. WIRQUIN, B. DUTHOIT, "Conception of an isothermal calorimeter for concrete – determination of the apparent activation energy", *Mater. Struct.,* vol. 35, pp. 389-394, 2002.

[BRU 92] J. BRUNNSTRØM-JENSEN, *Flyveaskes Puzzolane Reaktion*, Technical University of Denmark, 1992.

[BYF 80] J. BYFORS, *Plain Concrete at Early Ages*, CBI Research Report, vol. 80, Swedish Cement and Concrete Research Institute, Stockholm, 1980.

[CAR 82] N. J. CARINO, "Maturity Functions for Concrete", in *Colloque International sur le Béton Jeune*, Paris, pp. 123-128, 1982.

[CAR 84] J. CARINO, "The maturity method: theory and application", *Cement, Concrete Aggregates*, vol. 6, pp. 61-73, 1984.

[CHA 94] G. CHANVILLARD, L. D'ALOÏA, "Prévision de la résistance en compression au jeune âge du béton – application de la méthode du temps équivalent", *Bull. Liaison LPC*, vol. 193, pp. 39-51, 1994.

[COO 92] W. D. COOK, B. MIAO, P. C. AITCIN, D. MITCHELL, "Thermal stresses in large high strength concrete columns", *ACI Mater. J.*, vol. 8, pp. 63-68, 1992.

[EMB 98] M. EMBORG, "Development of mechanical behaviour at early ages", in R SPRINGENSCHMID, *Prevention of Thermal Cracking in Cement at Early Ages*, report 15, RILEM, pp. 76-1148, 1998.

[FRE 77] P. FREIESLEBEN HANSEN, E. J. PEDERSEN, "Maturity computer for controlled curing and hardening of concrete", *Nordisk Betong*, vol. 1, pp. 21-25, 1977.

[FRE 82] P. FREIESLEBEN HANSEN, J. H. HANSEN, V. KJAER, E. J. PEDERSEN, "Thermal Properties of Hardening Cement Paste", in *Colloque International sur le Béton Jeune*, vol. 1, pp. 23-26, Paris, 1982.

[GUE 96] I. GUENOT, J. M. TORRENTI, P. LAPLANTE, "Stresses in early-age concrete: comparison of different creep models", *ACI Mater. J.*, vol. 93, pp. 254-259, 1996.

[GUE 97] I. GUENOT-DELAHAIE, "Contribution à l'analyse physique et à la modélisation du fluage propre du béton", *Etudes et recherches des Laboratoires des Ponts et Chaussées*, "*Ouvrages d'Art*" series, vol. OA 25, April, LCPC, 1997.

[HAM 93] H. HAMFLER, S. RÖHLING, K. VAN BREUGEL, *Methods for Calculating Temperatures in Hardening Concrete Structures*, Chapter 6 in RILEM TC 119 TCE, Draft No. 4, private communication, 1993.

[JOL 94] C. JOLICOEUR, J. BILODEAU, M. A. SIMARD, "Etude calorimétrique de l'hydratation du ciment – détermination de l'énergie d'activation", Step report, LCPC unpublished document, 1994.

[JUS 92] H. JUSTNES, E. J. SELLEVOLD, G. LUNDEVALL, "High strength concrete binders – part a: composition of cement pastes with, and without condensed silica fume", in V.M. MALHOTRA, *4th CANMET/ACI International Conference on Fly Ash, Silica Fume, Slag and Natural Pozzolans in Concrete*, Istanbul, 2 SP132-47, pp. 873-889, 1992.

[JUS 94] H. JUSTNES, I. MELAND, J. O. BJOERGUM, J. KRANE, "The Mechanism of Silica Fume Action in Concrete Studied by Solid State 29Si NMR", in P. COLOMBEY, A. R. GRIMMER, *Seminar on Application of NMR Spectroscopy to Cement Science*, Guerville, 1992.

[KAD 00] H. KADA-BENAMEUR, E. WIRQUIN, B. DUTHOIT, "Determination of Apparent Activation Energy of Concrete by Isothermal Calorimetry", *Cement Concrete Res.*, vol. 30, pp. 301-305, 2000.

[KAD 02] H. KADA, M. LACHEMI, N. PETROV, O. BONNEAU, P. C. AITCIN, "Determination of the coefficient of thermal expansion of high performance concrete from initial setting", *Mater. Struct.*, vol. 35, pp. 35-41, 2002.

[KLI 58] P. KLIEGER, *Effect of Mixing and Curing Temperature on Concrete Strength*, Research Department Bulletin RX103, Portland Cement Association, 1958.

[LAP 93] P. LAPLANTE, "Propriétés mécaniques des bétons durcissants: analyse comparée des bétons classiques et à très hautes performances", *Etudes et recherches des Laboratoires des Ponts et Chaussées*, "*Ouvrages d'Art*" series, OA13, LCPC, 1993.

[LAR 00] F. DE LARRARD, "Structures granulaires et formulation des bétons", *Etudes et recherches des Laboratoires des Ponts et Chaussées*, "*Ouvrages d'Art*" series OA34, LCPC, Paris, 2000.

[LAR 92] F. DE LARRARD, R. LE ROY, "Relations entre formulation et quelques propriétés mécaniques des bétons à hautes performances", *Mater Struct.*, vol. 25, pp. 464-475, RILEM, 1992.

[LEA 70] F. M. LEA, *The Chemistry of Cement and Concrete*, Edward Arnold Ltd, 1970.

[LER 34] W. LERCH, R. H. BOGUE, "Heat of hydration of Portland cement pastes", *Bureau of Standards J Res*, vol. 12, pp. 645-664, 1934.

[LER 96] R. LE ROY, "Déformations instantanées et différées des bétons à hautes performances", *Etudes et recherches des Laboratoires des Ponts et Chaussées*, "*Ouvrages d'Art*" series, LCPC, 1996.

[MAM 70] M. MAMILLAN, "Recherches expérimentales sur l'accélération du durcissement du béton par le chauffage", *Ann. ITBTP*, vol. 267-268, pp. 133-203, 1970.

[MAM 82] M. MAMILLAN, "Traitement thermique des bétons", in *Le Béton Hydraulique*, pp. 261-269, ENPC editions, 1982.

[MEJ 90] O. MEJLHEDE-JENSEN, *Mikrosilicas Puzzolane Reaktion*, Technical University of Denmark, 1990.

[MIA 01] B. MIAO, *Risque de Fissuration dans les Applications Horizontals*, Research report, Laboratoire Central de Recherché, vol. 67, Lafarge, 2001.

[MIA 88] B. MIAO, Effets mécaniques dus au retrait de dessiccation du béton, PhD Thesis, ENPC, 1988.

[MIA 89] B. MIAO, *Comportement à Long-terme d'un Elément Massif en Béton*, Research report LCPC-ATILH, 1989.

[MIA 97] B. MIAO, "Analytical model for evaluating the mechanical compatibility of concrete repair materials", in *Int. Conf. on Civil Engineering Materials*, Ottawa, pp. 621-631, 1997.

[MOR 89] P. MORABITO, "Measurement of the thermal properties of different concretes", *High Temperatures – High Pressures*, vol. 21, pp. 51-59, 1989.

[MOU 03] P. MOUNANGA, Etude expérimentale du comportement de pâtes de ciment au très jeune âge: hydratation, retraits, propriétés thermophysiques, PhD Thesis, University of Nantes, 2003.

[NEV 00] A. M. NEVILLE, *Propriétés du Béton*, Eyrolles, 2000.

[OH 03] B. H. OH, S. W. CHA, "Non-linear analysis of temperature and moisture distributions in early-age concrete structures based on degree of hydration", *ACI Mater. J.*, vol. 100, pp. 361-370, 2003.

[OLU 91a] F. A. OLUOKUN, "Prediction of concrete tensile strength from its compressive strength : evaluation of existing relations for normal weight concrete", *ACI Mater. J.*, vol. 88, pp. 302-309, 1991.

[OLU 91b] F. A. OLUOKUN, E. G. BURDETTE, J. H. DEATHERAGE, "Splitting tensile strength and compressive strength relationship at early ages", *ACI Mater. J.*, vol. 88, pp. 115-121, 1991.

[PCA 95] PCA, *Dosage et Contrôle des Mélanges de Béton*, 6th edition, Portland Cement Association, 1995.

[POW 46-47] T. C. POWERS, T. L. BROWNYARD, "Studies of the physical properties of hardened portland cement paste", *J. Am. Concrete Inst.*, vol. 18, pp. 101-132, 249-336, 469-504, 549-602, 669-712, 845-880, and 933-992, 1946-1947.

[POW 47] J. C. POWERS, "A discussion of cement hydration in relation to the curing of concrete", in *Proc. Highway Research Board*, vol. 27, pp. 178-188, 1947.

[PRI 51] W. H. PRICE, *Factors Influencing Concrete Strength*, American Concrete Institute, vol. 47, pp. 417-432, 1951.

[REG 80] M. REGOURD, E. GAUTHIER, "Comportement des ciments soumis au durcissement accéléré", *ITBTP Annexes*, vol. 387, pp. 83-96, 1980.

[RIC 91] D. N. RICHARDSON, "Review of variables that influence measured concrete compressive strength", *J. Mater. Civ. Eng.,* vol. 3, 2, 95-112, 1991.

[ROS 98] F. S. ROSTASY, T. TANABE, M. LAUBE, "Assessment of external restraint", in R. SPRINGENSCHMIDT, *Prevention of Thermal Cracking in Concrete at Early Ages – State of the Art Report*, Report 15, RILEM, 1998.

[ROU 96] S. ROUSSEL, Détermination de l'énergie d'activation apparente des bétons – Application à la maturométrie, PhD thesis, ENS of Cachan, 1996.

[SCH 00] A. SCHWARTZENTRUBER, C. CATHERINE, "La méthode du mortier de béton équivalent (MBE). Un nouvel outil d'aide à la formulation des bétons adjuvantés", *Mater. Struct.*, vol. 33, pp. 475-482, 2000.

[SCH 00b] G. DE SCHUTTER, L. TAERWE, "Fictitious degree of hydration method for the basic creep of early age concrete", *Mater. Struct.*, vol. 33, pp. 370-380, 2000.

[SCH 02] A. SCHWARTZENTRUBER, M. PHILIPPE, MARCHESE G. MARCHESE, O. LAURENCE, "Cracking tendency of UHSC – influence of fibers and expansive admixtures", in G. KÖNIG, F. DEHN, T. FAUST, *High Strength/High Performance Concrete*, vol. 2, pp. 1391-1405, University of Leipzig, 2002.

[SCH 03a] G. DE SCHUTTER, "Thermal properties", in A. BENTUR, *Early Age Cracking in Cement Systems*, report 25, RILEM, 2003.

[SCH 03b] G. DE SCHUTTER, K. KOVLER, "Short term mechanical properties", in A. BENTUR, *Early Age Cracking in Cement Systems*, report 25, RILEM, 2003.

[SCH 95a] G. DE SCHUTTER, L. TAERWE, "General hydration model for portland cement and blast furnace slag cement", *Cement Concrete Res.*, vol. 25, pp. 593-604, 1995.

[SCH 95b] G. DE SCHUTTER, L. TAERWE, "Specific heat and thermal diffusivity of hardening concrete", *Mag. Concrete Res.*, vol. 47, pp. 203-208, 1995.

[SCH 96] G. DE SCHUTTER, L. TAERWE, "Degree of hydration-based description of mechanical properties of early age concrete", *Mater. Struct.*, vol. 29, pp. 335-344, 1996.

[SPR 78] J. H. SPROUSE, R. B. PEPPLER, *Setting Time, Significance of Tests and Properties of Concrete and Making of Concrete Materials*, STP169B, ASTM, pp. 105-121, 1978.

[TAZ 00] E. TAZAWA, R. SATO, E. SAKAI, S. MIYAZAWA, "Work of JCI Committee on autogeneous shrinkage", in V. BAROGHEL-BOUNY & P.C. AÏTCIN, *Shrinkage of Concrete – Shrinkage 2000*, Paris, pp. 21-40, 2000.

[TAZ 95] E. TAZAWA, S. MIYAZAWA, "Influence of cement and admixture on autogenous shrinkage of cement paste", *Cement Concrete Res.*, vol. 25, pp. 281-287, 1995.

[THO 30] T. THORVALDSON, W. G. BROWN, C. R. PEAKER, "Studies on the thermochemistry of the compounds occurring in the system CaO-Al2O3-SiO2 – IV. The heat of solution of tricalcium aluminate and its hydrates in hydrochloric acid", *J. Am. Chem. Soc.*, vol. 52, pp. 3927-3936, 1930.

[TOR 92] J. M. TORRENTI, "La Résistance du béton au très jeune âge", *Bull. Liaison LPC*, vol. 179, pp. 31-41, 1992.

[TOR 95] J. M. TORRENTI, P. ARISTAGHES, K. DOMBERNOVSKI, F. EDON, I. GUENOT, P. MONACHON, "La simulation des effets thermiques dans le béton au jeune âge: exemples d'applications", *Annales de l'Institut Technique du Bâtiment et des Travaux Publics*, vol. 535, concrete series, pp. 322, 1995.

[TOR 96] J. M. TORRENTI, "Comportement mécanique du béton – Bilan de six années de recherche", *Etudes et recherches des Laboratoires des Ponts et Chaussées*, "*Ouvrages d'Art*", vol. OA 23, LCPC, 1996.

[TOR 99] J. M. TORRENTI, L. GRANGHER, M. DIRUY, P. GENIN, "Modeling concrete shrinkage under variable ambient conditions", *ACI Mater. J.*, vol. 96, pp. 35-39, 1999.

[TRI 82] H. W. TRINHZTFY, J. BLAAUWENDRAAD, J. JONGENDIJK, "Temperature development in concrete structures taking account of state dependent properties", in *Colloque International sur le Béton Jeune*, Paris, pp. 211-218, 1982.

[TUR 04] P. TURCRY, Retrait et fissuration des bétons autoplaçants – influence de la formulation, PhD thesis, University of Nantes, 2004.

[VER 50] G. J. VERBECK, C. W. FOSTER, "Long-time study of cement performance in concrete: chapter 6 – the heat of hydration of the cements", in *Proceedings of the American Society for Testing and Materials*, vol. 50, pp. 1235-1257, 1950.

[VER 60] G. VERBECK, "Chemistry of hydration of portland cement – III. energetics of the hydration of Portland cement", in *4th International Symposium on the Chemistry of Cements*, Washington, pp. 453-465, 1960.

[VER 68] G. J. VERBECK, R. H. HELMUTH, "Structures and physical properties of cement paste", in *5th Int. Symp. on the Chemistry of Cement*, Tokyo, pp. 1-32, 1968.

[WAL 00] V. WALLER, "Relations entre composition des bétons, exothermie en cours de prise et résistance en compression", *Etudes et recherches des Laboratoires des Ponts et Chaussées*, "*Ouvrages d'Art*" series, OA35 – LCPC, 2000.

[WHI 78] D. WHITING, A. LITVIN, S. E. Goodwin, "Specific heat of selected concretes", *Journal of American Concrete Institute*, vol. 75, pp. 299-305, 1978.

[WHI 88] D. WHITING, Permeability of selected concretes, in *Permeability of Concrete*, pp. 195-222, ACI SP-108, ACI, 1988.

[WIR 02] E. WIRQUIN, M. BRODA, B. DUTHOIT, "Determination of the apparent activation energy of one concrete by calorimetric and mechanical means: influence of a superplasticizer", *Cement Concrete Res.*, vol. 32, pp. 1207-1213, 2002.

[YSS 95] M. P. YSSORCHE, Microfissuration et durabilité des bétons à hautes performances, PhD Thesis, INSA of Toulouse, 1995.

Chapter 8

Modeling Concrete at an Early Age

8.1. Introduction

In this chapter we present an example of the modeling of concrete at an early age. Modeling is achieved by the experimental determination of the parameters of the behavior of concrete at an early age.

8.2. The coupled thermo-chemo-mechanical problem

8.2.1. *Macroscopic modeling of hydration*

Although hydration is, in fact, the result of several complex reactions that could be modeled at a micro-level (see, for instance, the NIST approach [BEN, 00]), for the modeling of early age behavior we generally choose to consider only one global reaction, and the modeling is developed within the frame of poromechanics of reactive media [COU 95]. Of course other approaches are possible (see [THE 87; BRE 91; BOR 94; NIU 95; EMB 98; BER 01; CER 02; SCH 02; BAZ 03; CHA 03; BEN 08]).

Chapter written by Franz-Josef ULM, Jean-Michel TORRENTI, Benoît BISSONETTE and Jacques MARCHAND.

8.2.1.1. *Hydration degree*

At a macroscopic level concrete at early age can be considered as a porous medium composed of a solid skeleton and a porous space that contains fluids (water, vapor, air). In the following we will consider a closed system without exchange with the exterior[1]. The conservation of the water mass could be written as:

$$\frac{dm_e}{dt} = -\frac{dm_{se}}{dt} \qquad [8.1]$$

where m_e is the water mass per unit volume and m_{se} the water mass per unit volume which has reacted with the cement. If $m_{se}(\infty)$ is the water mass we need for complete hydration of the cement, we can define the degree of hydration ξ with the following equation [POW 48]:

$$\xi(t) = \frac{m_{se}(t)}{m_{se}(\infty)};0 \leq \xi \leq 1 \qquad [8.2]$$

The intrinsic dissipation of the closed system in non-isothermal conditions reads [COU 95]:

$$\varphi dt = \sigma_{ij} d\varepsilon_{ij} - SdT - d\psi \geq 0 \qquad [8.3]$$

where S is the entropy of the system and ψ the Helmholtz free energy, function of external and internal variables, which allow the description of all the sates of the system. Strains ε_{ij} and temperature T are the external state variables. Internal variables are m_{se} (i.e. not controlled from the exterior, for a detailed derivation from mass-conservation and thermodynamics of open porous continua, see [COU 95]), irreversible strains ε_{ij}^p resulting, for instance, from microcracking, ε_{ij}^f creep strains and χ the hardening/softening variable. Free energy is assumed formally as $\psi = \psi\left(\varepsilon_{ij}, T, m_{se}, \varepsilon_{ij}{}^p, \varepsilon_{ij}{}^f, \chi\right)$.

In the case of a chemically inert material and when no plastic evolutions occur, the dissipation is zero, which leads to the standard state equations:

$$\sigma_{ij} = \frac{\partial \psi}{\partial \varepsilon_{ij}} \quad S = -\frac{\partial \psi}{\partial T} \qquad [8.4]$$

1 In fact a coupling between drying and hydration could exist. It is particularly important for the skin of concrete structures (see an example of such a coupling in [OH 03]).

8.2.1.2. *Chemo-elastic cross-effects: aging elasticity and autogeneous shrinkage*

The creation of hydration products leads to stiffening of the matter, which-at the macro-level of material description-appears as a variation of the elasticity modulus in time. At a micro-level of material description, this aging elasticity cannot be regarded as an actual change of the elastic material properties of the matter constituting concrete, but rather as a change in the concentration of the non-aging constituents, i.e. the hardened cement gel ([BAZ 95]). Furthermore, this change in concentration of the hardened cement gel is accompanied by a volume reduction (Le Chatelier contraction), coupled with a capillary shrinkage related to the formation of menisci due to water consumption through hydration ([WIT 82; ACK 88; HUA 95]). Both aging elasticity and shrinkage related to the hydration reaction are chemo-mechanical cross-effects. Aging elasticity could be expressed in an incremental form:

$$d\sigma_{ij} = \left(K - \frac{2}{3}G\right)d\varepsilon_{kk}\delta_{ij} + 2Gd\varepsilon_{ij} + 3Bd\xi\delta_{ij} \qquad [8.5]$$

where the bulk and the tangent shear moduli K and G depend on the degree of hydration ξ as suggested from experiment [BYF 80; LAU 90]. Several authors give relations for the evolution of the Young's modulus with hydration, generally using a relation between the Young's modulus and the compressive strength. We will discuss this point later. Modulus B expresses the cross-effect between the elastic strain and the hydration degree. Indeed, if there is no stress, equation [8.5] could be written as:

$$d\varepsilon_{kk} = -\frac{3B}{K}d\xi = -3bd\xi \qquad [8.6]$$

b has a positive value and is the tensor of chemical dilatation coefficient, which relates the increase of the degree of hydration to the strains of chemical origin. In the isotropic case *b* is a constant. This relation has already been experimentally demonstrated [BUI 79; PAR 90; MOU 03].

Figure 8.1 gives an example of dependence of chemical shrinkage strains on the degree of hydration.

8.2.1.3. *Thermo-mechanical and thermo-chemical cross-effects*

From equation [8.4], for a non-isothermal evolution around temperature $T = T_0$ and without plastic or creep strains, we obtain the following relations:

$$d\sigma_{ij} = \left(K - \frac{2}{3}G \right) d\varepsilon_{kk}\delta_{ij} + 2Gd\varepsilon_{ij} - 3\alpha KdT\delta_{ij} + 3Bd\xi\delta_{ij} \qquad [8.7]$$

$$T_0 dS = C_{th}d\theta - 3T_0\alpha Kd\varepsilon_{kk} - Ld\xi \qquad [8.8]$$

α is the thermal dilation coefficient and accounts for the thermo-mechanical cross-effect. It depends on the degree of hydration, but experimental results show that the variation becomes rapidly negligible (see Chapter 7). Nevertheless, it is possible to take this variation into account (see [BOG 87] for an example).

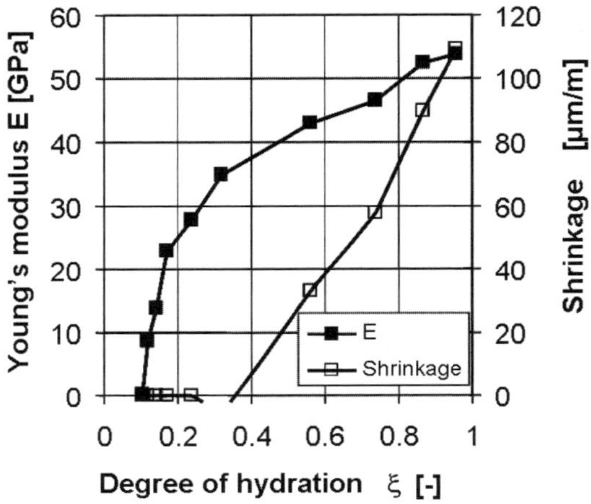

Figure 8.1. *Chemo-mechanical couplings: aging elasticity E(ξ) and dependence of chemical shrinkage strains on degree of hydration [ULM 99]*

The entropy balance equation of the closed porous medium around a reference temperature T_0, neglecting the dissipation, φ, as a source of internal heat, reads:

$$T_0 \frac{dS}{dt} = r - \nabla \cdot q \, ; \qquad [8.9]$$

where r represents the rate of external volumic heat sources (for instance, cooling pipes) and $-\nabla \cdot q$ the rate of external heat due conduction (for instance, heating forms). We adopt here an isotropic linear conduction relation, $q = -k\nabla T$, heat flux vector and k the isotropic thermal conductivity. k is very soon independent of the degree of hydration [BAS 95; MOU 03]. Combining equations [8.8] and [8.9] we obtain:

$$C_{th}\frac{dT}{dt} - 3T_0\alpha K\frac{d\varepsilon}{dt} - L\frac{d\xi}{dt} = r - k\nabla^2 T \qquad [8.10]$$

The physical significance of the terms on the left-hand side of [8.10] are [ULM 95]:

– C_{th} the volume heat capacity per unit of volume in an isodeformation experiment carried out at constant hydration degree (i.e. the heat quantity in order to increase of 1° a unit of volume). As for the thermal dilation coefficient, experimental results have shown that this term could be considered as constant and independent of ξ (see Chapter 7). Nevertheless, a slight decrease in the thermal conductivity and in the heat capacity can be noted with the degree of hydration [SCH 95];

$-3T_0\alpha K$ is the latent heat due to the deformation $d\varepsilon_{kk}$ per unit of volume in an isothermal experiment carried out at a constant degree of hydration; it can be considered as negligible with respect to the latent heat of hydration;

– L is the latent heat due to hydration; $L > 0$ corresponds to an exothermic reaction.

Therefore, assuming the latent heat due to deformation is negligible, equation [8.10] becomes:

$$C_{th}\frac{dT}{dt} = r - k\nabla^2 T + L\frac{d\xi}{dt} \qquad [8.11]$$

8.2.1.4. *Chemo-plastic hardening*

We now take into account plastic strains. Free energy Ψ could be written as:

$$\psi = W(T, \varepsilon_{ij}, \varepsilon_{ij}{}^{p}, \xi) + U(\chi, \xi) \qquad [8.12]$$

where $U(\chi, \xi)$ is the part of free energy associated with plastic hardening/softening phenomena in the material, as detailed in [COU 95; COU 96]. Using equations [8.12] and [8.3] and [8.4] we obtain:

$$\varphi dt = \sigma_{ij} d\varepsilon_{ij}{}^{p} + \zeta d\chi + A d\xi \geq 0 \qquad [8.13]$$

$$\text{with: } \sigma_{ij} = -\frac{\partial W}{\partial \varepsilon_{ij}{}^{p}}\ ;\ \zeta = -\frac{\partial U}{\partial \chi};\ A = -\frac{\partial (W+U)}{\partial \xi} \qquad [8.14]$$

$$d\sigma_{ij} = \left(K - \frac{2}{3}G\right)\left(d\varepsilon_{ij} - d\varepsilon_{ij}{}^{p}\right)\delta_{ij} + 2G\left(d\varepsilon_{ij} - d\varepsilon_{ij}{}^{p}\right) - 3\alpha K dT \delta_{ij} + 3B d\xi \delta_{ij} \qquad [8.15]$$

The first two terms of equation [8.13] represent the dissipation associated with the plastic strains of the skeleton. It is the classical expression obtained for an elastoplastic solid. ζ denotes the hardening force. A is the affinity of the chemical reaction, which expresses the imbalance between the free water and the water combined in solid phase in the macroscopic modeling: it is the driving force of the process of the hydration reaction.

In contrast to standard hardening/softening plasticity, where the plastic threshold depends only on the plastic evolutions [i.e. $\zeta = \zeta(\chi)$], the material threshold may evolve due to the extent of the chemical reaction. This coupling between plastic evolutions and chemical reaction is called chemical hardening [COU 96], and is expressed by a dependence of hardening force, ζ, on both the plastic variable, χ, and degree of hydration, ξ, i.e. $\zeta = \zeta(\chi,\xi)$. This accounts for strength growth as chemo-plastic hardening in early-age concrete [ULM 96]. Concerning the development of the compressive strength of concrete with the degree of hydration, a linear variation [BYF 80; MIN 78] can be adopted, with a threshold corresponding to a mechanical percolation [ACK 88; TOR 05]. De Schutter has proposed a more general relationship [SCH 96]:

$$\frac{f_c(\xi)}{f_c(\xi = 1)} = \left(\frac{\xi - \xi_0}{1 - \xi_0}\right)^a \qquad [8.16]$$

where f_c is the compressive strength, ξ_0 the percolation threshold and a a constant (if a=1, we have a linear relation). A similar relation could be considered for the Young's modulus if we assume that $E = kf_c{}^n$ [ROS 93; SCH 96]. Concerning Poisson's ratio, only a few of results exist. Bernard has shown the influence of water in its evolution [BER 03]. Nevertheless the influence water can be kept constant because the main part of the evolution occurs for very low values of the degree of hydration.

Concerning the elasto-plastic behavior, we have few results. De Schutter [SCH 95] proposed that Sargin's law be used with parameters depending on the degree of hydration, but he did not consider the softening behavior. Regarding elastoplasticity, the strength growth could be associated to the hardening force (see [8.16]). If we assume, in a first approach, that the chemoplastic coupling is linear (a=1), we obtain:

$$\zeta = -\frac{\partial U}{\partial \chi} = \xi \zeta_\infty - \xi \frac{\partial U_0}{\partial \chi} \qquad [8.17]$$

where $\xi\zeta_\infty$ = the strength growth as pure chemical hardening and $-\xi\dfrac{\partial U_0}{\partial \chi}$ its evolution due to plastic evolutions, both defined with respect to their values at complete hydration.

As in the standard plastic model, the plastic admissible stress states are defined by a loading function $f = f(\sigma_{ij}, \zeta) \leq 0$ and only plastic evolutions occur, if a loading point σ is at the boundary of the elasticity domain and where it remains $f = df = 0$. Therefore, standard expressions can be applied for both the flow rule and the hardening rule:

$$d\varepsilon^p{}_{ij} = d\lambda \frac{\partial f}{\partial \sigma_{ij}}; \qquad d\chi = d\lambda \frac{\partial f}{\partial \zeta} \tag{8.18}$$

The plastic multiplier, $d\lambda$, and the hardening modulus, H, verify the following relation [COU 96]:

$$\frac{\partial f}{\partial \zeta}\frac{\partial \zeta}{\partial \chi} d\chi = -d\lambda H \tag{8.19}$$

The constituency condition df=0 used with equations [8.17] and [8.18] in equation [8.19] is written in the form:

$$d\lambda = \frac{1}{H}\left[\frac{\partial f}{\partial \sigma_{ij}} d\sigma_{ij} + \frac{\partial f}{\partial \zeta}\frac{\partial \zeta}{\partial \xi} d\xi\right] \tag{8.20}$$

$$H = \xi H_0 = \xi\left(\frac{\partial f}{\partial \zeta}\right)^2 \frac{\partial^2 U_0}{\partial \chi^2} \tag{8.21}$$

The term $\left[\dfrac{\partial f}{\partial \sigma_{ij}} d\sigma_{ij}\right]$ in equation [8.20] corresponds to the classical term of the plastic multiplier when the second term $\left[\dfrac{\partial f}{\partial \zeta}\dfrac{\partial \zeta}{\partial \xi} d\xi\right]$ represents the chemo-plastic coupling. Finally, the hardening modulus, H, is only associated to the plastic behavior. We can find in [ULM 99] an example of the application of the Willam-Warnke criterion [WIL 73] in the frame of chemoplasticity and in [HEL 00] an extension of chemoplasticity to the case of a multi-surfaces criterion.

8.2.1.5. *Creep at early age*

It is very important to take into account creep at early age if we wants a quantitative prediction of the stresses generated within concrete structures and microcracking to these stresses [GUE 94; TOR 96; SCH 03]. For instance, in the case of totally restrained strains, if creep is not taken into account, we can underestimate tensile stresses at the beginning of the cooling of the structure. We will consider here only basic creep (the effect of desiccation at an early age being limited to the skin of concrete structures).

Several authors have proposed modeling for basic creep at early age. De Schutter has used one Kelvin-Voigt unit to link directly basic creep to the degree of hydration [SCH 99; SCH 00]. This could be improved using several Kelvin-Voigt units [BEN 08]. Here we present an approach proposed by Sercombe *et al.* [SER 00]. This approach is developed in the framework of closed reactive porous media.

Creep in concrete can be divided into short-term creep and long-term creep. Short-term creep is largely affected by stress-induced water movement and redistribution in the capillary porous space [WIT 74]. The underlying mechanism of long-term creep seems, rather, situated in to the relaxation of micro-pre-stress [BAZ 97]. The creep properties are strongly age dependent, concerning both creep amplitude and creep kinetics. Short-term aging during hydration is explained by the volume of growth of hydration products in the capillary pores, which stiffens the overall material rigidity and decreases the strain rate.

Here we consider only the short-term creep. This creep is reversible and is taken into account using a scalar variable, w (the water mass). The free energy of the system can be written as a function of the previously defined variables:

$$\psi = W(\theta, \varepsilon_{ij} - \varepsilon_{ij}^{p}, \varepsilon^{ij^{f}}, \xi, w) + U(\chi, \xi, w) \qquad [8.22]$$

Use of equation [8.22] in the intrinsic dissipation of the closed system (equation [8.3]) reads:

$$\varphi dt = \sigma_{ij} d\varepsilon_{ij}^{p} + \zeta d\chi + A d\xi + \sigma_{ij} d\varepsilon_{ij}^{f} + A_w dw + S_\gamma d\gamma \geq 0 \qquad [8.23]$$

The supplementary state equations are the following:

$$\sigma_{ij} = -\frac{\partial W}{\partial \varepsilon_{ij}^{f}}; \; A_w = -\frac{\partial (W+U)}{\partial w}; \; S_\gamma = -\frac{\partial U}{\partial \gamma} \qquad [8.24]$$

For the evolution of the process at the origin of short-term creep, the simple case of a linear law can be considered [ULM 98]. It leads to expression of the rate of water mass w in the following manner:

$$\eta_w \dot{w} = g_{ij}\sigma_{ij} - kw \qquad [8.25]$$

where η_w is the viscosity associated to the short-term creep, $g_{ij} = \partial^2 W / \partial w \partial \sigma_{ij}$ accounts for the influence of stress σ on the diffusion of water mass w and $k = -\partial^2 (W+U)/\partial w^2 = g_{ij} C_{ijkl} g_{kl} + \partial^2 U / \partial w^2$ a constant of equilibrium for the diffusion process, C_{ijkl} is the elasticity tensor.

Multiplying equation [8.25] by the tensor $g_{ij} = g_{0ij} / f(\xi)$ and assuming that $d\varepsilon_{ij}^{ve} = g_{ij} dw$ where ε_{ij}^{ve} is the visco-elastic strain associated with the reversible creep (i.e. the short-term creep), leads to an aging visco-elastic law of the form:

$$\tau \dot{\varepsilon}^{ve}{}_{ij} = \frac{1}{kf(\xi)^2} g_{0ij} g_{0kl} \sigma_{kl} - \varepsilon_{ij}^{ve} \qquad [8.26]$$

where $\tau = \eta_w / k$ is the characteristic time of short-term creep (or of the diffusion process). Equation [8.26] is a linear function of the stress and the magnitude of the visco-elastic strains ε_{ij}^{ve} (short-term creep) depends on the state of hydration of the material, via the coupling with the state variable ξ. The kinetics of short-term creep could also be made dependent on the state of hydration of the material, by considering a non-constant characteristic time $\tau = \tau(\xi)$. Note also that equation [8.26] corresponds to the behavior of an aging Kelvin Voigt unit.

In the case of a uniaxial creep test, with a constant stress, σ_0, the short-term creep compliance could be defined as $J^{ve} = \varepsilon^{ve} / \sigma_0$. Equation [8.26] leads then to:

$$\tau(\xi) \frac{dJ^{ve}}{dt} = \frac{1}{kf(\xi)^2} - J^{ve} = J_\infty^{ve}(\xi) - J^{ve} \qquad [8.27]$$

$J_\infty^{ve} = 1 / \left[kf(\xi)^2 \right]$ can be regarded as the asymptotic (final) value of the viscous creep compliance. The relations $\tau(\xi)$ and $J_\infty^{ve}(\xi)$ could be determined from creep tests at early age (see, for instance, [SER 00]).

8.2.2. *Identification of the hydration kinetics*

To simplify this problem, we neglect here the influence of plastic and viscous strains on the kinetics (we assume a chemo-elastic evolution).

The hydration reaction occurs when water (as well as dissolved ions) diffuses through the layers of hydrates already formed as unhydrated cement. Once they meet, new hydrates are formed according to a thermally activated reaction; then the water is chemically and/or physically combined. Hence, the diffusion of water through the layers of hydrates can be considered as the rate-determining process of the hydration kinetics (the chemical reaction itself is very fast and its kinetics could be neglected).

At a macroscopic level this could be expressed by the following relation:

$$A = \eta(\xi)\frac{d\xi}{dt}\exp\left(\frac{E_a}{RT}\right) \quad [8.28]$$

where E_a is the activation energy of the hydration and R the ideal gas constant. The factor $\exp\left(\frac{E_a}{RT}\right)$ accounts for the thermally activated character of the reaction according to the Arrhenius concept, while the factor $\eta(\xi)$ accounts for the increase of the thickness of the hydrate layers already formed, which increases the micro-diffusion time of the free water to reach the unhydrated cement. So it is an increasing function of ξ because the micro-diffusion rate decreases with the degree of hydration.

From the expression of the free energy and the state equations we obtain the incremental form of the affinity A:

$$dA = -Bd\varepsilon_{ii} - L\frac{dT}{T_0} \quad [8.29]$$

This equation shows couplings that are symmetrical to chemo-mechanical and thermo-chemical couplings. It shows that the affinity depends of the volumetric strain and of the temperature, T. These couplings indeed exist: to obtain very high strength concrete we know that heating and mechanical pressure allow improvement of the performance [RIC 95]. But, in classical conditions these couplings are weak and could be neglected. In this case, the affinity depends only on the degree of hydration:

$$A = A_0 - a(\xi) \quad [8.30]$$

with $A_0 = a(\xi = 1)$ because $A(\xi = 1) = 0$ (thermodynamic equilibrium).

For technical reasons (to describe the effects of the hydration kinetics on the evolution of the mechanical properties of concrete), we can use an equivalent time t_{eq}, or maturity, such that:

$$t_{eq}(t) = \int_{S=0}^{S=t} \exp\left[\frac{E_a}{R}\left(\frac{1}{T_0} - \frac{1}{T(s)}\right)\right] ds \qquad [8.31]$$

When $T=T_0$, the reference temperature, $t_{eq}=t$. Using this maturity as a variable in equation [8.29], we obtain:

$$\tilde{A}(\xi) = \frac{A}{\eta(\xi)} = \frac{d\xi}{dt}\exp\left(\frac{E_a}{RT}\right) = \exp\left(\frac{E_a}{RT_0}\right)\frac{d\xi}{dt_{eq}} \qquad [8.32]$$

where $\tilde{A}(\xi)$ can be considered as normalized affinity. The solution of the differential equation [8.32] is a relation between the maturity and degree of hydration $\xi=\xi(t_{eq})$. Thus, two samples made of the same concrete that has the same maturity, have undergone equivalent thermal histories (i.e. curing) with respect to the hydration process: their hydration degree is the same.

8.2.3. *Identification of the normalized affinity*

The identification of the normalized affinity $\tilde{A}(\xi)$ implies, for a given temperature history $\theta(t)$, a specific evolution of the hydration kinetics (or the degree of hydration $\xi(t)$). Regarding the macroscopic scale of a concrete sample, the hydration kinetics can be followed only indirectly through the effects the phenomenon induces, i.e. through its cross-effects with heat generation, strength evolution, or autogenous shrinkage. In terms of the theory, the kinetics function $A(\xi)$ can be determined by exploring the Maxwell-symmetries of the modeling, provided that the order of the coupling is known [ULM 96]. This is illustrated here for adiabatic calorimetric experiments and isothermal strength growth evolution tests. The test-data used hereafter are taken from Laplante [LAP 93].

8.2.3.1. *Adiabatic calorimetric experiment*

Under adiabatic conditions, equation [8.11] becomes:

$$C\frac{d\theta}{dt} = L\frac{d\xi}{dt} \qquad [8.33]$$

Assuming the volume heat capacity, C, and the latent hydration heat, L, are constant, equation [8.33] leads to:

$$\xi(t)=\frac{\theta^{ad}(t)-\theta_0}{\theta_\infty-\theta_0} \quad [8.34]$$

where $\theta^{ad}(t)$ is the temperature in the adiabatic test. Finally, using [8.34] in [8.32] gives:

$$\tilde{A}(t)=\frac{d\theta^{ad}(t)/dt}{\theta_\infty-\theta_0}\exp\left(\frac{E_a/R}{\theta^{ad}(t)}\right) \quad [8.35]$$

Equations [8.34] and [8.35] finally gives the relation between $\tilde{A}$ and ξ.

8.2.3.2. *Isothermal strength evolution*

Taking equation [8.16], in the linear case, the compressive strength f_c evolves according to:

$$\frac{df_c}{dt}=\frac{f_c(\xi=1)}{(1-\xi_0)}\frac{d\xi}{dt} \quad [8.36]$$

Equation [8.36], included in equation [8.32], assuming an isothermal conservation of the samples at a constant temperature θ_0 gives:

$$\tilde{A}(t)=\frac{(1-\xi_0)}{f_c(\xi=1)}\frac{df_c(t)}{dt}\exp\left(\frac{E_a}{R\theta_0}\right) \quad [8.37]$$

Figure 8.2 presents the evolutions of the temperature in an adiabatic test and of the strength in isothermal conditions. From these results, we can compare the obtained affinities. In Figure 8.3 we can see that the values for $\tilde{A}(\xi)$ determined for this concrete from two independent experiments, a calorimetric and a mechanical, coincide and confirm the intrinsic nature of the kinetics function $\tilde{A}(\xi)$. This agreement must be seen in perspective with the timescale of the observation of the adiabatic calorimetric experiment and the isothermal strength evolution tests (see Figure 8.2): due to the thermal activation of the hydration reaction, the hydration rate is much higher under adiabatic conditions than under isothermal conditions. The obtained result thus confirms the insignificant influence of temperature, T, on affinity, $\tilde{A}(\xi)$, and thus, the validity of the weak coupling hypothesis on temperature, T [ULM 96].

8.3. Data collection and experimental methods

Consistent with the theoretical approach discussed in the previous section, an inventory of the main experimental methods developed to study early-age concrete is proposed hereafter. Determination of the degree of hydration, monitoring of volume changes and characterization of mechanical properties are addressed.

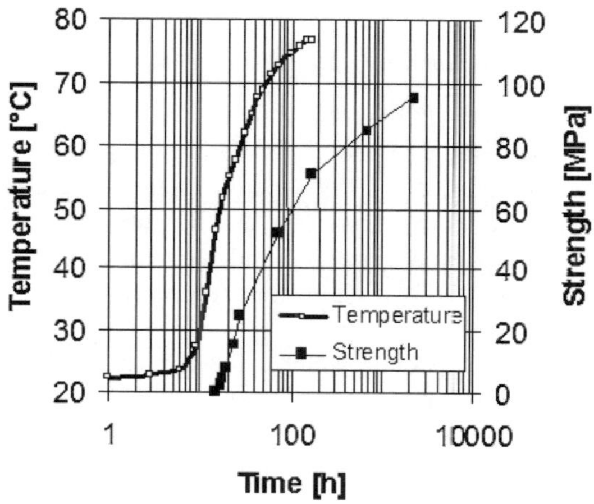

Figure 8.2. *Adiabatic temperature evolution and isothermal strength evolution as a function of time for a high performance concrete (experimental values from Laplante) [ULM 99]*

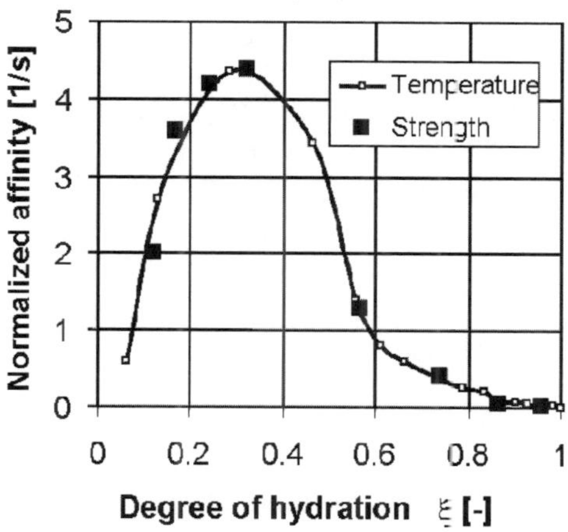

Figure 8.3. *Kinetic function, $\tilde{A}(\xi)$, determined experimentally from an adiabatic calorimetric experiment ("temperature"), and from the isothermal strength evolution ("strength") [ULM 99]*

8.3.1. *Degree of hydration*

Direct determination of the degree of hydration of cementitious materials is practically impossible due to the extremely complex nature of the reaction products [LAP 93]. Therefore, indirect methods must be used to quantify the reaction progress. Existing methods are based on either determining the amount of hydrated cement (by weight) or measuring the heat release during hydration.

8.3.1.1. *Determination of the amount of hydrated cement*

Powers and Brownyard [POW 48] showed that the amount of chemically combined water in a hardening cementitious material is proportional to the amount of products formed, regardless of the reaction stage. Therefore, the degree of hydration (ξ) can be indirectly expressed as the ratio between the amount of chemically bound water at time t ($m_{se}(t)$) and the ultimate amount of chemically bound water ($m_{se}(t = \infty)$) when hydration stops (equation [8.2]).

The amount of chemically bound water can be quantified by successive dryings of powdered cement paste samples at 105°C and 1,000°C. Drying at 105°C allows the interstitial and adsorbed waters to evaporate. Although still under debate, it is generally considered that chemically bound water remains intact at this temperature, and that the hydrates disintegrate only at approximately 390°C [NEV 00]. With subsequent drying at 1,000°C, the hydrates collapse and allow the chemically bound water to evaporate.

The amount of water that is chemically bound to the cement paste at time t is thus determined by the weight loss of the sample between successive dryings at 105°C and 1,000°C, after subtracting the cement loss-on-ignition [BAL 74].

Each mineral phase of the cement clinker (C_3S, C_2S, C_3A, and C_4AF) reacts with water to form hydrates. The volume of water required for the hydration reaction differs for each phase. Molina [MOL 92] identified the amount of water required to hydrate each clinker component ($g_{water}/g_{component}$). The volume of chemically bound water in the material at the end of hydration can be theoretically determined by considering the volume fraction of each clinker phase in the cement and the coefficients determined by Molina [COP 60].

Because of the crystalline structure of clinker, X-ray diffraction (XRD) analysis may also be used to determine the amount of hydrated cement at a given time, t. Comparative results show good correlation with recorded amounts of chemically bound water [COP 60]. However, Byfors [BYF 80] reported discrepancies, which he attributed to the difficulty in stopping hydration at an early stage, when it normally progresses very rapidly.

8.3.1.2. *Heat release determination*

The cement hydration reaction is highly exothermal. Hence, the heat released during hydration provides a parameter to describe the evolution of the degree of hydration. The degree of hydration is determined based on the ratio between the cumulative heat released at time t ($Q(t)$) and the cumulative heat released at the end of hydration ($Q(\infty)$), expressed as follows:

$$\xi(t) \cong \frac{Q(t)}{Q(\infty)} \qquad [8.38]$$

Three different calorimetric methods are used to measure the heat of hydration of cement [LAP 93; MOR 98]: the adiabatic test, the isothermal test, and dissolution heat determination. In the adiabatic test, heat loss is prevented and the hydration heat release can be calculated by measuring the temperature rise. The isothermal calorimetry test consists of maintaining the test sample at constant temperature throughout hydration and measuring the radiant energy flux, based upon which the heat release can be calculated. Dissolution heat determination is an indirect method that allows the determination of the heat of hydration at time t by the difference between the dissolution heat of anhydrous cement and the dissolution heat of hydrated cement paste at time t.

The results of Copeland and colleagues [COP 60] and Danielsson [DAN 66] show a linear relation between heat release and bound water content.

The choice between calorimetry testing and determination of the chemically bound water content in the material depends on the particular context and needs. In addition to generating fundamental data for calculations, adiabatic and isothermal calorimetry tests have the advantage of providing a continuous evaluation of the degree of hydration. Nevertheless, they are not necessarily representative of the actual thermal conditions inside a structure or within a laboratory specimen intended for mechanical or dilatometric testing. While the determination of chemically bound water content requires periodic recordings, it allows representative samples to be taken directly from the structure or test specimen at selected times.

8.3.2. ***Volume changes***

Measuring strains in cementitious materials at early age is a real challenge due to the large number of parameters that can affect the experimental data reliability [BOU 93]. Disturbances induced by the measuring device itself, thermal effects, and measuring device precision are just a few examples of potential error sources.

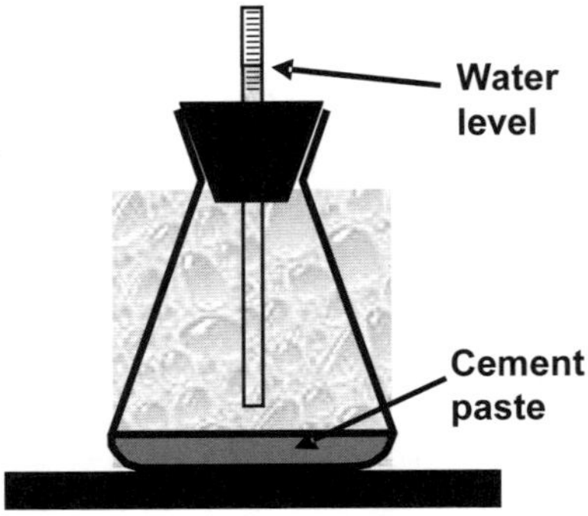

Figure 8.4. *Typical volumetric test method for the measurement of Le Chatelier contraction in cement paste [BIS 02]*

Two types of methods are used to measure free strains in cementitious materials. First, linear methods involve simply measuring the relative displacement of two reference points in the specimen, usually along its main axis. Second, volumetric methods use direct evaluation (displaced fluid volume) or indirect evolution (buoyant weight) of volume changes in representative samples.

8.3.2.1. *Cementitious material shrinkage during hydration*

The shrinkage of cementitious materials during hydration actually comprises two phenomena, the Le Chatelier contraction and self-desiccation shrinkage. The first one, a purely chemical phenomenon, is the volume deficit that characterizes the cement hydration reaction. The second, a physicochemical phenomenon, is a direct consequence of the Le Chatelier contraction occurring in a closed hardening system (i.e. without water intake from the exterior), which results from the progressive decrease in internal relative humidity in the hydrating material porous network.

8.3.2.1.1. Le Chatelier contraction (*internal* or chemical shrinkage)

The Le Chatelier contraction is generally measured using a volumetric test method [LEC 04]. Typically, the test consists of immersing a small amount of cement paste in an Erlenmeyer flask filled with water and sealed with a plug incorporating a capillary tube (Figure 8.4). The volume of liquid displaced during hydration is then measured directly through the water level in the capillary tube. The observed volume change initially corresponds to the plastic material subsidence, and gradually shifts to the filling of the air voids created in the hardening structure [BOI 99; JUS 94; GEI 83].

Albeit simple, this procedure exhibits a few problems and limitations. First of all, the chemical shrinkage results can be affected by the bleeding that occurs in some cementitious materials during the first hours of hydration [BAR 82]. As the results can be significantly affected by the resulting parasitic water flow, the investigated materials should be formulated to prevent bleeding. When bleeding cannot be avoided, the initial readings ("time zero") must be delayed.

Moreover, the test is very sensitive to temperature variations. At 20°C, for example, the coefficient of thermal expansion (CTE) of water is roughly 17 times higher than that of cement paste [LAP 94]. To overcome this problem, experimental devices to measure the Le Chatelier contraction are generally placed in a thermal bath to maintain the material at constant temperature throughout testing.

A further difficulty in volumetric determination of the Le Chatelier contraction is the sensitivity of the results to the material surface to volume ratio and permeability [JUS 94]. The material layer placed in the test flask must be sufficiently shallow to allow water to fully penetrate the cement paste and continuously fill the created voids; otherwise, the Le Chatelier contraction can be only partially recorded. Therefore, a lower material thickness limit must be specified [BOI 99; GEI 83].

Chemical shrinkage can also be determined in a gravimetric fashion. The method uses a device similar to the one described above, except that the water level in the capillary tube is kept constant throughout testing. This condition is achieved by adding water through the capillary tube at regular time intervals or by keeping the unit immersed in water. The device buoyant weight (filled container, material, and capillary tube) is recorded periodically, and the material contraction is calculated accordingly [JUS 94]. The experimental difficulties encountered with this method are similar to those described previously for the volumetric evaluation of the Le Chatelier contraction.

Given the inherent difficulties in measuring the Le Chatelier contraction, the available methods are almost exclusively applicable to cement pastes. However, the test described subsequently in section 8.3.2.1.3 could enable a quantitative evaluation of the Le Chatelier contraction in mortars and concretes.

8.3.2.1.2. Self-desiccation shrinkage (*external* or autogenous shrinkage)

Self-desiccation shrinkage of mortar and concrete is most typically measured with a length change method similar to that used for drying shrinkage determination. After maturing for approximately 18 to 24 hours, specimens are withdrawn from the mold and immediately coated with a thin impermeable film (usually self-adhesive aluminum). The test method simply consists of recording the relative displacement of the reference plugs placed on the side or end faces of the specimen at different time periods, "time zero" coinciding with completion of the sealing operation.

Obviously, strain monitoring that begins 24 hours after water-cement contact neglects a potentially significant portion of self-desiccation shrinkage, particularly in materials with low water/binding ratio. Along with the increasing use of these materials and the early cracking tendency they sometimes exhibit, the need for a more accurate strain assessment commencing immediately after casting has grown considerably. This has led to the development of customized dilatometers such as the one illustrated in Figure 8.5. The main difficulty in designing these devices was to eliminate or, at least, to reduce below a satisfactory level the movement restriction caused by the strain sensors, and if applicable, the friction with the mold walls [PAI 76].

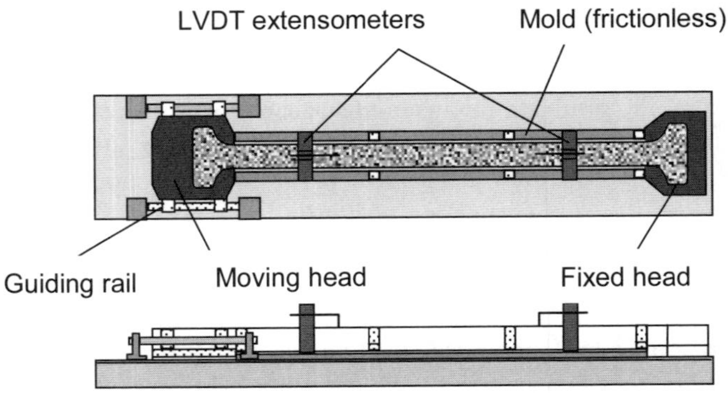

Figure 8.5. *Horizontal dilatometer for the measurement of early age deformations in cementitious materials: extensometers connected with embedded anchorages [PIG 99]*

Accordingly, most of the devices that were developed do not require the specimen to be released from the mold [TOM 00; BJØ 99; KOV 94; PAI 76], the latter being part of the testing system. In addition, great care is taken to minimize friction between the specimen and the mold walls. For instance, some molds are built with materials such as polystyrene [TAW 97]. Alternatively or as a supplementary measure, a thin film with a low friction coefficient can be laid in between the mold and material prior to casting [TOM 00; BJØ 99].

Generally, these devices are equipped with very precise strain gages or sensors connected to an automatic data acquisition system. When inductive displacement transformers (IDT) or linear variable differential transducers (LVDT) are used as external displacement sensors, the specimen movement is transmitted to the moving part of the measuring devices through custom anchoring systems (small rigid parts embedded in the specimen). The use of contactless wave reflective sensors has also been reported in some studies [MOR 99; HOL 99]. A variety of tests have been developed, according to the displacement or strain-measuring procedure. They can be classified into six main categories [HAM 02], as follows:

– nail-type axial anchoring, with the nail head embedded in the concrete and extending through the end plates (Figure 8.6) [BJØ 99];

– end plates anchored in the concrete (Figure 8.7) [MOR 99];

– transverse bars embedded in the test specimen: beam-type (Figure 8.8) [TAK 99];

– vertical supports embedded in the test specimen: slab-type (Figure 8.9) [HOL 99];

– embedded strain gages [HAN 99];

– metal plates sitting on top of upright cylindrical specimens (Figure 8.10) [JEN 97].

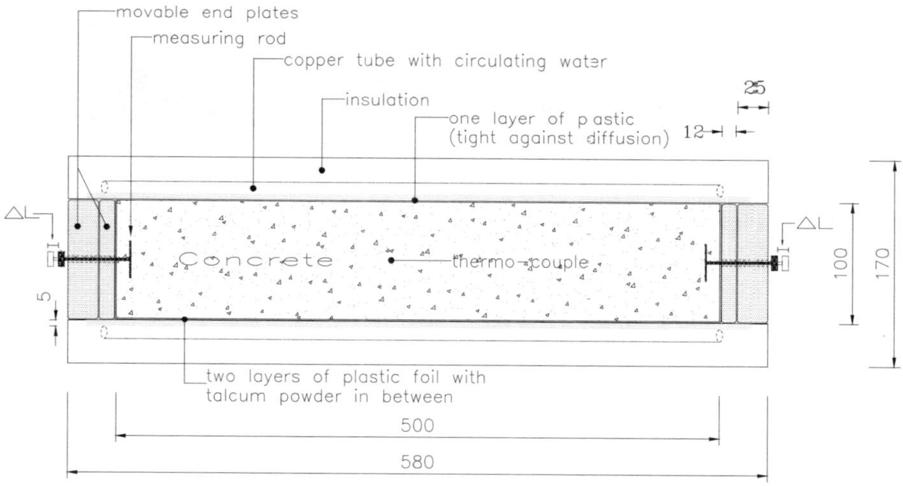

Figure 8.6. *Horizontal dilatometer for the measurement of early age deformations in cementitious materials: extensometers connected to the specimen with embedded nail-type anchorages through the mould end plates [BJØ 99]*

Self-desiccation shrinkage can also be measured using a volumetric method, again almost immediately after water-cement contact. Such a method was first proposed by Justnes and colleagues [JUS 94] and later refined by Boivin [BOI 99]. It consists of filling a latex membrane with cement paste and recording the weight of the sample immersed in water over time (Figure 8.11). Based on Archimedes' principle, the external volume change of the specimen is readily calculated from the buoyant weight change. Because, theoretically, there is no external water ingress, the test measures the overall macroscopic volume change due to self-desiccation in a hydrating system (also called *external shrinkage*). This type of test is usually performed under isothermal conditions to eliminate the potential influence of temperature on the recorded data (influence upon self-desiccation and thermal deformations).

Test samples must be carefully prepared to avoid capturing air in the latex membrane and to ensure that the membrane is properly sealed. Samples must also be rotated while immersed in water to minimize bleeding [HAM 02]. The procedure and main error sources for this type of test are described in details by Justnes and colleagues [JUS 94].

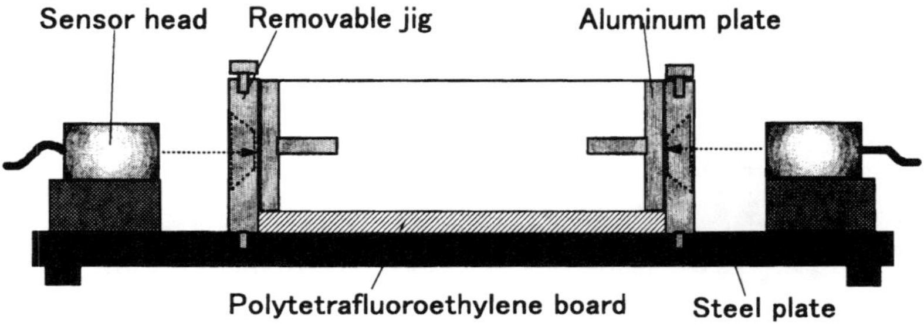

Figure 8.7. *Horizontal dilatometer for the measurement of early age deformations in cementitious materials: extensometers connected to anchored end plates [MOR 99]*

Some limitations and inconsistencies regarding self-desiccation shrinkage determination are found in the literature. First, some tests, such as the volumetric method proposed by Justnes and colleagues [JUS 94], are essentially designed for cement paste. Although they can be extended successfully in some cases to mortar, they are definitely not suitable for concrete. Besides, test results reported by Hammer and colleagues [HAM 02] indicate that concrete self-desiccation shrinkage cannot be reliably predicted based on that of its constitutive cement paste. According to these authors, in the absence of a satisfactory explanation for such discrepancy, doubts on the validity of test results will remain. Conversely, LeRoy [LER 95] previously observed an apparently satisfactory correlation between final self-desiccation shrinkage in paste and corresponding self-desiccation shrinkage in concrete (using the length change test method). However, strains were not monitored during the first 18 to 24 hours of the tests and, unfortunately, the comparative dynamic shrinkage curves for paste and concrete were neither presented nor discussed in the report.

In a different study, linear and volumetric measurements of self-desiccation shrinkage in cement pastes were systematically compared by Barcelo and colleagues

[BAR 97]. It was found that length change data converted into equivalent volumetric values were roughly three-times smaller than the actual volumetric data. This finding raised questions on the validity of strain determination using a latex membrane in the volumetric external shrinkage test. Boivin [BOI 99] identified two parameters potentially contributing to alter volumetric shrinkage experimental data: gradual depression induced by the material self-desiccation inside the membrane (the authors concluded that due to the sealing, a greater depression was induced in the volumetric shrinkage sample than in length change specimens) and pressure exerted by the stretched membrane upon the specimen during setting.

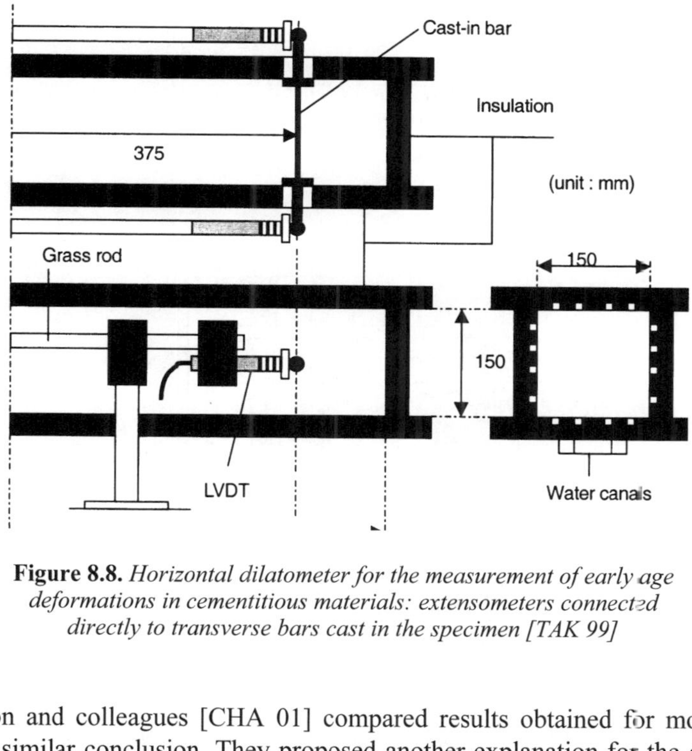

Figure 8.8. *Horizontal dilatometer for the measurement of early age deformations in cementitious materials: extensometers connected directly to transverse bars cast in the specimen [TAK 99]*

Charron and colleagues [CHA 01] compared results obtained for mortars and reached a similar conclusion. They proposed another explanation for the difference between the two types of measurements (length change versus volumetric), which in that study reached approximately one order of magnitude. They asserted that the membrane was being sucked into the specimen surface pores and bubbles under the effect of capillary depression, thus altering the test results.

Furthermore, the results of simple tests performed in Norway cast doubt on the effectiveness of latex membranes to prevent water movement [HAM 02]. In a specimen consisting of a latex membrane filled with water only and exposed to a 30% relative humidity atmosphere, a water loss of 0.5% by weight was recorded over a 24-hour period. Tests are underway at Laval University to determine whether the capillary depression built-up in a hydrating cementitious material is sufficient to cause water to penetrate the membrane and, if so, whether an alternate fluid could be used in lieu of water.

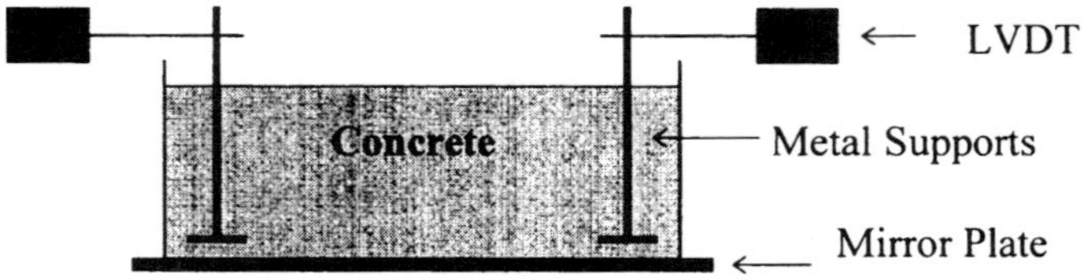

Figure 8.9. *Horizontal dilatometer for the measurement of plastic shrinkage deformations in cementitious materials: extensometers connected to vertical stems embedded in the slab specimen [HOL 99]*

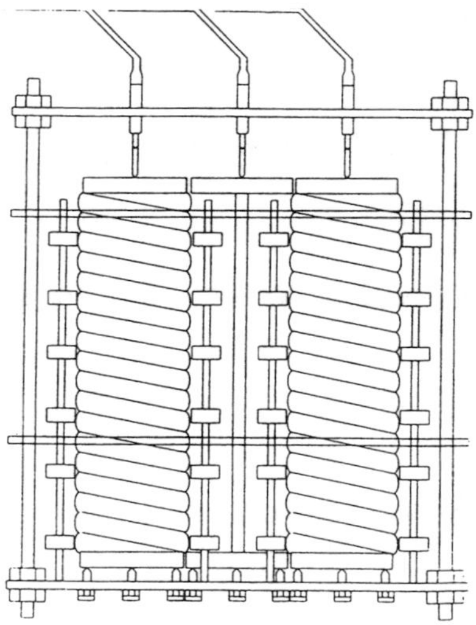

Figure 8.10. *Vertical dilatometer for the measurement of shrinkage deformations in cementitious materials: extensometers connected to metal plates sitting on top of upright specimens cast in cylindrical flexible tubes [JEN 97]*

8.3.2.1.3. Combined determination of *internal* (chemical) and *external* (self-desiccation) shrinkage

A promising test developed by Gagné and colleagues [GAG 99] for studying the early-age behavior of cementitious material allows simultaneous determination, with a single device, of both internal (chemical) and external (self-desiccation) shrinkage almost immediately after water-cement contact. *External* shrinkage, as defined by the authors, is the apparent (external) volume loss of a cementitious material in the absence of hygrometric exchange with the surrounding medium. *Internal* shrinkage corresponds to the cumulative volume of internal voids (more or less the chemical shrinkage having occurred in the hardened system) being created in the material under the same conditions. The combined volume changes must approach the value of the total Le Chatelier contraction, regarding the degree of hydration.

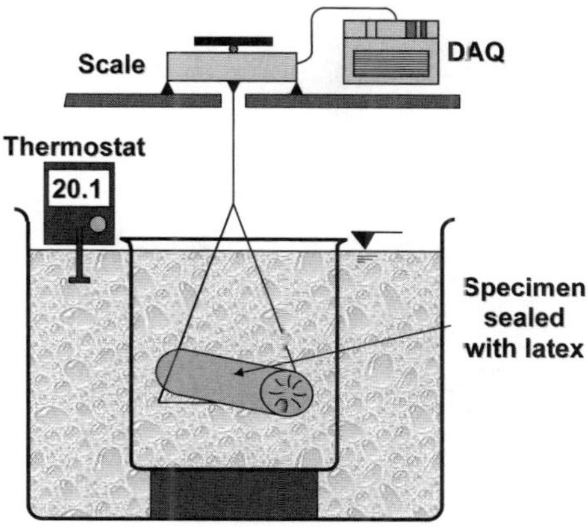

Figure 8.11. *Volumetric test method for the measurement chemical shrinkage in cementitious materials [BIS 02]*

The test setup consists of a dual-compartment cell, as illustrated in Figure 8.12. The test material is placed in the inner chamber, consisting of a flexible cylindrical membrane (latex) axially crossed by three perforated coaxial tubes. The external chamber, which has rigid metallic walls, is filled with water. During testing, the system is immersed in a thermal bath. Self-desiccation shrinkage, or external shrinkage, is measured by monitoring the water displacement in the graduated capillary tube connected to the external chamber. Airflow penetrating the internal chamber via the perforated tubes is measured to determine the volume of internal voids generated during hydration. This device yields quite satisfactory results on cement pastes, with combined volume change values (corrected for the degree of hydration) approaching corresponding theoretical values for the Le Chatelier contraction. The authors believe that it can be adapted to characterize mortars and concretes.

8.3.2.2. *Thermal deformation*

Obviously, there are a number of similarities between the test methods for the determination of thermal and shrinkage strains. This section focuses on the specific problems in determining thermal strain, which are primarily associated with the simultaneous occurrence of autogenous shrinkage and the cross-effects between temperature and self-desiccation (potentially nonlinear effects).

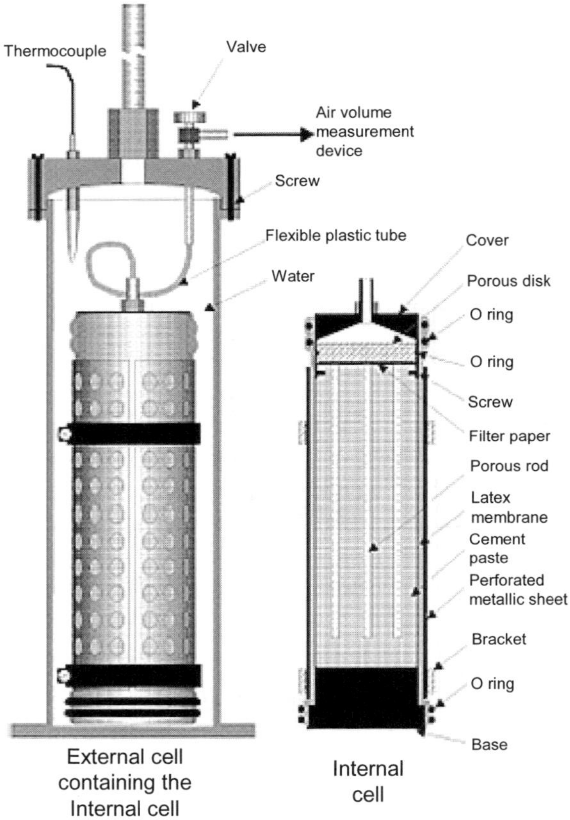

Figure 8.12. *Combined determination of chemical shrinkage and external self-desiccation shrinkage in cementitious materials using a dual-compartment cell [GAG 99]*

Like shrinkage, thermal deformations can be assessed either through volume or length change measurements. However, while volumetric measurements are suitable for use in isothermal conditions, they become quite challenging under variable temperatures [BOU 03], as the container and fluid used are themselves subject to strains that must be precisely determined.

Linear measurement systems are more customary, as they offer greater flexibility, particularly with regards to the type of displacement sensor used. When monitoring is to be undertaken immediately after placement, specimens can be cast horizontally or vertically in the testing devices, with the inner faces of the mold treated to minimize friction.

An innovative approach makes use of flexible tubes into which the concrete is directly placed for length change monitoring. Boulay and colleagues [BOU 93] and Jensen and colleagues [JEN 95] successfully used plastic accordion-shape tubes. This type of system is readily immersed in a thermal bath.

Various methods have been developed to measure autogenous thermal strains in concrete. Weigler and colleagues [WEI 74] developed a method where the autogenous shrinkage strain component is eliminated by differentiation between two specimens subjected to different temperature regimes. In that study, thermal strain curves and the corresponding coefficients of thermal expansion (CTE) were deduced by performing dilatometric tests at different temperature ranges (15-20°C and 20-25°C). Although this approach might seem attractive, it is unfortunately not accurate, because in real time, it does not account for the potentially large variation in maturity between concrete specimens compared for identification purposes.

Moreover, Bjøntegaard's [BJØ 99] results show that, even if equivalent maturity is considered, the thermal component cannot be satisfactorily isolated in tests performed at different temperatures. The same conclusion was drawn from a series of tests conducted under variable temperature regimes. The alternate test strategy proposed by Bjøntegaard to directly determine the thermal dilation coefficient – and hence the thermal component only – at various times during hydration consists of superimposing a low amplitude thermal cycle to the intended thermal regime, for example, a thermal history simulating *in situ* conditions. The cycle amplitude must be sufficiently low so that CTE remains constant in the range considered, while sufficiently large to allow accurate measurement of the corresponding strain. In addition, the thermal cycling period must be short enough to minimize the effects on the autogenous shrinkage component, but still long enough to establish a uniform temperature field. Figure 8.13 illustrates the principle of the test.

8.3.3. *Mechanical behavior*

The mechanical properties of concrete evolve considerably at early age [DES 03]. The material response to the various loads that come along these changes depends on its mechanical strength and visco-elastic properties.

8.3.3.1. *Mechanical strength*

There are no specifically adapted tests to characterize the mechanical properties of concrete (compressive, tensile, and flexural strengths; elastic modulus; Poisson's ratio) in the first 18 to 24 hours after water-cement contact. Generally, ordinary test procedures intended for hardened concrete are carried out as soon as the material can be stripped from the mold and that testing can be performed without altering the specimens. These characterization tests are not described in detail here, as they are well known and abundantly documented elsewhere.

Among the testing devices developed in recent years to characterize early age concrete behavior in restrained movement condition, some can yield information on the evolution of tensile strength and elastic modulus shortly after casting. This is the case for instance with the discretized restrained shrinkage (DRS) tests described in the following subsection (section 8.3.3.2). As the specimen is cast directly in the testing device, tests up to failure can be performed to determine material tensile strength of very early age concrete. However, when multiple results are required, the procedure is tedious, as it requires repeated testing. For the elastic modulus, the stepwise (discretized) test procedure allows monitoring changes in elastic modulus over time immediately from the time of setting.

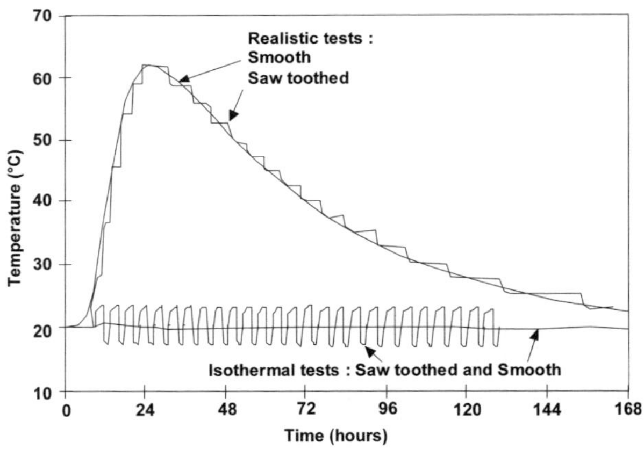

Figure 8.13. *Schematic diagram of the experimental procedure for evaluating the thermal dilation coefficient of cementitious materials during hydration: superimposed low-amplitude/short-period saw-toothed thermal cycling [BJØ 99]*

8.3.3.2. *Viscoelastic properties*

Similar to the determination of mechanical strength and elastic properties, the viscoelastic properties of young concrete can be determined with the usual test methods intended for hardened concrete, as soon as the specimens can be stripped from their molds without damage. To obtain information on the viscoelastic nature of early age concrete (0-24 hours), specimens must be cast directly in the testing apparatus, as in the methods designed to assess early age concrete shrinkage and cracking sensitivity.

8.3.3.2.1. Creep tests for hardened concrete

The most widely documented and commonly used test is the *compressive creep* test, for which standardized procedures and devices have been adopted (for example, the RILEM CPC 12 method in Europe and the ASTM C512 Standard Test Method in North America). These test methods are not discussed further here, except to mention that it is quite difficult to perform on concrete in the first 24 to 30 hours.

With the aim of limiting or eliminating cracking in concrete structures, increasing emphasis has been put on the tensile properties of concrete, especially the visco-elastic behavior. Although no creep test in tension has been standardized yet, a number of testing devices have been developed in recent years to better characterize the phenomenon [UME 94; BIS 95; POS 98; BIS 97, DEL 99; PAN 01; VAY 01; MOR 03], usually in connection with drying shrinkage cracking mitigation.

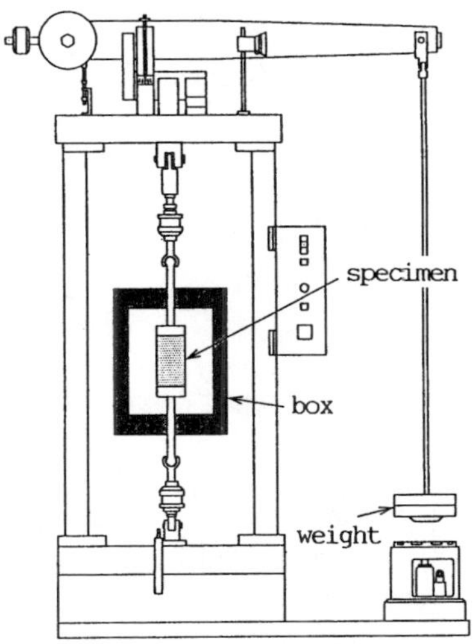

Figure 8.14. *Direct tensile creep rig for cementitious materials using a lever-arm loading system [UME 94]*

Creep test rigs typically rely on a rather simple design, as illustrated in Figure 8.14, using lever-arm systems to apply the load and, in some cases, to measure the specimen strain. With a proper mechanical anchoring system, testing can be undertaken as early 24 to 30 hours after casting, provided that stringent test procedures are followed [BIS 95; PAN 01].

Conducting direct tensile creep tests is difficult (parasitic effects of eccentricity, low amplitude strain measurement, etc.) and cumbersome. A promising alternative, particularly for industrial applications, is the *flexural creep test*. Data on this topic are unfortunately scarce, but a recent study showed that flexural creep results correlate well with direct tensile creep test results [MOR 03]. In this study, 50×100×1,200 mm beams were subjected to a sustained four-point gravity loading using a simple and inexpensive test setup (Figure 8.15). Such flexural creep test specimens can be loaded within the first 24 to 30 hours after casting more easily than direct tensile creep specimens, though a strict procedure must still be obeyed.

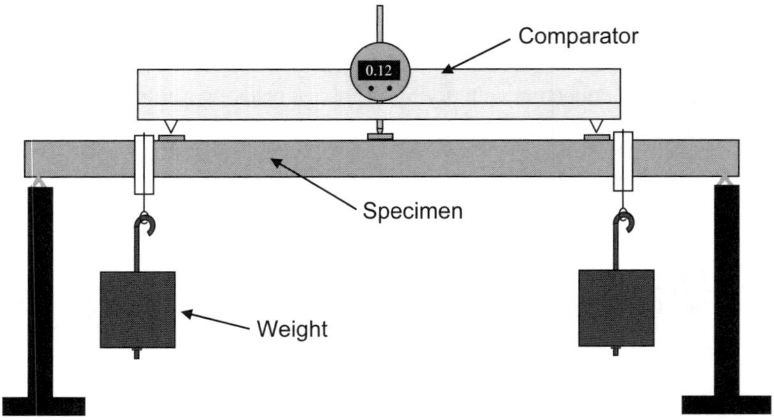

Figure 8.15. *Flexural creep test rig for concrete and mortar [MOR 03]*

Although it is commonly accepted that creep and relaxation are dual aspects of the same phenomenon, this is not supported by a great deal of experimental data, mostly because of the inherent complexity of the related test methods. Morimoto and Koyanagi [MOR 94] developed two different devices (Figure 8.16a and b) in a comparative study of concrete relaxation in compression and tension at early age (starting at t=1 day). In the case of compressive tests, 100×100×400 mm prismatic specimens are loaded in a closed-loop hydraulic testing machine allowing strain control. The specimens, initially loaded up to a given stress level, are maintained thereafter at the corresponding strain throughout testing. For tensile testing, 100×100×860 mm prismatic specimens are subjected to direct loading in a high-rigidity frame. Unfortunately, the anchoring procedure and the loading system were not described in the article.

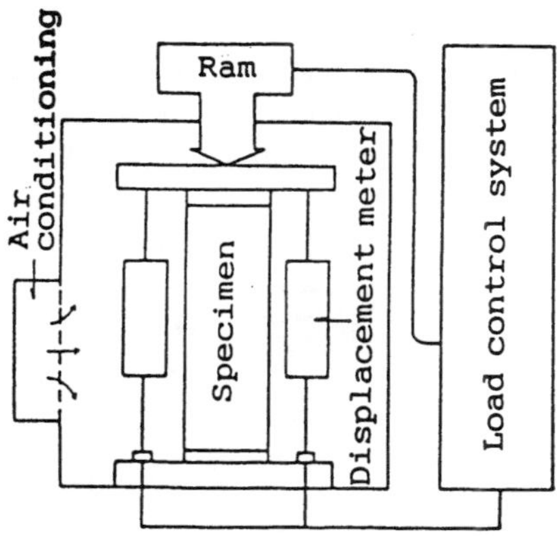

a)

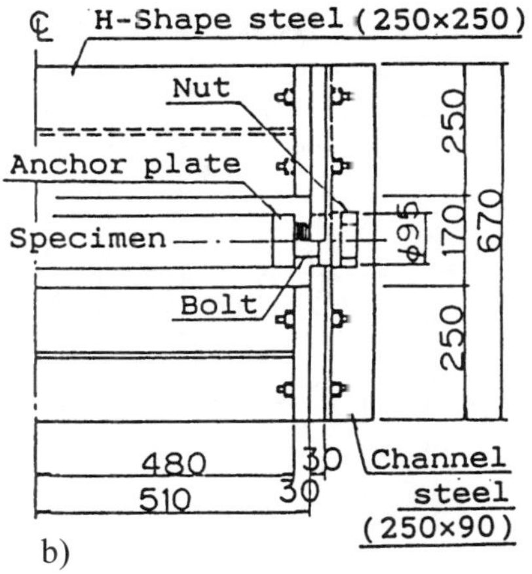

b)

Figure 8.16. *Relaxation test setups for cementitious materials: a) compressive loading, b) tensile loading [MOR 94]*

8.3.3.2.2 Creep tests in hardening concrete

In the wake of recent research related to the study and characterization of early volume changes in concrete, test methods have been developed to measure the resulting stresses induced under conditions that simulate real structures (restrained movements). With well-designed test conditions and appropriate treatment of the experimental data, concrete visco-elastic behavior can be quantitatively assessed from the time of setting.

Basically, the restrained strain tests developed to study early-age concrete fall into two distinct categories: *passive-restraint* and *active-restraint* tests. The first category includes the *cracking frame test*, the length change test performed on prismatic specimens reinforced with an instrumented rebar, and the *restrained shrinkage ring test*. The second category, inspired by the *cracking frame test*, includes different versions of the DRS test.

The *cracking frame test*, first developed at the Federal Institute for Materials Research and Testing in Germany, involves a prismatic concrete cast horizontally and having a straight prismatic central part with flared ends fitting into rigid grips [SPR 94]. As shown in Figure 8.17, the restraint is provided by stiff lateral steel members intended to keep the distance between the grips as constant as possible. The load induced by the restrained thermal and shrinkage strains is measured through strain gages mounted on the lateral members. The cracking frame was built mainly to study thermal cracking in concrete, so the apparatus formwork was insulated in order to reproduce the temperature development in a thick cross-section.

The output of the test is the thermal and autogenous shrinkage stress curve. The visco-elastic component is obtained by simultaneously measuring the total free strain (thermal strain and hygrometric shrinkage) in an identical control specimen subjected to the same exposure conditions (temperature and relative humidity).

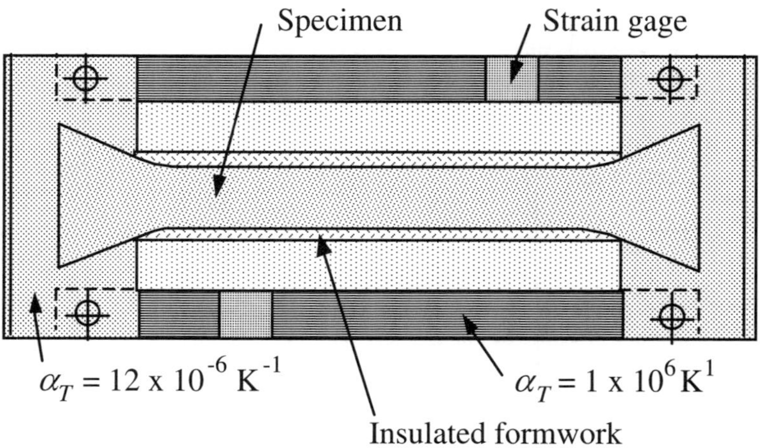

Figure 8.17. *Restrained strain apparatus for concrete (passive restraint): the cracking frame [SPR 94]*

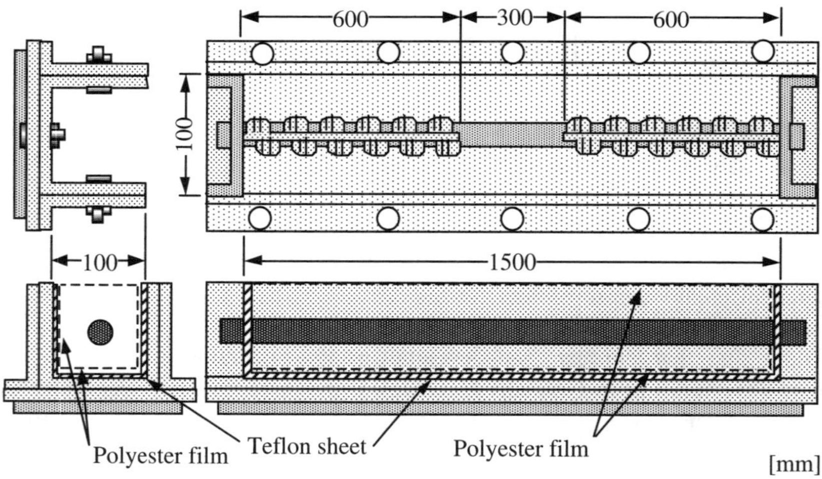

Figure 8.18. *Schematic diagrams of the mold and sample used in the restrained shrinkage test (passive restraint) adopted by the Japan Concrete Institute [JCI 96]*

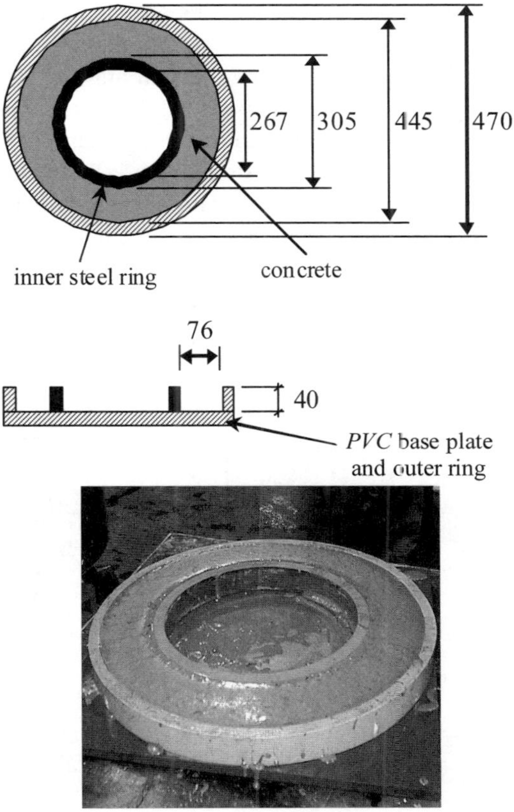

Figure 8.19. *Example of restrained shrinkage ring test for cementitious materials (passive restraint)*

To obtain information on the viscoelastic properties in any of the three aforementioned passive restraint tests, the free strain must be measured simultaneously in unrestrained control specimens similar in shape and size to the restrained specimens and submitted to the same exposure conditions (in the case of the rings, free shrinkage can be measured on prismatic control specimens having equivalent volume/surface ratio [ATT 01]). It is then possible to calculate the visco-elastic strain component (elastic strain + creep), which corresponds to the difference between the strain curves obtained for the restrained and control specimens. Knowing the stress evolution in the concrete specimen over time, the experimental compliance curve (*J*) can be determined.

Tests performed on prismatic specimens reinforced with an instrumented steel rebar were mainly developed in Japan [SAT 98; MAT 98; OHN 98]. Figure 8.18

illustrates the mold and specimen used in the procedure adopted by the Japan Concrete Institute [JCI 96]. The specimen simply consists of a prismatic beam with an embedded steel rebar that restrains its longitudinal movement. The surface of the central section of the rebar is ground, polished, and covered with Teflon® along a sufficient length to prevent bond and ensure uniform stress distribution in both materials (steel, concrete) in that area. A strain gage is installed on the steel rebar beneath the Teflon® coating to record the strain induced during the test and allow determination of the corresponding stress as soon as the material has set.

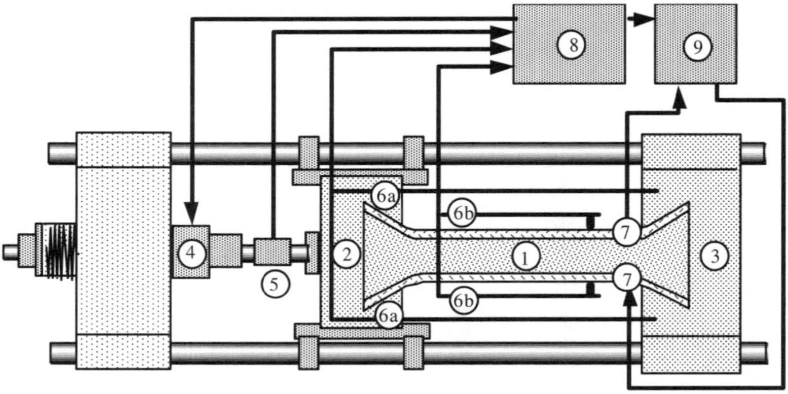

(1) Specimen
(2) Adjustable cross-head
(3) Fixed cross-head
(4) Step motor
(5) Load cell
(6a) Measurement of cross-head movements
(6b) Length measurement with carbon fiber bars
(7) Formwork with heating/cooling system
(8) PC for controlling and recording
(9) Cryostal for cooling/heating of the formwork

Figure 8.20. *Restrained strain apparatus for concrete (active restraint): the Temperature-Stress Testing Machine (TSTM) [SPR 94]*

The restrained shrinkage ring test is another passive-restraint test used to study early age behavior, although it was developed and used primarily to study drying shrinkage cracking. The principle is to cast an annulus of concrete around an inner steel ring, which restrains concrete contraction. The relative dimensions of the two components must be carefully selected to limit the stress gradient in the concrete, which is readily evaluated [WEI 01]. Besides its simplicity, the advantage of this approach is that it eliminates localized stress transfer areas, where stress concentrations invariably develop, and the overall volume of the specimen is stressed rather uniformly. A typical test setup is illustrated in Figure 8.19. In addition to the inner steel ring, the mold is made of two removable parts, an external cylindrical wall and a bottom plate. Similar to the two other passive tests, the restraining steel ring is instrumented (on its inner face) with strain gages to monitor strains in the steel and determine the evolution of stresses in both the steel and concrete rings right from the time the material has set.

As in the passive-restraint tests, the stress variation is not controlled in any way; it is not possible, based on the sole results of the tests, to distinguish between elastic strain and component creep. To do so, the change in elastic modulus must be independently determined under exposure conditions similar to those of the restrained shrinkage test. This potentially significant limitation can be overcome by using active-restraint tests.

Moreover, it is important to realize that 100% restraint is hardly ever achieved in passive-restraint tests unless disproportionately restraining elements are used, that the effective restraint depends on the relative rigidity of the concrete and the restraining element, and that it can change significantly during testing due to ageing and creep.

The first testing device with active control specifically designed to study early age restrained autogenous shrinkage was built at the *Laboratoire Central des Ponts et Chaussées* (LCPC) in France about 20 years ago [PAI 89]. This modified version of the *cracking frame*, called by the authors the *self-cracking apparatus*, fully counteracts the volume changes. The lateral restraining members are replaced by a loading device attached to one of the two end grips, which is mobile, and equipped with a dynamometer. From the time of casting and throughout the test, the length of the specimen is maintained constant by means of a closed-loop system that monitors the strain and controls the loading device. During about the same period, a similar device called the *Temperature-Stress Testing Machine* (TSTM) was built at IBM in Germany for the study of thermal cracking (see Figure 8.20). Subsequently, some investigators [KOV 94; BLO 95; PIG 99] have built sophisticated systems on the basis of those developed at LCPC and IBM. Basically, they consist of a horizontal mold with flared ends in which the fresh concrete is placed directly after mixing. The mold is designed in such a way to minimize friction between concrete and the mold walls in the central section of the specimen, the restraint being exclusively provided by the bearing forces generated at the ends. One of the gripping ends is fixed, while the other is mobile and connected in series to a load cell and to a closed-loop control system. Contrary to the two active tests described previously (*self-cracking apparatus* and TSTM), load adjustment to prevent strain in the specimen is not performed continuously, but rather discretely. When the shrinkage-induced deformation has reached a selected threshold value (ranging typically between ± 1 to ± 5 μm/m), a load increment is applied to bring the specimen back to its original position. The operating principle is illustrated in Figure 8.21. As the threshold strain value is reduced, the test approaches the conditions found in a purely restrained shrinkage test (Figure 8.21b). The load being increased step by step (instead of evolving in a quasi-continuous fashion), this procedure is also referred to as the DRS test.

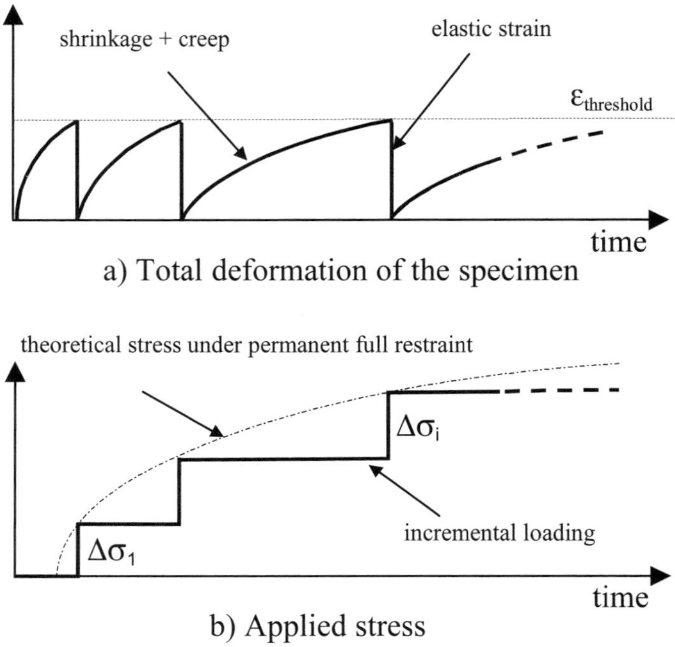

Figure 8.21. *Principle of the discretized restrained shrinkage test (active restraint) [BIS 01]*

The raw test results are the evolution of stress and total strain over time. By monitoring simultaneously the free strain on a similar, unrestrained specimen exposed to identical exposure conditions, the evolution of both elastic and creep strains can be determined. This is because the load increment is applied over a relatively short duration at the beginning of each cycle, allowing the tangent elastic modulus to be determined at the age and stress level considered.

Some of these systems are equipped with devices to control precisely the specimen temperature [SPR 94; BJØ 99; CHA 02] during testing so that various thermal histories can be imposed, for instance, to simulate real conditions in a structure and/or characterize cross-effects between phenomena.

One of the great advantages of the DRS test is its great versatility. Depending on the system, control procedures can be modified in a variety of ways to suit various experimental needs: restrained volume change testing (with or without discretization), creep testing (under constant or variable stress), relaxation testing (under constant or variable strain), etc. When the system features include thermal control, the possibilities increase considerably.

8.4. Conclusion

Along with the development of models, the test methods to characterize early age concrete behavior and properties made substantial progress in recent years. In particular, sophisticated techniques have been developed to determine material properties under conditions simulating those of actual structures. Nevertheless, much remains to be done to meet all identification needs. As emphasized in section 8.2, current models rely on assumptions about the evolution or cross-effects of certain parameters and phenomena. One challenging aspect that calls for immediate attention is the study and characterization of the coupled effects between self-desiccation shrinkage, thermal strain, and mechanical response of the material in the first hours and days of hydration. The evolution of material damage in this period also needs to be investigated. Of course, beyond the mechanical considerations, it is necessary to continue performing exhaustive physicochemical characterization of cementitious material in the very early stages of hydration.

8.5. Bibliography

[ACK 88] ACKER P., *Comportement mécanique des bétons. apports de l'approche physico-chimique*, research report of the LPC, no. 152, 1988.

[ATT 01] ATTIOGBE E.K., SEE H.T., MILTENBERGER M.A., "Tensile creep in restrained shrinkage", in ULM, F. J., BAZANT, Z. P., WITTMAN, F. H., *RILEM 6th International Conference on Creep, Shrinkage & Durability Mechanics of Concrete and other Quasi-Brittle Materials*, Boston, Elsevier, pp. 651-656, 2001.

[BAL 74] BALLANDER U., *Accelerated provning av betongs hallfasthet - Bestämning av hydrationsgrad*, Swedish Cement and Concrete Research Institute, Stockholm, 1974.

[BAR 82] BARON J., BUIL M., "Le retrait autogène de la pâte de ciment durcissante", *Bull. Liaison LPC*, no. 117, pp. 65-70, 1982.

[BAR 97] BARCELO L., BOIVIN S., RIGAUD S., ACKER P., CLAVAUD B., "Linear *vs* volumetric autogeneous shrinkage measurement: Material behavior or experimental artifact?", in *Proceedings of Second International Research Seminar on Self-Desiccation and its Importance in Concrete Technology*, Lund, pp. 109-126, 1997.

[BAS 95] BASTIAN G., KHELIDJ A., "Propriétés thermophysiques d'un béton fraîchement coulé", *Bull. Liaison LPC*, no. 200, pp. 25-35, 1995.

[BAZ 03] BAZANT Z.P., KIM J.K., JEON S.E., "Cohesive fracturing and stresses caused by hydration heat in massive concrete wall", *J. Eng. Mech.*, ASCE, vol. 129, pp. 21-30, 2003.

[BAZ 97] BAZANT Z.P., HAUGGARD A., BAWEJA S., ULM F.J., "Microprestress solidification theory for concrete creep. I: Aging and drying effects", *J. Eng. Mech.*, ASCE, vol. 123, pp. 1188-1194, 1997.

[BEN 00] BENTZ D., "CEMHYD3D: a three dimensional cement hydration and microstructure development modeling package. Version 2.0", *NISTIR 6485*, US Department of Commerce, April 2000.

[BEN 08] BENBOUDJEMA F., TORRENTI J.M., "Early-age behavior of concrete nuclear containments", *Nucl. Eng. Des.*, vol. 238, pp. 2495–2506, 2008

[BER 01] O. BERNARD, E. DÉNARIÉ, E. BRÜHWILER, Comportement au jeune âge du béton et limitation de la fissuration traversante des structures hybrides, research report no. 563, Office fédéral des routes, 2001.

[BER 03] BERNARD O., ULM F.J., LEMARCHAND E., "A multiscale micromechanics-hydration model for the early-age elastic properties of cement-based materials", *Cement Concrete Res.*, vol. 33, pp 1279-1295, 2003.

[BIS 01] BISSONNETTE B., MARCHAND J., DELAGRAVE A., BARCELO L., "Early-age behavior of cement-based materials", in J.P. SKALNY, *Materials Science of Concrete – VI,* American Ceramic Society, 2001.

[BIS 02] BISSONNETTE B., MARCHAND J., MARTEL C., PIGEON M., "Influence of superplasticizer on the volume stability of hydrating cement pastes at early age", in *Concrete: Material Science to Applications – A Tribute to Surendra P. Shah*, ACI Spring Convention, Detroit, pp. 167-188, 2002.

[BIS 95] BISSONNETTE B., PIGEON M., "Tensile creep at early ages of ordinary, silica fume and fiber reinforced concretes", *Cement Concrete Res.*, vol. 25, pp. 1075-1085, 1995.

[BIS 97] BISSONNETTE B, Le fluage en traction: un aspect important de la compatibilité des réparations minces en béton, PhD thesis, Laval University, 1997.

[BJØ 99] BJØNTEGAARD Ø., Thermal dilation and autogenous deformation as driving forces to self-induced stresses in high performance concrete, PhD thesis, Norwegian University of Science and Technology, 1999.

[BLO 95] BLOOM R., BENTUR A., "Free and restrained shrinkage of normal and high-strength concretes", *ACI Mater. J.*, vol. 92, pp. 211-217, 1995.

[BOG 87] VAN DEN BOGERT P.A.J., DE BORST R., NAUTA P., "Simulation of the mechanical behavior of young concrete", in *Proceedings IABSE Colloquium on Computational Mechanics of Reinforced Concrete Structures,* Delft, 1987.

[BOI 99] BOIVIN S., Retrait au jeune âge du béton – Développement d'une méthode expérimentale et contribution à l'analyse physique du retrait endogène, PhD thesis, École Nationale des Ponts et Chaussées, 1999.

[BOR 94] DE BORST R., VAN DEN BOOGARD A., "Finite element modeling of deformation and cracking in concrete at early ages", *J. Eng. Mech.,* ASCE, vol. 120, 1994.

[BOU 03] BOULAY C., "Determination of the coefficient of thermal expansion", Early age cracking in cementitious systems", in A. BENTUR, RILEM report 25, *Early Age Cracking in Cementitious Systems*, RILEM Publications, pp. 217-224, 2003.

[BOU 93] BOULAY C., PATIES C., "Mesure des déformations du béton au jeune âge", *Mater. Struct.*, vol. 26, pp. 307-311, 1993.

[BRE 91] VAN BREUGEL K., Simulation of hydration and formation of structure in hardening cement-based materials, PhD thesis, Delft University of Technology, 1991.

[BUI 79] BUIL M., Contribution à l'étude du retrait de la pâte de ciment durcissante, research report of the LPC, no. 92, 1979.

[BYF 80] BYFORS J., Plain concrete at early ages, CBI research report 3:80, Swedish Cement and Concrete Research Institute, 1980.

[CER 02] CERVERA M., FARIA R., OLIVER J., PRATO T., "Numerical modeling of concrete curing, regarding hydration and temperature phenomena", *Comp. Struct.*, vol. 80, pp. 1511-1521, 2002.

[CHA 01] CHARRON J.-P., MARCHAND J., BISSONNETTE B., "Early-age deformations of hydrating cement systems, systematic comparison of linear and volumetric shrinkage measurements", *Concrete Sci. Eng.*, vol. 3, pp. 168-173, 2001.

[CHA 02] CHARRON J.-P., Contribution à l'étude du comportement au jeune âge des matériaux cimentaires en conditions de déformations libre et restreinte, PhD thesis, Laval University, 2002.

[CHA 03] CHARRON J.P., MARCHAND J., BISSONNETTE B., PIGEON M., GÉRARD G., "Empirical models describing the behavior of concrete at early age", in A. BENTUR, RILEM report 25, *Early Age Cracking in Cementitious Systems*, RILEM Publications, 2003.

[COP 60] COPELAND L.E., KANTRO D.L., VERBECK G.J., "Chemistry of hydration of Portland cement", in *Fourth International Symposium on Chemistry of Cement*, paper IV-3, Washington, 1960.

[COU 95] COUSSY O., *Mechanics of Porous Continua*, J. Wiley and Sons, 1995.

[COU 96] COUSSY O., ULM F.J., "Creep and plasticity due to chemo-mechanical couplings", in *Archive of Applied Mechanics*, vol. 66, Springer Verlag, pp. 523-535, 1996.

[DAN 66] DANIELSSON U., Conducting Calorimeter Studies of the Heat of Hydration of a Portland Cement, CBI research report 38, Swedish Cement and Concrete Research Institute, 1966. (after [BYF 80])

[DEL 99] DELA B.F., Eigenstresses in hardening concrete, PhD thesis, Technical University of Denmark, 1999.

[DES 03] DE SHUTTER G., KOVLER K., "Short term mechanical properties", in A. BENTUR, RILEM report 25, *Early Age Cracking in Cementitious Systems*, RILEM Publications, pp. 101-110, 2003.

[EMB 98] EMBORG M., "Models and methods for computation of thermal stresses and cracking risks", in R. SPRINGENSCHMIDT, RILEM report 15, *Prevention of Thermal Cracking in Concrete at Early Ages*, edited by 1998.

[GAG 99] GAGNÉ R., AOUAD I., SHEN J., POULIN C., "Development of a new experimental technique for the study of the autogenous shrinkage of cement paste", *Mater. Struct.*, vol. 32, pp. 635-642, 1999.

[GEI 83] GEIKER M., Studies of Portland cement hydration by measurements of chemical shrinkage and a systematic evaluation of hydration curves by means of the dispersion model, PhD thesis, Institute of Mineral Industry Technical University of Denmark, 1983.

[GUE 94] GUENOT I., TORRENTI J.M., LAPLANTE P., "Stresses in concrete at early ages: comparison of different creep models", *Proceedings of the RILEM Symposium on Avoidance of Thermal Cracking in Concrete at Early Ages*, Munich, 1994.

[HAM 02] HAMMER E.E., BJØNTEGAARD Ø., SELLEVOLD E.J., "Measurement methods for testing of early autogenous strain", in A. BENTUR, RILEM report 25, *Early Age Cracking in Cementitious Systems*, RILEM Publications, pp. 207-216, 2002.

[HAN 99] HANEHARA S., HIRAO H., UCHIKAWA H., "Relationships between autogenous shrinkage, the microstructure and humidity changes at inner part of hardened cement paste at early ages", in E.-I. TAZAWA *Autogenous Shrinkage of Concrete, Proceedings Int. Workshop* 1998, E & FN Spon, pp. 89-100, 1999.

[HEL 00] HELLMICH C., SERCOMBE J., ULM F.J., MANG H., "Modeling of early-age creep of shotcrete. II: application to tunneling", *J. Eng. Mech.,* vol. 126, pp. 292-299, 2000.

[HOL 99] HOLT E.E., LEIVO M.T., "Autogenous shrinkage at very early ages", in E.-I. TAZAWA *Autogenous Shrinkage of Concrete, Proceedings Int. Workshop* 1998, E & FN Spon, pp. 133-140, 1999.

[HUA 95] HUA C., *Analysis and Modeling of Autodessiccation Shrinkage of Hardening Cement Paste.* Research report of the LPC, no. 0A15. LPC, 1995.

[JCI 96] JAPAN CONCRETE INSTITUTE, *JCI Research Report of Autogenous Shrinkage,* pp. 199-201, 1996.

[JEN 95] JENSEN O.M., HANSEN P.F., "A dilatometer for measuring autogenous deformation in hardening Portland cement paste", *Mater. Struct.*, vol. 28, pp. 406-409, 1995.

[JEN 97] JENSEN O.M., CHRISTENSEN S.L., DELA B.F., HANSEN J.H., HANSEN P.F., NIELSEN A., *HETEK – Control of Early Age Cracking in Concrete – Phase 2: Shrinkage of Mortar and Concrete*, Report 110, Road Directorate, Denmark Ministry of Transportation, 1997.

[JUS 94] JUSTNES H., REYNIERS B., SELLEVOLD E.J., "An evaluation of methods for measuring chemical shrinkage of cementitious pastes", *Nordic Concrete Res.*, vol. 25, pp. 45-61., 1994.

[KOV 94] KOVLER K., "Testing system for determining the mechanical behavior of early age concrete under restrained and free uniaxial shrinkage", *Mater. Struct.*, 27, pp. 324-330, 1994.

[LAP 93] LAPLANTE P., "Propriétés mécaniques des bétons durcissants: analyse comparée des bétons classiques et à très hautes performances", *Etudes et recherches des Laboratoires des Ponts et Chaussées*, "*Ouvrages d'Art*", OA13, LCPC, 1993

[LAP 94] LAPLANTE P., BOULAY C., "Évolution du coefficient de dilatation thermique du béton en fonction de sa maturité aux tous premiers âges", *Mater. Struct.*, vol. 27, pp. 596-605, 1994.

[LAP 94] LAPLANTE P., BOULAY C., "Evolution du coefficient de dilatation thermique du béton en fonction de sa maturité", *Mater. Struct.*, vol. 27, 596-605, 1994.

[LAU 90] LAUBE, M. (1990) Constitutive model for the analysis of temperature-stresses in massive structures. PhD thesis, Braunschweig University of Technology.

[LEC 04] LE CHATELIER H., *Recherches expérimentales sur la constitution des mortiers hydrauliques*, Dunod, 1904.

[LER 95] LE ROY R., "Déformations instantanées et différées des bétons à hautes performances", *Études et recherches des laboratoires des ponts et chaussées, Série "Ouvrages d'Art*", OA22, LCPC, 1996.

[MAT 98] MATSUHITA H., TSURUTA H., "The influence of quality of coarse aggregate on the autogenous shrinkage stress in high-fluidity concrete", in *Autogeneous Shrinkage of Concrete*, E & FN Spon, pp. 339-350, 1998.

[MIN 78] MINDESS S., YOUNG J.F., LAWRENCE F.V., "Creep and drying shrinkage of calcium silicates pastes. I: specimen preparation and mechanical properties", *Cement Concrete Res.*, vol. 8, pp. 591-600, 1978.

[MOL 92] MOLINA L., On Predicting the Influence of Curing Conditions on the Degree of Hydration, CBI research report, 5:92, Swedish Cement and Concrete Research Institute, Stockholm, 1992.

[MOR 03] MORENCY M., VAYSBURD A.M., VON FAY K., BISSONNETTE B., Development of a Test Method to Evaluate Cracking Tendency of Repair Materials, Phase I Report, U.S. Bureau of Reclamation, Research Report 2004-1, March 2005, 45 pp.

[MOR 94] MORIMOTO H., KOYANAGI W., "Estimation of Stress Relaxation in Concrete at Early Ages", in R. SPRINGENSCHMID, RILEM report 25, *Thermal Cracking in Concrete at Early Ages*, E& FN Spon, pp. 95-102, 1994.

[MOR 98] MORABITO P., "Methods to determine the heat of hydration of concrete", in R. SPRINGENSCHMID, RILEM report 15, *Prevention of Thermal Cracking in Concrete at Early Ages*, RILEM publications, 1998.

[MOR 99] MORIOKA M., HORI A., HAGIWARA H., SAKAI E., DAIMON M., "Measurement of autogenous length changes by laser sensors equipped with digital computer systems", in EI-ICHI TAZAWA, *Autogenous Shrinkage of Concrete, Proceedings Int. Workshop 1998*, editor, E & FN Spon, pp. 191-200, 1999.

[MOU 03] MOUNANGA P., Etude expérimentale du comportement de pâtes de ciment au très jeune âge: hydratation, retraits, propriétés thermophysiques, PhD thesis, University of Nantes, 2003.

[NEV 00] NEVILLE A.M., *Propriétés des bétons*, Éditions Eyrolles, 2000.

[NIU 95] NIU Y.Z., TU C.L., LIANG R.Y., ZHANG S.W., "Modeling of thermomechanical damage of early-age concrete", *J. Struct. Eng., ASCE*, vol. 121, pp. 717-726, 1995.

[OH 03] OH B.H., CHA S.W., "Non linear analysis of temperature and moisture distributions in early-age concrete structures based on degree of hydration", *ACI Mater. J.*, vol. 100, pp. 361-370, 2003.

[OHN 98] OHNO Y., NAKAGAWA T., "Research of test method for autogenous shrinkage stress in concrete", in *Autogenous Shrinkage of Concrete*, E & FN Spon, pp. 351-358, 1998.

[PAI 76] PAILLÈRE A.M., SERRANO J.J., "Appareil d'étude de la fissuration du béton", *Bull. Liaison LPC*, vol. 83, pp. 29-38, 1976.

[PAI 89] PAILLÈRE A.M., BUIL M., SERRANO J.J., "Effect of fiber addition on the autogenous shrinkage of silica fume concrete", *ACI Mater. J.*, vol. 86, pp. 139-144, 1989.

[PAN 01] PANE I., HANSEN W., "Early age tensile creep and stress relaxation in Portland cement concrete-Effects of cement replacement with mineral additives", in F.J. ULM, Z.P. BAZANT, F.H. WITTMANN *Creep, Shrinkage and Durability Mechanics of Concrete and Other Quasi-Brittle Materials, Proceedings of the Sixth International Conference, CONCREEP-6*, , Elsevier, pp. 605-611, 2001.

[PAR 90] PARROTT L.J., GEIKER M., GUTTERIDGE W.A., KILLOH D., "Monitoring Portland cement hydration: comparison of methods", *Cement Concrete Res.*, vol. 20, pp. 919-926, 1990.

[PIG 99] PIGEON M., TOMA G., DELAGRAVE A., BISSONNETTE B., MARCHAND J., PRINCE J.-C., "Equipment for the analysis of the behavior of concrete under restrained shrinkage at early ages", *Mag. Concrete Res.*, vol. 52, pp. 297-302., 1999.

[POS 98] POSTON R.W., KESNER K.E., EMMONS P.H., VAYSBURD A.M., "Performance criteria for concrete repair materials, Phase II Laboratory results properties of hardened cement paste", Report REMR-CS-57, USACoE WES, 1998.

[POW 48] POWERS T.C., BROWNYARD T.L., "Studies of the physical properties of hardened cement paste", Research Laboratories of the Portland Cement Association, Bulletin 22, Chicago, IL, 1948.

[RIC 95] RICHARD P., CHEYREZY M., "Les bétons de poudres réactives", *Ann. l'ITBTP,* vol. 532, 1995.

[ROS 93] ROSTASY F.S., GUTSCH A., LAUBE M., "Creep and relaxation of concrete at early ages – experiments and mathematical modeling", in *Proceedings of the 5th International RILEM Symposium on Creep and Shrinkage of Concrete*, Barcelona, 1993.

[SAT 98] SATO R., XU M., YANG Y., "Stresses due to autogenous shrinkage in high-strength concrete and its prediction", in *Autogeneous Shrinkage of Concrete*, E & FN Spon, pp. 327-338, 1998.

[SCH 00] DE SCHUTTER G., TAERWE L., "Fictitious degree of hydration method for the basic creep of early age concrete", *Mater. Struct.*, vol. 33, pp. 370-380, 2000.

[SCH 02] DE SCHUTTER G., "Finite element simulation of thermal cracking in massive hardening concrete elements using degree of hydration based material laws", *Comp. Struct.*, vol. 80, pp. 2035-2042, 2002.

[SCH 03] DE SCHUTTER G., KOVLER K., "Visco elastic response", in A. BENTUR, RILEM report 25, *Early Age Cracking in Cementitious Systems*, RILEM Publications, pp. 111-120, 2003.

[SCH 95] DE SCHUTTER G., TAERWE L., "Specific heat and thermal diffusivity of hardening concrete", *Mag. Conc. Res.*, vol. 47, pp. 203-208, 1995.

[SCH 96] DE SCHUTTER G., TAERWE L., "Degree of hydration-based description of mechanical properties of early age concrete", *Mater. Struct.*, vol. 29, pp. 335-344, 1996.

[SCH 99] DE SCHUTTER, G., "Degree of hydration based Kelvin model for the basic creep of early age concrete". *Mater. Struct.*, vol. 32, pp. 260-265, 1999.

[SER 00] SERCOMBE J., HELLMICH C., ULM F.J., MANG H., "Modeling of early-age creep of shotcrete. I: model and model parameters", *J. Eng. Mech.*, vol. 126, pp. 289-291, 2000.

[SPR 94] SPRINGENSCHMID R., BREITENBUCHER R., MANGOLD M., "Thermal cracking in concrete at early-ages", in R. SPRINGENSCHMID, RILEM Proceedings 25, *Thermal Cracking in Concrete at Early Ages*, E& FN Spon, pp 137-144, 1994.

[TAK 99] TAKADA K., VAN BREUGEL K., KOENDERS A.A.B., KAPTJIN, N., "Experimental evaluation of autogenous shrinkage of lightweight aggregate concrete", *Autogenous Shrinkage of Concrete, Proceedings Int. Workshop 1998*, Ei-ichi TAZAWA (editor), E & FN Spon, London, pp. 221-232, 1999.

[TAW 97] TAZAWA E. I., MIYAZAWA S., "Autogeneous shrinkage of concrete and its importance in concrete technology", in Z.P. BAZANT, I. CAROL, *Creep and Shrinkage of Concrete*, E & FN Spon, pp. 159-168, 1997.

[THE 87] THELANDERSSON S., "Modeling of combined thermal and mechanical action in concrete", *J. Eng. Mech.*, ASCE, vol. 113, pp. 893-906, 1987.

[TOM 00] TOMA G., Comportement des bétons au jeune âge, PhD thesis, Laval University, 2000.

[TOR 05] TORRENTI J.M., BENBOUDJEMA F., "Mechanical threshold of concrete at an early age", *Mater. Struct.*, vol. 38, pp. 299-304, 2005.

[TOR 96] TORRENTI J.M., "Comportement mécanique du béton – bilan de six années de recherche", *Etudes et recherches des Laboratoires des Ponts et Chaussées*, "*Ouvrages d'Art*", OA 23, LCPC, 1996

[ULM 95] ULM F.J., COUSSY O., "Modeling of thermo-chemo-mechanical couplings of concrete at early ages", *J. Eng. Mech.*, ASCE, vol. 121, pp. 785-794, 1995.

[ULM 96] ULM, F.J., COUSSY O., "Strength growth as chemo-plastic hardening in early age concrete", *J. Eng. Mech.*, ASCE, vol. 122, 1996.

[ULM 98] ULM F., COUSSY O., HELMICH C., "Chemoplasticity: a review of evidence", *Proc. EURO-C Conference*, pp. 421-439, 1998.

[ULM 99] ULM F. "Couplages thermochemomécaniques dans les bétons – un premier bilan", *OA31*, LCPC, 1999.

[UME 94] UMEHARA H., UEHARA T., IISAKA T., SUGIYAMA A., "Effect of creep in concrete at early ages on thermal stress", in R. SPRINGENSCHMID, RILEM report 25, *Thermal Cracking in Concrete at Early Ages*, E& FN Spon, pp 79-86, 1994.

[VAY 01] VAYSBURD A.M., EMMONS P.H., BISSONNETTE B., "Some aspects of evaluating cracking sensitivity of repair materials", *RILEM International Conference on Early Age Cracking in Cementitious Systems,* Haïfa (Israel), pp.169-186, 2001.

[WEI 01] WEISS W.J., FERGUSON S., "Restrained shrinkage testing: the impact of specimen geometry on quality control testing for material performance assessment", in ULM, F. J., BAZANT, Z. P., WITTMAN, F. H., *RILEM 6th International Conference on Creep, Shrinkage & Durability Mechanics of Concrete and other Quasi-Brittle Materials*, Boston, Elsevier, pp. 645-651, 2001.

[WEI 74] WEIGLER H., KARL S., Junger beton, beanspruchung – festigkeit – verformung (teil 1), *Betonw. Fertigteil Tech.* vol. 6, pp. 392-401, 1974.

[WIL 73] WILLAM K.J., WARNKE E.P., "Constitutive model for the triaxial behavior of concrete", *IABSE Proc.*, vol. 19, Bergame, 1973.

[WIT 82] WITTMANN F.H., "Creep and shrinkage mechanisms", in Z.P. BAZANT, F.H. WITTMANN, *Creep and Shrinkage of Concrete Structures*, J Wiley and Sons, 1982.

Chapter 9

Delayed Effects – Creep and Shrinkage

9.1. Introduction

Concrete is an aging material: its internal structure and mechanical characteristics evolve with time. Under a given stress, concrete reacts instantaneously but undergoes delayed strains too (under prolonged loading: constant loading, pre-stressing, etc.). The increasing diversification of concretes and the development of high-performance concretes (HPC, ultra-high performance fibre-reinforced concrete (UHPFRC), etc.) have produced diverse long-term behaviors, and thus the requirement of delayed phenomena quantification. The success of concrete structures, especially those undergoing significant constant strains, such as the containment of nuclear power plants, pre-stressed bridges, or high-rise structures, built to last around 50 to 100 years, involves knowledge of the long-term behavior of the concrete.

Two different types of tools are used to design and handle such types of structures:

– models, generally based on the linear viscosity, used in construction codes or, for a more precise approach, used in finite element code calculation;

– tests, whose results allow the engineers to control the design and to achieve the structure in good conditions.

Chapter written by Farid BENBOUDJEMA, Fékri MEFTAH, Grégory HEINFLING, Fabrice LEMAOU and Jean-Michel TORRENTI.

Regarding fundamental research, the tests enable us to better understand the physical mechanisms occurring within the material. Practical work has then to be carried out relying on modeling. Indeed, identifying the different parts of delayed strains requires a reverse reasoning that strongly relies on modeling. This reverse identification is essential as structural effects (cracking) affect the practical measurement of delayed strains. Isolating the intrinsic behavior of the material to model it depends then upon:

– models of cracking behavior approved to quantify the structural effects;

– conventional decomposition of those deformations in different components in order to experimentally quantify them, whereas the driving mechanisms of delayed deformations seem to be coupled.

This cognitive research contributes to precise the limits of the models and thus improve it. The purpose of this chapter is to illustrate that approach of delayed strains in concrete.

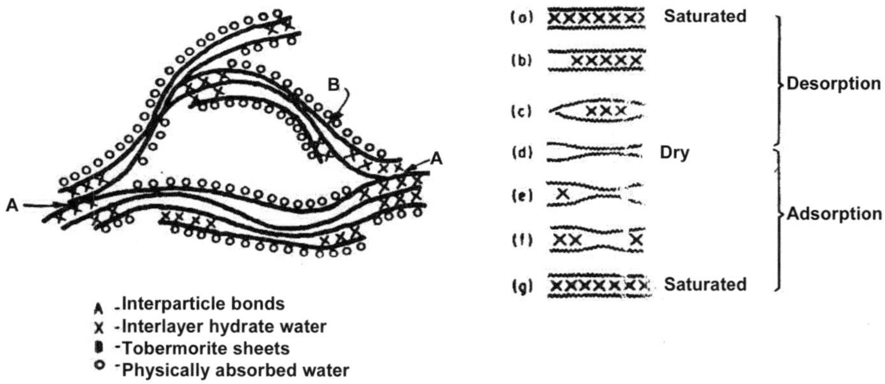

Figure 9.1. *C-S-H gel [FEL 68] microstructure*

9.2. Definitions and mechanisms

Conventionally, delayed strains of concrete are split into two components:

– shrinkage strains;

– creep strains.

These two main components can also be divided into subcomponents, depending on the different conditions and stress the material is experiencing.

Although it is becoming increasingly obvious that the driving mechanisms of dimensional variations are linked to the microscopic characteristics of the material [ISH 68; YOU 74; JEN 92; ACK 01], this broad division of delayed strains is still used. In particular, every subcomponent can be independently identified relying on a relatively simple experimental reasoning. Before continuing with the subject of strains, the microstructure of cement paste has to be briefly presented in order to describe the different mechanisms that control these delayed deformations.

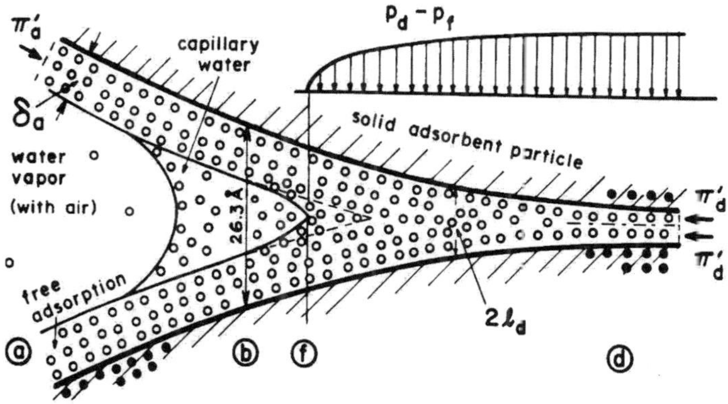

Figure 9.2. *Idealized sketch of the hindered adsorption zone and of the transition into the macropores [BAŽ 72]*

9.2.1. *Microstructure of cement paste*

Numerous models can be found in the scientific literature, nevertheless, the model proposed by Feldman and Sereda (Figure 9.1) appears superior to explain the mechanical behavior of cement paste (creep in particular) [GUE 97]. Hydrates can be modeled on C-S-H particles as simple, badly crystallized sheets, which create interlayer spaces during their joining. The sheets can have a relative movement, and either penetration or departure of water from the interlayer spaces is possible. The interlayer bonds are of solid-solid type, linked via weak Van der Walls forces and high ionic/covalent forces.

We are then within the hindered adsorption zone (Figure 9.2). Within those zones, strongly adsorbed water is under a given pressure of about 130 MPa [BAZ 70], known as disjunction pressure, in contrast to the attraction forces, which exist in between C-S-H particles and which maintain the skeleton's structure. The water itself is, therefore, a structural component of the material and is capable of sustaining local stresses.

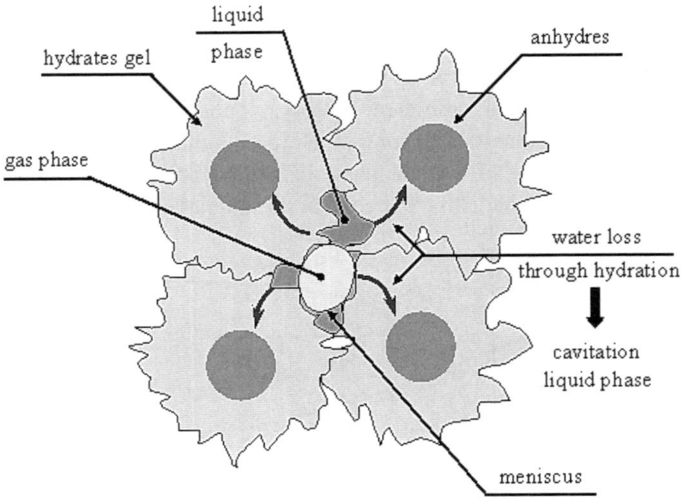

Figure 9.3. *Liquid phase cavitation and menisci appearance due to the ongoing hydration during the setting process*

Recent results [ULM 07] concerning concrete at the nanoscopic scale, have shown that the particles of C-S-H are organized in an isotropic form despite the anisotropic shape of the constitutive sheets. Two types of particles exist: low and high-density C-S-H [CON 04]. These different types have different behaviors under instantaneous loading and also in the case of delayed strains.

9.2.2. Shrinkage

Shrinkage is the dimensional variation in time generated by different driving mechanisms, mainly chemical reactions during hydration, variation of water content in the cement paste, but also temperature variation. These variations can be due to the hydration reaction, or due to climatic or industrial (removing formwork, thermal treatment) factors. They are split into five main mechanisms [AIT 98] and can occur either after one another or simultaneously.

9.2.2.1. *Plastic shrinkage*

Chronologically, plastic shrinkage occurs first and develops in concretes that lose water when in a plastic state. In general, this water loss takes place through evaporation, but can also result from the adsorption by an adjacent concrete or by dry ground. Plastic shrinkage develops before and during the concrete's setting process.

9.2.2.2. *Chemical shrinkage*

The hydration reaction involves a decrease of 8 to 12% of the initial volume; this phenomenon is called Le Chatelier's contraction. It corresponds to a potential linear shrinkage of the matrix of 3 to 4%. Indeed, the produced hydrates have a smaller volume compared to the sum of the anhydrous cement and water volumes. Therefore, a free contraction process initiates as soon as hydration begins. At this stage, the bridges formed between the hydrates doesn't make up a skeleton rigid enough to prevent a relative joining of the hydrated grains. This shrinkage also develops before and during the concrete's setting process.

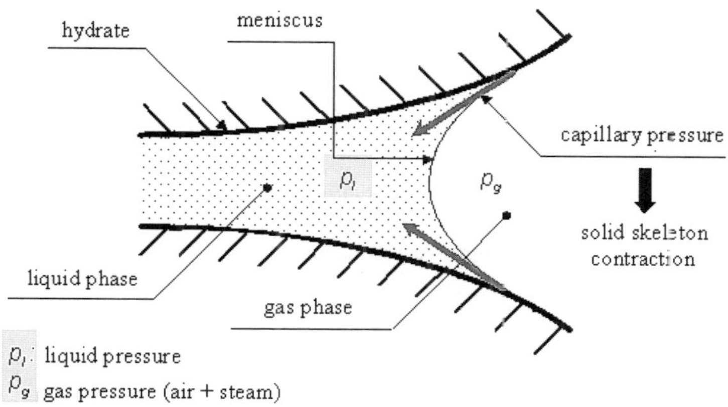

Figure 9.4. *Joining of two sheets of hydrates under the effect of capillary pressure*

9.2.2.3. *Autogenous shrinkage or self-desiccation*

As soon as concrete sets, hydrates, making up the new structure, percolate and form a rigid connected network [TOR 05]. The free contraction due to le Chatelier's contraction becomes impossible. Therefore, the liquid phase, which keeps on decreasing because of hydration, no longer fills all the available space. Cavitation is initiated: the spreading liquid water is vaporized and a gaseous network is rapidly formed (Figure 9.3). The coexistence of the two phases leads to the formation of menisci at the liquid/gas interface, and to the formation of capillary pressures [WIT 76; BAR 83; BOU 07], significant enough to trigger an autogenous shrinkage of the hardened structure (Figure 9.4). Nevertheless, the only physical action of the capillary pressures doesn't explain the observed effect on the shrinkage mechanism due to the presence of ionic elements in the liquid phase [BEL 01]. Therefore, other chemical forces are taken into account. This is why the distance between two opposed surfaces of hydrate solid particles is supposed to be obtained as a result of the equilibrium between the Van der Waals and capillary pressure attractive forces, and conversely, the disjunction pressure repulsive forces [WIT 01]. The pressures are likely to be due to the overlap of polarized layers of adsorbed water or to the chemical potential variation of the adsorbed water layers near the solid walls [PES 00].

Therefore the evolution of autogenous shrinkage deformation is linked to the kinetics of cement hydration. This evolution is very rapid during the first days and its progress is about 60 to 90 % after 28 days. The parameters that influence hydration kinetics, such as the characteristics and fineness of the cement and the water to cement ratio W/C (W being the amount of water (l/m^3) and C the amount of cement (kg/m^3)), play a major part on the amplitude of the autogenous shrinkage. This deformation does not exceed 100×10^{-6} for concretes whose ratio W/C is greater than 0.45, but rapidly increases when this ratio is below 0.4 and then can reach 300×10^{-6}.

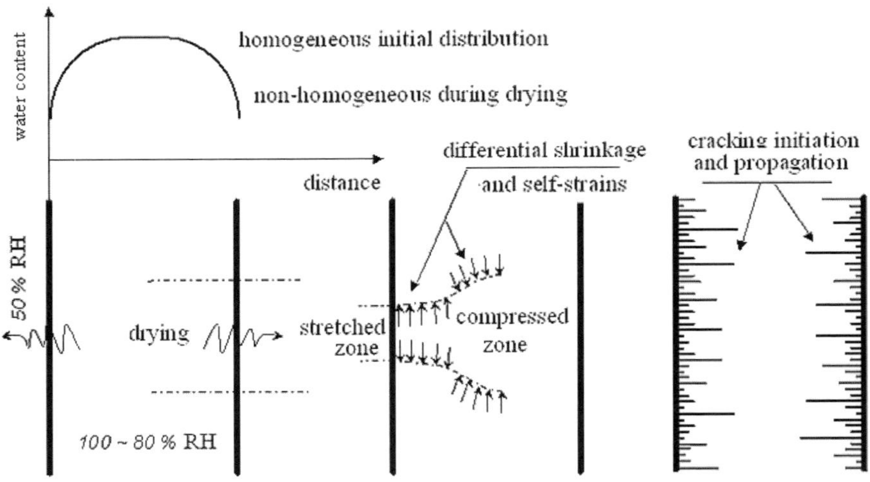

Figure 9.5. *Structural effect of the desiccation restraint induced by drying*

It is worth noting that autogenous shrinkage occurs in a homogenous way in a structure, when this structure is being built during the same concreting operation and when it is not being blocked by its supports or by its formwork. In that case, the kinetics of hydration, and therefore, kinetics of water consumption act uniformly on the structure volume. When a fresh concrete layer is placed on a hardened concrete layer, the difference in the kinetics often leads to an early cracking in the upper layer where strains are restrained.

In addition, and despite its isotropic characteristic, increase in autogenous shrinkage also causes a diffuse degradation of the concrete material in early age. Indeed, dimensional variations of the hardening cement paste are incompatible with the presence of rigid aggregates, leading to the initiation of microcracking within the material, particularly at the paste/aggregates interface.

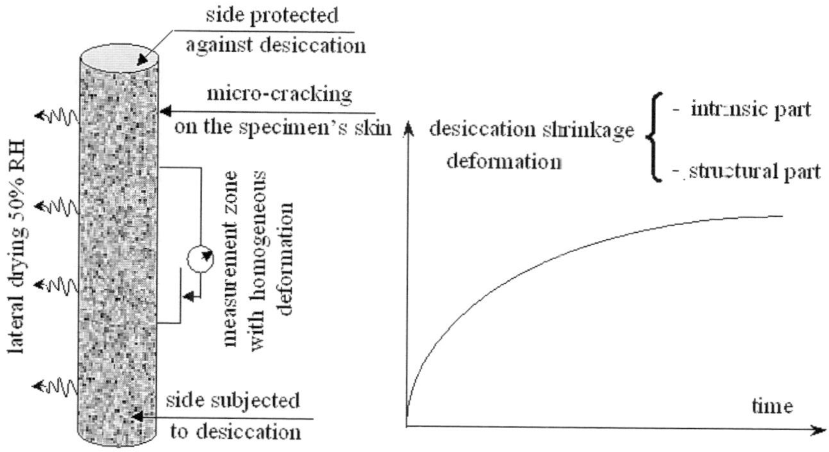

Figure 9.6. *Desiccation shrinkage strain: unloaded specimen subjected to drying*

9.2.2.4. *Desiccation shrinkage*

The driving mechanism of this component of delayed strain is intrinsically similar to autogenous shrinkage as it is also due to the variation in water content, leading to matrix contraction. Nevertheless, this mechanism, unlike the former, presents a heterogenous characteristic. In this case, variation in the level of water occurs through loss of water outwards of the material when subjected to drying: this loss being governed by the initial hygroscopic degree of the material (100 to 80% relative humidity) and the outer surroundings (depending of the country considered) which are not balanced. Desiccation shrinkage has a slower kinetics process compared with autogenous shrinkage and describes the progression in time of the drying front of the sample. This process is extremely slow: 10 years for a 16 cm diameter specimen, its duration being proportional to the square of its dimension. Thus, regarding massive structures, drying time, and therefore, desiccation shrinkage time, often exceeds the given lifetime of the structure. Therefore, the final amplitude becomes is much less important to the engineer in comparison to the evolution rate.

Non-uniform distribution of drying conditions lead to differential deformations of shrinkage between the sample inner zones and its outer zones, especially during the transitional phase of the process (Figure 9.5). The consequence of the strong gradients is the generation of a self-stresses system (compression on the core and tension at the edges), which leads to the initiation of a cracking process within the material [BAŽ 94, GRA 95]. This structural effect strongly depends on the geometry of the sample or the structure and on the boundary conditions.

It is worth noting that the experiment, which must allow the identification of this component, consists of measuring the deformation evolution in time on a mechanically unloaded specimen bearing a single lateral desiccation (Figure 9.6). The shrinkage deformation is measured regarding the longitudinal direction. This measurement must be performed within a zone of the specimen far enough from the edges to avoid the additional effect of the boundary conditions. The measuring zone has also to be stretched enough for the cracking distribution to be sufficiently homogenous and the representation of an average strain to be justified.

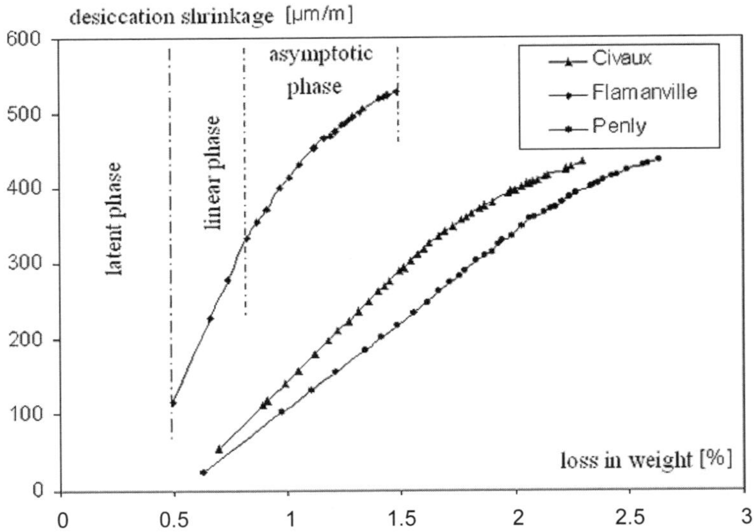

Figure 9.7. *Evolution of desiccation shrinkage strain with water loss in mass for different types of concretes [GRA 95]*

However, this measurement remains global. It is made of the simultaneous effect of two contrasting components: the contraction due to shrinkage and the extension due to cracking of the material. Therefore, intrinsic shrinkage, which is essential to the identification of an objective behavior law, can not be separated from its structural characteristics. The separation of these two parts requires the use of modeling to enable reverse identification of the apparent shrinkage measured experimentally.

As an example, the analysis of experimental results, obtained on different concrete formulations [GRA 95], shows that three distinct phases can be observed when shrinkage strain is represented as a function of the relative mass loss (Figure 9.7):

– a first phase, where drying does not induce any apparent shrinkage;

– a second phase, where shrinkage is proportional to mass loss;

– a third phase, where shrinkage becomes asymptotic.

A part of this behavior is due to an intrinsic phenomenon and another part to a structural effect due to cracking. The use of a cracking model will be useful to identify the limit between the material's intrinsic behavior and the structural part. We will try to bring some answers to those questions in section 9.4.

9.2.2.5. *Thermal shrinkage*

The reaction of hydration of cement is strongly exothermic: from 150 to 450 J/g of cement (Figure 9.8). This release of heat generates a rise in temperature compared to the initial temperature. This elevation may reach 50°C, sometimes more when the object is a massive one. When concrete is cooling down, thermal shrinkage deformation can then reach 500×10^{-6} depending on the cement's proportions and nature. As the skin cools faster than the core, a thermal gradient is set up, which generates self-stresses and, therefore, diffuse cracking of the skin.

The bigger the element, the more important the effect. Moreover, the material's rigidity evolves in value to one closer to the final value during the rise in temperature.

If the deformations are restrained, this dissymmetry leads to small compressive stresses when the temperature is rising and to important tensile stresses when concrete is cooling down. These stresses can be such that they lead to a located cracking that can cross all sections [TOR 96].

9.2.2.6. *Conclusion*

The different processes of delayed strain exhibit very different kinetics due to their chemical, hydric, or thermal nature. In a 16 cm diameter cylinder, hydration heat diffuses during 1 or 2 days, whereas drying lasts more than 10 years. Knowledge of these processes enables them to be modeled satisfactorily: in the short term, drying remains very superficial and barely affects hydration; in the long term, climatic temperature variations remain cyclic and do not really modify the drying kinetics.

9.2.3. *Creep*

Creep is the delayed dimensional variation due to the application of a constant external mechanical loading. This phenomenon seems to be highly related to the presence of water in the material. Indeed, different studies show that a (partial or total) preliminary drying of the loading leads to lower measured creep deformations [NEV 57; GLU 62; MUL 64; RUE 68; WIT 70; BRO 76]. Nevertheless, the hypotheses regarding fundamental mechanisms controlling creep do not have unanimous support within the scientific community.

Thus Powers [POW 68] suggested that creep is governed by diffusion of the adsorbed water. Feldman [FEL 72] considered that creep of a cement paste is due to a progressive crystallization of amorphous silicates. This crystallization is accelerated as a result of the drying or the use of a stress. The role of water is limited to the redistribution in the porous spaces leading to a densification of the solid's skeleton and, therefore, to an irreversible apparent creep.

Another mechanism, suggested by Ruetz [RUE 68], considers that creep is due to a relative sliding of hydrates sheet by shearing. This sliding process concerns the adsorbed layers of water in between sheets where water acts as a lubricant. Ruetz considered that this water is affected by drying, which disturbs the ordered structure of the adsorbed layers. Their rigidity then decreases, which amplifies the creep process.

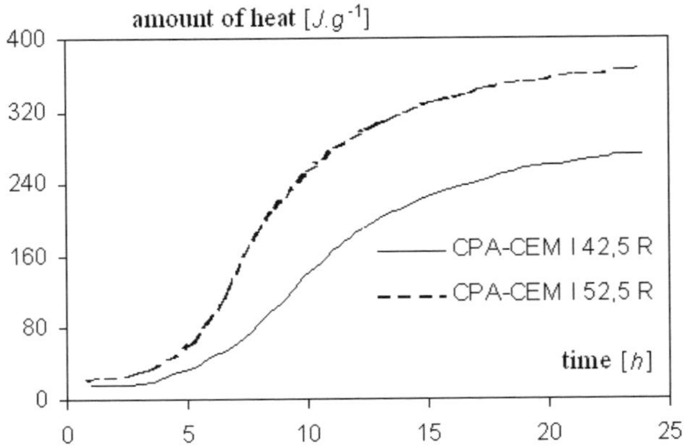

Figure 9.8. *Evolution of the hydration heat of a CEM I cement measured according to the NF P 15-436 norm [BAR 97]*

Starting with the same reasoning, Ishai [ISH 68] considered at the starting point of creep water diffusion occurs on an observation scale different from that observed in hydric transfer due to desiccation. Thus, creep originates from the diffusion of adsorbed water in the inter- and intra-crystalline space. The latter is modified thanks to the use of mechanical loading.

Ageing through recrystallization under stress or through C-S-H polymerization, which is induced by drying or the use of a stress, has other hypotheses proposed in the literature [ALI 4; BEN 79].

More recently, another creep mechanism, based on the theory of solidification, has been proposed by Bazant *et al.* [BAŽ 97a]. The mechanism seems then to be of a mechanical nature on the scale of C-S-H nanopores. It consists of a progressive rearrangement of the bonds between the adjacent surfaces of hydrate sheets during the sliding of C-S-H sheets. This rearrangement is made possible as a result of the unstable and unordered nature of the over-stretched bonds due to disjunction pressures within the hindered adsorption zones.

Recent results [VAN 09] investigating concrete on a nanoscale seem to indicate that creep could also be explained by rearrangement of particles of C-S-H. In this case the packing density of these particles would be a major parameter to control the creep process of concrete.

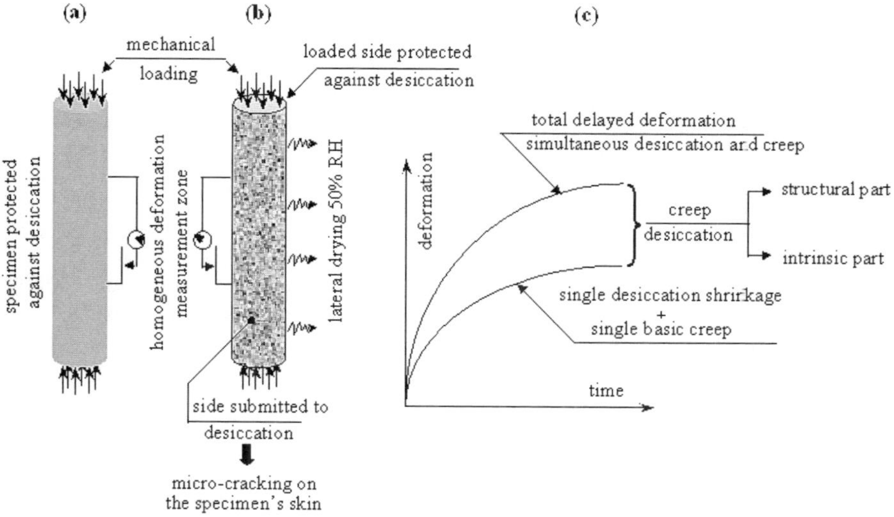

Figure 9.9. *Experimental identification of creep components*

Regarding experimental identification, creep is defined as equal to the total delayed strain minus the shrinkage. The scientific community generally agrees that [ACK 83] creep can be split into two distinct components, which correspond to at least two different physico-chemical mechanisms:

– basic creep: defined as the measured deformation of a concrete specimen in hydric equilibrium and thus without humidity exchange with its surroundings (Figure 9.9a);

– desiccation creep: defined as the additional part that can be measured when the loaded specimen is submitted to a simultaneous desiccation (Figures 9.9b and 9.9c).

9.2.3.1. *Basic creep*

The kinetics analysis of this delayed strain [RUE 68; ULM 98] for different concretes highlights two separate kinetic regimes. The first, short-term creep, rapidly evolves during the few days after loading. The other, long-term creep, is characterized by significant ageing kinetics, but is considered to be non-asymptotic [ULM 98].

9.2.3.1.1. Short-term creep

Short-term creep could be associated with a process of capillary water microdiffusion due to external applied forces [WIT 82; ULM 98]. The measured value of 25 kJ/mol of the activation energy associated with short-term creep [DAY 83], close to the one of 15 kJ/mol associated with a phenomenon of free water migration (capillary and physically adsorbed water) [DIAL 87], seems to confirm this hypothesis. Moreover, based on experimental tests [ULM 98], short-term creep kinetics is influenced by the W/C ratio. However, the main difference between concretes bearing different W/C ratios comes from the capillary pores [BAR 94], which would be related to the microdiffusion process in question.

Diffusion is initiated under the action of external forces when stresses are relayed, on a microscopic scale, to the hydration compounds surrounding capillary pores (Figure 9.10). Locally, this microscopic stress transfer results in a thermodynamic unbalance between the free adsorption molecules of water in the transmission zones and molecules further away. In order to restore the equilibrium, the water molecules spread into the adsorbed water layers (surface diffusion) towards free zones, leading to the deformation of the solid skeleton.

9.2.3.1.2. Long-term creep

It appears that the kinetics of long-term basic creep depend upon the type of concrete [ULM 98], and thus explain the origin of long-term creep from the hydrate pores (nanopores). Regarding the considerations on the role played by disjunction pressure considerations, the mechanism cannot be linked to water microdiffusion (short-time and asymptotic phenomenon), and seem to have a mechanical origin. Within the hindered adsorption zones, solid adjacent surfaces are linked via mutual attraction or direct atomic bonds. This bonding forces balance the induced disjunction pressure. Locally super-stretched and unstable bonds are likely to break, and are reformed in the adjacent zones of lesser pressure due to C-S-H sheet sliding (Figure 9.11). Other potential ruptures will then occur at other sites as bonding forces become relaxed [BAŽ 97a].

This chain reaction process successively consumes creep sites that have been activated by mechanical loading. This exhaustion leads to an observed ageing of creep kinetics as a function of the material's age. It is worth noting that the macroscopic stress, due to its much lower order of magnitude compared with the disjunction pressure, does not really modify the amplitude of the bonding forces at nanopore level [BAŽ 70]. Therefore the dislocation process results in the intrinsic instability of the hydrates.

It should be added that the mechanisms and the explanation provided regarding their origin remain hypothetical. The rate of their occurrence cannot be easily determined using the usual techniques available; their experimental confirmation is non-existent. Moreover, this approach seems to contradict some experimental observations as irreversible basic creep is considered deviatoric. However, tests also show an irreversible part on the spherical component of basic creep, which presents an asymptotic characteristic highlighting the occurrence of an associate diffusive process [GOP 69; JOR 69; GLU 72; BEN 00a].

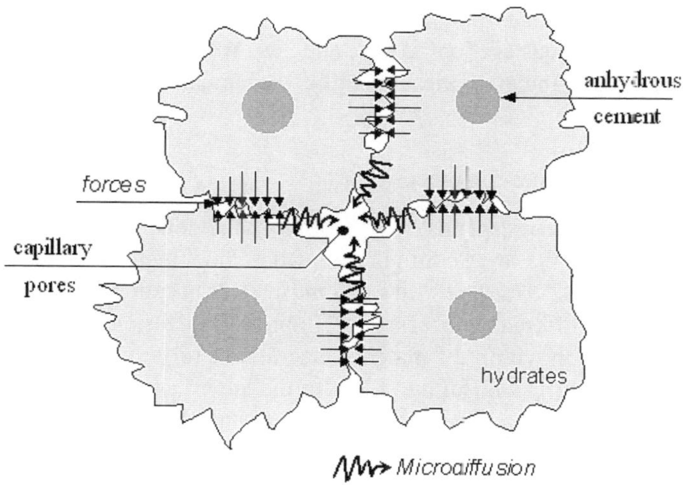

Figure 9.10. *Short-term creep mechanism [ULM 98]*

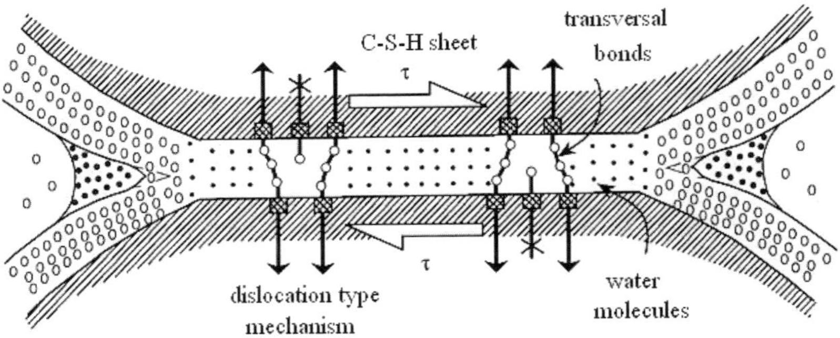

Figure 9.11. *Long-term creep mechanism [ULM 98]*

9.2.3.2. *Desiccation creep*

When a specimen is simultaneously submitted to desiccation and a mechanical loading (Figure 9.9b), the occurring creep mechanism is affected by the variation in time of the contents, in addition to its non-homogenous distribution (due to drying) within the sample. The shrinkage mechanism is modified due to the presence of a mechanical loading affecting, in particular, the microcracking process and, therefore, also affecting the shrinkage structural element. A much more significant delayed strain is measured when compared with the obtained sum of basic creep and shrinkage strains, measured on similar specimens (Figure 9.9c). This phenomenon is called the Pickett effect. Desiccation creep deformation is then part of the delayed strain obtained by removing the delayed global strain, the basic creep, and shrinkage (of desiccation and endogenous if the test is carried out at an early age).

The origins of desiccation creep and its role in global creep are still controversial. Nevertheless, the scientific community agrees on two main mechanisms.

9.2.3.2.1. Intrinsic desiccation creep

This component corresponds to the basic behaviour of the material at the microscopic scale in the ideal configuration where no structural effect is occurring. Bazant [BAŽ 85, BAŽ 97ab] specifies that intrinsic desiccation creep is linked to the presence of two different processes of humidity diffusion within concrete: a macroscopic diffusion within the material macropores (drying and rehumidification) and a microscopic diffusion (Figure 9.12). In the latter case, the local flow of water molecules between the hindered adsorption zones and the capillary pores in cement paste increase the rate of rupture of the C-S-H atomic bonds, leading to the formation of an additional creep.

This additional strain is also observed when the specimen is rehumidified. This is because the stain remains unchanged but the direction of the relative movement (inward diffusion) of water molecules changes during rehumidification. This observation can also be confirmed thanks to experimental results where a cumulative strain is noticed during humidity cycles [HAN 60].

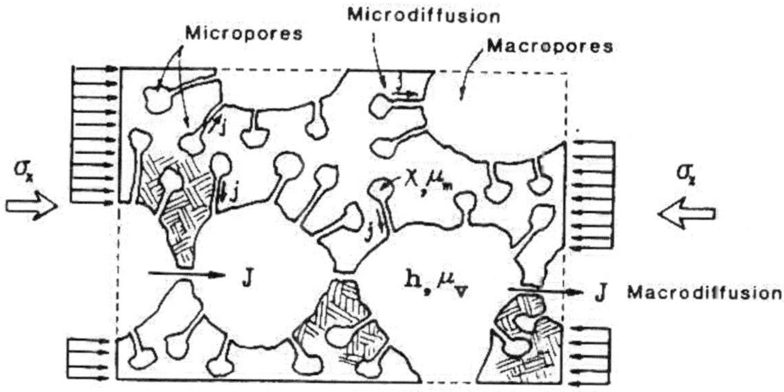

Figure 9.12. *Diffusion process of water into cement paste under stress resulting in the intrinsic part of desiccation creep [BAŽ 85]*

9.2.3.2.2. Structural desiccation creep

Because of desiccation, skin cracking cannot be avoided. The measured apparent shrinkage on a cracked specimen is lower than the material's potential shrinkage [ACK 88, GRA 95, MEF 00] obtained on a loaded specimen. Indeed, loading the compressed specimen limits cracking and thus involves an additional part of apparent shrinkage. This type of shrinkage is usually called structural desiccation creep as shrinkage deformation is always measured on a non-loaded specimen during a conventional experimental approach.

Granger [GRA 95] claimed that structural desiccation creep results in about 20% of desiccation creep, reaching its asymptotic value after 1 year. In addition, Bazant [BAŽ 94] considered that strain increases when drying starts in order to reach its maximum after about 10 days, to then decrease. Nevertheless, when the specimen is totally dried, cracks tend to reform and the structural element can also have a sizeable contribution during this phase [MEF 00, BEN 01a].

Moreover, as was the case for desiccation shrinkage, identifying the intrinsic behavior of desiccation creep requires the use of modeling to quantify the structural element. Reverse identification will rely, in particular, on a cracking model allowing the limit between the intrinsic behavior and the structural effect to be located.

9.3. Experimental approach

Specific tests evaluate every occurring strain. The total strain, ε, at every instant and derived from a measurable displacement field, is by definition the sum of the elastic deformation and the different delayed strains [LER 96a; ACK 98]:

$$\varepsilon_e = \varepsilon_{th} + \varepsilon_{es} + \varepsilon_{ds} + \varepsilon_{bc} + \varepsilon_{dc} \quad [9.1]$$

with ε_e being the instantaneous elastic strain, ε_{th} the thermal shrinkage strain (non-measurable with the following tests), ε_{es} the endogenous shrinkage strain, ε_{ds} the desiccation shrinkage, ε_{bc} the basic creep, and ε_{dc} the desiccation creep.

The terms of this equation, based on linear viscosity hypotheses, justify the total number of concrete specimens that will be used during the tests. To characterize the delayed behavior of concrete, it is necessary to have four series of specimens ready to be used:

– endogenous shrinkage is directly measured on a specimen being non-loaded and protected from desiccation;

– desiccation shrinkage is deduced from the measurement of the dimensional variation of a second specimen that is non-loaded and not protected against desiccation. By hypothesis, at a given time, the measurement being carried out is the sum of an endogenous shrinkage strain and a desiccation shrinkage strain;

– the performed measurements on protected and loaded specimens provide the concrete basic creep once endogenous shrinkage strain is deduced;

– the same type of measurement on a fourth non-protected and loaded specimen provides the desiccation creep, once shrinkage is measured in drying conditions and basic creep is deduced.

The tests have twin aims: to validate models used in codes and more complex models introduced in finite element codes, and in the case of structures sensitive to delayed strains, to characterize the delayed behavior of specific concretes. The tests, and, in particular, the creep test, cannot be easily performed. Indeed, their duration, ranging from a few months to a few years, demands particular precautions: regulation of the temperature and the relative humidity of the test chambers, efficient protection against specimen desiccation, and a measuring system that does not derivate in time. Moreover, performed measurements are frequently subject to inherent dispersions regarding the setting of different specimens, as well as the measuring system. These factors have then to be taken into account when interpreting the results to predict the behavior of the material over longer periods of time, from a dozen to a hundred years.

9.3.1. *Tests*

Under compressive stress maintained constant in time, concrete strains increase as a function of time. This raise can reach, as a relative value, two to three times the value of instantaneous strain. In the same way, without any stress, the material undergoes significant shrinkage. Regarding the types of concrete, the order of magnitude of creep and shrinkage strains measured are as follows:

– longitudinal creep strains: 500 to $1{,}500\times10^{-6}$;

– longitudinal shrinkage strain: 50 to 500×10^{-6}.

The evolution of longitudinal strains with regards to time is dependent of the time t, the age of the concrete at loading, tc, the applied stress σ_{app} and the boundary conditions (temperature θ and relative humidity, RH).

Generally, creep test consists of loading a cylindrical specimen of concrete, of age t_c, under an uniaxial compressive stress σ_{app} kept constant. The conditions of temperature and hygrometry are also constant (air-conditioned room: θ=20±1°C and RH=50±5%). Longitudinal strains as a function of time, $\varepsilon(t)$, are then measured. To those creep specimens are associated non-loaded shrinkage specimens placed in the same test chamber. The tests (creep and shrinkage) can last from 6 to 12 months, or even several years.

Assuming that creep is proportional to the applied stress (hypothesis of linear visco-elasticity), measuring those strains allow the calculation of the shrinkage or compliance function $J(t)$ defined by:

$$J(t) = \frac{\varepsilon(t) - \varepsilon_s(t)}{\sigma_{app}} \qquad [9.2]$$

where $\varepsilon_s = \varepsilon_{es} + \varepsilon_{ds}$ is the global strain measured on a specimen being non-loaded and submitted to desiccation. Strains of thermal or chemical origins are not relevant in this case. The strain of endogenous shrinkage, ε_{es}, is measured on a specimen being non-loaded and protected against desiccation.

The endogenous shrinkage measured when the specimen is protected against desiccation is actually not the same as that measured when the specimen is drying. Indeed, endogenous shrinkage is linked to the reaction rate of hydration, whereas this reaction depends on the relative humidity: the hydration reaction stops if the internal relative humidity falls under about 80% [XI 94]. Nevertheless, this distribution of strains stands if, when the test of desiccation shrinkage starts, the hydration reaction does not significantly evolve. This is generally the case when the specimen is older than 28 days.

The loading of the creep specimen is about 30% of the compressive strength at the day of the loading. This stress value corresponds to the plausible value of the stress in concrete during the lifetime of the structure. In addition, this allows concrete to stay within the validity range of linear visco-elasticity.

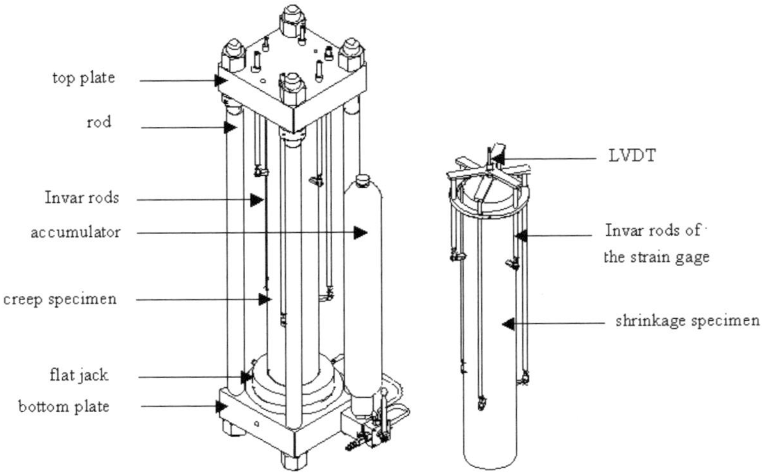

Figure 9.13. *Framed creep and shrinkage specimens*

9.3.1.1. *Control of the boundary conditions*

The control of the boundary conditions (θ, RH), in order for the gradients of temperature and relative humidity of the specimen to be zero, requires resolution of the problems of specimen conservation [ACK 98] and the air-conditioning system. An air-conditioning system allows the temperature and relative humidity of the test chambers to be regulated, following the rules θ =20±1°C and RH=50±5%, within the whole volume and without any stratification of the air layers.

The choice of a height (100 cm) of the creep and shrinkage specimens is important as it avoids the parasite effects of the water content boundary conditions and achieves a better stress and displacement repartition within the measured zone.

9.3.1.2. *Specimen conservation*

Numerous studies have highlighted the main role of water within concrete and the importance of hydric unbalance between the core and the surroundings of the material [ACK 88]. In order to make the specimens waterproof, a new technique has been created, based on aluminum foil, which consists of protecting tested specimens under endogenous conditions with auto-adhesive aluminum foil (acrylic adhesive) [TOU 96].

Desiccation-free tests (basic creep, endogenous shrinkage) are performed on specimens, covered with auto-adhesive aluminum foil creating a double thickness, after release from the mould. Regarding desiccation tests, specimens are directly stocked in the test chamber.

9.3.2. *Testing equipment*

9.3.2.1. *Frame and specimens*

Tests are performed on molded cylindrical specimens with a 16 cm diameter and a 100 cm height. In general, a shrinkage specimen (non-loaded and with the same dimensions and conservation conditions) is associated with the creep specimen (as a result of conservation conditions).

The creep specimen is framed (Figure 9.13) as a result of:

– a frame bearing two extremely rigid metallic plates [SER 76], linked by four rods, associated with a generator of hydraulic force. This latter is made of a flat hydraulic jack, linked to a four-way valve and an oleopneumatic accumulator (containing nitrogen under high pressure), the aim of which is to limit variations of the pressure of the hydraulic jack. The pressure drop in the oil in the jack is due to specimen shortening, which generates a variation in the volume in the accumulator, causing a small drop in pressure of the nitrogen. This pressure drop becomes lower as the relative variation of volume becomes lower, therefore, the accumulator volume is important;

– a load cell attached to the top plate: a hydraulic dynamometer, with the aim of measuring the applied force on the specimen at any time and to periodically check that the applied load remains constant;

– an extensometer measuring the relative displacement of two sections 50 cm apart in the main part of the specimen (the same type of extensometer is used to measure shrinkage). It is made of two sets of three vertical rods leaning on sealed inserts at an angle of 120° in concrete and 25 cm away from each end. This relative displacement is sent to the top section thanks to the metallic rods and is measured by a displacement sensor (differential transformer with a stroke of ±1 mm and a resolution of 0.12 µm over their global stroke). The choice of a material used to make measuring rods has to be able to withstand the retained strain (2×10^{-6}) and thermal variations; Invar, which has a thermal dilatation coefficient in the order of 0.5×10^{-6} /°C is used.

9.3.2.2. *Loading and measuring techniques*

9.3.2.2.1. Loading

The creep specimen is hardened by its management conditions, and is stored in the test chamber. Preceding its loading, its sides are adjusted beforehand by a stone cutter.

The preparation of the frame includes several phases: maintenance, inflation of the accumulator, calibration of the load cell. Calibration consists of comparing the load cell response to the response of the calibrated force sensor during the preparation of the frame for the test. The cell surface is calculated by graphical extrapolation of the recording of a hydraulic pressure generated by an external pump (measured by a pressure sensor) and the applied force on a metallic wedge, also measured by a reference force sensor [ACK 98]. At the end of this phase, the specimen is put in place inside the frame and also centered. The extensometer is turned on and data acquisition configured.

The fact that the behavior of the material at a certain age depends on the whole stress history imposes an accurate definition of stress. The four-way valves, located at the bottom of the accumulator in the generator circuit of pressure, enables the specimen to be isolated during its set up into the frame and an instantaneous loading to be performed, the loading being applied within 0.4 s.

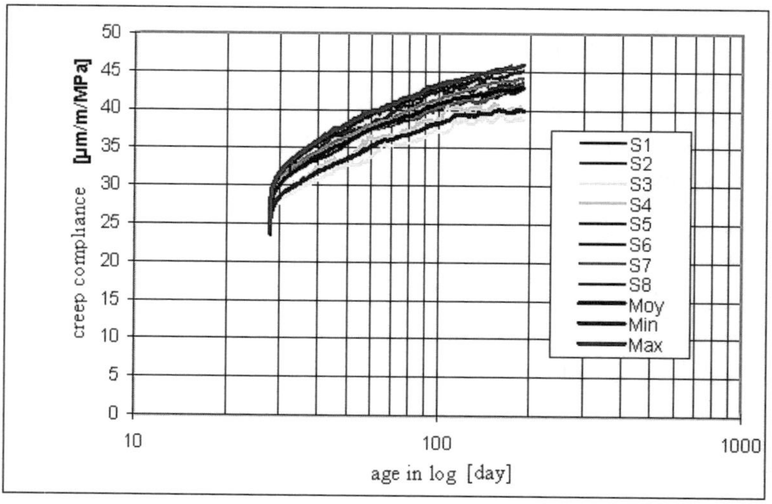

Figure 9.14. *Experimental curves of basic creep compliance*

9.3.2.2.2. Measurement of the applied stress

The quality of the creep test depends on the capacity of the loading device to maintain a constant stress: the accumulator provides this function. The use of a load cell with a previously measured effective section allows the real stress applied to the specimen to be determined by a graphical method. The latter is measured just after having performed loading and about every 3 months thereafter.

9.3.2.2.3. Measurement of longitudinal strains

The measurement of shrinkage, ε_s, is carried out as soon as possible after releasing the specimens and their conditioning for storage.

The specimens for the creep test are loaded at different ages, depending on the objective: 3, 7, 14, or 28 days. The loading system and data acquisition system allow a reliable description to be obtained at the start of the test, particularly, the determination of the instantaneous elastic strain, ε_e (the acquisition frequency is higher during the loading).

9.3.3. *Data processing*

Once the tests have been carried out, with recorded measures over an adequate period of time (from a few months to a year), the test results are compared to models [LER 96a; MEL 99]. In the case where the industrial problem under investigate concerns the already performed design of a structure, we will make sure the experimental curve is located within a domain set by the contractor.

In the case where tests are useful to the design of a structure, in particular to feasibility studies, the model is better adjusted by identification, in order to predict the delayed strains of the structure over the longer term: the coefficients of the model are then calculated based on the experimental results, and then the observed behavior on a shorter period is extrapolated, for real, on-site conditions (temperature, relative humidity, mean radius, etc.).

9.3.3.1. *Experimental data and statistical processing*

Another analysis of test results consists of the extrapolation of the evolution of delayed strains over a period of time (from a few years to a hundred years) using measurements only, regardless of a phenomenological model. This technique relies on a heavy experimental campaign [CLE 00], and considers the accuracy of the measurements and the causes of test dispersion:

– related to the material: concrete is heterogenous, and global dispersion of creep test is dependant of dispersion of its instantaneous mechanical characteristics. The weight of this heterogeneity (measured by its compressive strength) on the global dispersion of the creep test ranges from 6 to 8%;

– related to the test and depending on the application of the test process: the test material and its accuracy, or the operator. Compliance of basic creep is the parameter being studied.

The main results of the repeatability study are presented here as a compliance curve of basic creep for eight specimen pairs of creep/shrinkage for endogenous conditions. Each compliance curve corresponds to a batch (Figure 9.14).

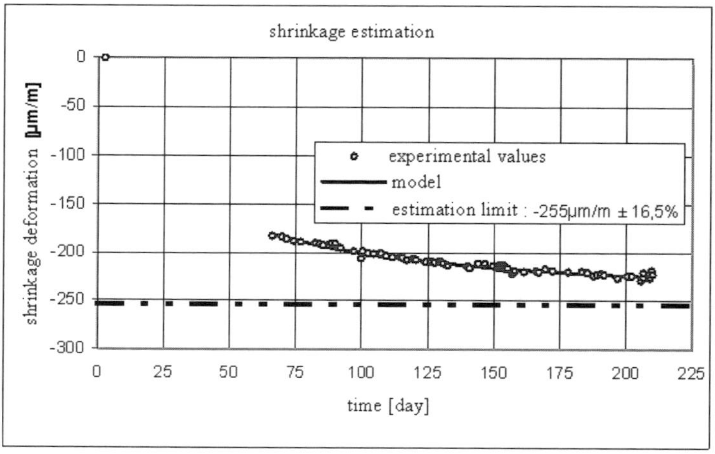

Figure 9.15. *Estimation example of infinite shrinkage*

With intial data exploitation, we obtain a confidence interval for the mean equal to ±7%. The development of a technique for the statistical processing of experimental data, based on a linear regression with weight and variable change with threshold, allows an asymptotic variable of the studied parameters (e.g. shrinkage) to be given, along with its tolerance (in between two limits).

This technique requires the choice of a variable change function and a temporal threshold under which processing is applied.

For the measured values x_i (time) and y_i (studied parameter: compliance or shrinkage) with a selection of the pair of values (x_i, y_i) for x_i≥threshold, the following variable change is applied:

$$z_i = (x_i - x_0)/(y_i - y_0) \quad [9.3]$$

with the terms x_0 and y_0, which correspond to the initial instantaneous values. The processing of the former data, allowing a compliance or long-term shrinkage value to be obtained at every test, leads to a confidence interval of ±16.5%, which can be applied to a single test. As we cannot repeat the same experiments for every type of concrete, we chose a constant confidence interval regarding both creep and shrinkage for a single test result.

9.3.3.2. *Creep coefficient determination*

Correct resolution of the measurements, particularly during instantaneous change, allows the initial compliance value, J_0, to be determined. The asymptotic value of compliance, determined by a statistical process (linear regression after changing the variables to weight and shrinkage) provides the infinite compliance value, J_∞. Creep coefficient, ϕ, is then given by the following equation:

$$\phi = \frac{J_\infty - J_0}{J_0} \tag{9.4}$$

If the creep coefficient, ϕ, and the longitudinal elasticity modulus, E, of a given concrete are known, we can deduce the global infinite creep strains.

9.4. Delayed response modeling

The prediction of the delayed strains of concrete for the study of the long-term performance of concrete structures (bridges, containments, etc.) inevitably requires the elaboration of identified predictive models by experimental studies.

Nevertheless, there are problems related to the exploitation and the extrapolation of the laboratory studies, which are due to the nature of the concrete itself: a heterogenous material in which the microstructure has a direct impact on the dispersion of the macroscopic measures obtained using the tests. The direct exploitation of the test results is also made difficult because of the occurrence of the structural effects due to cracking. It becomes harder to experimentally identify the intrinsic behavior of the material and, therefore, to propose some constituent rules for the phenomena.

Some recurrent issues also appear for the transition from a small specimen to a large, real structure. It is then essential to consider:

– the size effect related to desiccation: drying kinetics strongly depend on the structure's size and then modifies the delayed strain evolutions. Creep effect, insignificant for small structures, can become predominant for large structures where drying kinetics are slow [BAŽ 01];

– the size effect related to the mechanical behavior: tensile and compressive strengths and cracking energy vary as a function of the dimensions [VLI 00] and of the test boundary conditions [KOT 83]. The same applies for both creep and shrinkage strains [NEV 83].

Moreover, the tested specimens are generally studied during, at most, a few years, whereas the civil engineering structures can last around 50 years. As a result, some issues related to the difference between the concerned time scales appear:

– extrapolation of the results obtained in laboratory (weight loss and delayed strains) with much more important time intervals;

– the mechanical behavior of concrete becomes more fragile as the loading time increases [BAŽ 97c];

– the scale effect becomes more important as the considered structure ages [BAŽ 01].

Modeling is the only way to enable direct extrapolation of the obtained results with to geometry or a different environment. In particular, resorting to modeling enables the structural element to be isolated from the intrinsic element in order to identify every conventional component thanks to the reverse analysis process. This is what we will focus on in the following section.

9.4.1. *Modeling drying*

Water content strongly influences delayed strains of concrete, thus, the objective it to quantify those strains. Therefore, the drying process has to be modeled in order to obtain the evolution of the water content.

The initiating factor of concrete drying is the relative humidity gradient existing between the external environment and the material. This unbalance results in the outwards flow of concrete humidity following two coupled modes of transport [DAI 01; MAI 01]:

– humidity transport in a liquid phase is driven by the pressure gradient (controlled by the permeability of the porous network). Darcy's law describes this flow:

$$\boldsymbol{v} = -\frac{K}{\eta} \cdot \boldsymbol{grad}\left(p_l\right) \qquad [9.5]$$

where $\boldsymbol{v}$ (m/s) is the rate of liquid water filtration, K (m^2) the material intrinsic permeability (independent from the saturating liquid), η (kg/m/s) the dynamic viscosity of the saturating liquid (here water), and p_l (Pa) the saturating liquid pressure;

– humidity transport in a gas phase is driven by the concentration gradient (controlled by the water vapor diffusivity). Fick's first law describes this flow:

$$\boldsymbol{\Phi} = -D_{diff} \cdot \boldsymbol{grad}(w) \qquad [9.6]$$

where w (kg/m^3) is the water vapor concentration, D_{diff} (m^2/s) the water vapor diffusivity, and Φ_{diff} ($kg/m^2/s$) the water vapor flux.

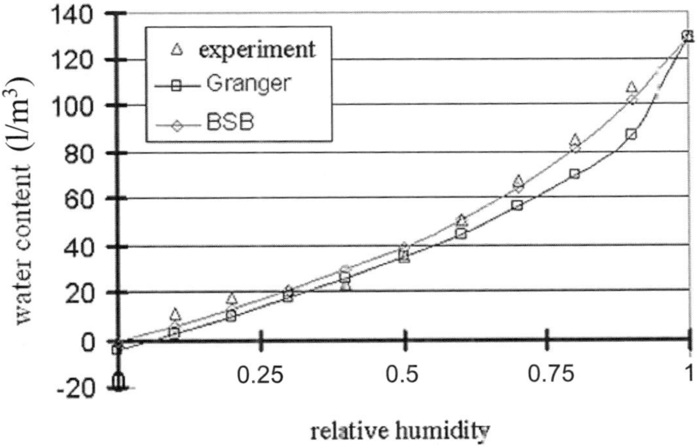

Figure 9.16. *Comparison between experimental desorption and simulated isotherms*

The existing coupling between the two transport phenomena results from the existing interaction between the vapor and liquid phases, through the condensation/evaporation phenomena.

It is difficult to separate the two transport phenomena when humidity flows are experimentally measured. To resolve this problem, equations [9.5] and [9.6] can be brought together as a single and simple equation. This approach has been used here where the equation to be resolved is similar to the Fick's second law [WIR 00]:

$$\frac{\partial C}{\partial t} = div\left(D(C) \cdot \boldsymbol{grad}(C)\right) \qquad [9.7]$$

where C (l/m^3) is the water content, $D(C)$ (m^2/s) the diffusivity including all the transport modes of humidity, strongly dependent of the water content.

Equation [9.7] considers both humidity flows due to diffusion and permeability. It is worth noting that this entry means that numerous hypotheses have to be respected [BEN 00b; COU 01].

We can use the relation proposed by Xi *et al.* [XI 94] to evaluate diffusity:

$$D(h) = D_0 \cdot \left[1 + a \cdot \left(1 - 2^{-10^{b(h-1)}}\right)\right] \quad [9.8]$$

where D_0, a, and b are material parameters depending on the W/C ratio [XI 94].

Equation [9.7] is formulated as a function of water content. The diffusion coefficient (equation [9.8]) can be expressed as a function of the same parameter, thanks to a desorption isotherm. The expression of the desorption isotherm is determined via the BSB model [BRU 69]:

$$C = \frac{A \cdot k \cdot V_m \cdot h}{(1 - k \cdot h) \cdot \left[1 + (A-1) \cdot k \cdot h\right]} \quad [9.9]$$

where h is the relative humidity, A, k, and V_m the model BSB constants.

The parameters are identified by the experimental plot proposed by Pihlajavaara [PIH 82]. It is worth noting that Granger [GRA 95] already suggested the approach via the desorption curve thanks to a second-degree equation, with a humidity range of 50 to 100%:

$$C = C_0 + \left(C_{eq} - C_0\right) \cdot \sqrt{2 \cdot (1-h)} \quad [9.10]$$

where C_0 and C_{eq} are the water contents corresponding to the initial relative humidity in the concrete and the relative humidity of the surroundings. The results obtained with two models are compared to the experimental results (Figure 9.16).

We can notice that the the BSB model provides the best superposition. We can also observe that equation [9.10] proposed by Granger [GRA 95] leads to negative water contents for a low relative humidity. This is not a problem as the concerned relative humidity is unusual.

The boundary condition at the drying surfaces level is of convective type. The water flow, $\vec{J}$ (l/m^2/s), between the concrete's exposed surface and the surroundings is calculated from the following equation [TOR 97]:

$$\vec{J} = \left[\beta \cdot \left(\left(2 \cdot C_0 - C_{eq}\right) - C_s\right)\right] \times \left(C_s - C_{eq}\right) \cdot \vec{n} \qquad [9.11]$$

where β is a material parameter equal to 5×10^{-10} m^4/s, C_0 the initial water content, C_s the water content on the drying surface, C_{eq} the water content corresponding to the relative humidity of the surroundings, and $\vec{n}$ the normal vector at the drying surface (outwards).

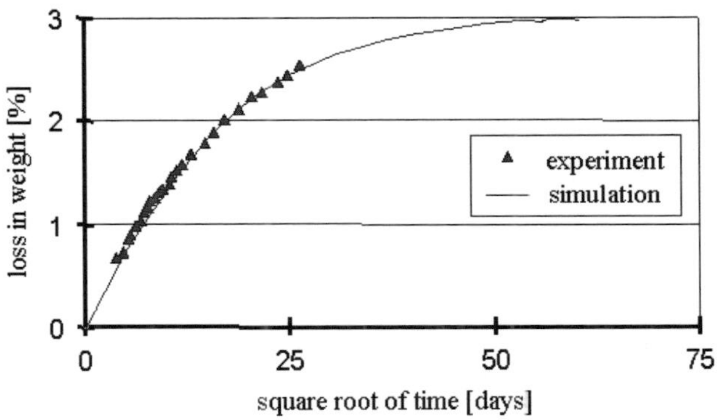

Figure 9.17. *Comparison between numerical and experimental weight loss*

The drying process commonly induces cracking of the skin of the concrete [ACK 83]. We could then be expecting the creation of cracks to accelerate the drying process. Experimental studies on the influence of cracking on the characteristics of humidity transport (diffusivity and permeability) generally examine the cracks generated by a mechanical loading [GER 00]. Nevertheless, experimental results show that the application of a compressive loading does not significantly modify the evolution of the specimen weight loss [MAN 41; HAN 60; NEV 60; LAS 97], although compressive loading strongly decreases the microcracking of the specimen. As a consequence, microcracking due to drying should not significantly modify the properties of humidity transport. The mechanical properties of humidity transport could be considered to be independent of the mechanical state of the material.

Resolution of the equations system [9.7]-[9.11] by the finite elements method allows the water content evolution, within the specimen, to be obtained. The parameters of the drying model must then be identified, by superposition, in order to reproduce the experimental results. This enables us to have water content distributions sufficiently representative of reality to be used in the modeling of delayed strains. As an example, the experimental and numeric evolutions of the water loss as a function of time are compared in Figure 9.17, for a cylindrical specimen, tested at the LCPC (*Laboratoire Central des Ponts et Chaussées*) of a 100 cm height and a 16 cm diameter, with a given concrete formulation [GRA 95; BEN 05].

The evolution associated with water content, along the radius of the specimen, is given at different times as seen in Figure 9.18. In particular, it highlights the existence of a strong gradient in the specimen when drying commences. This gradient then gradually tapers off. This shows that the drying process occurs in a non-homogenous way within the specimen.

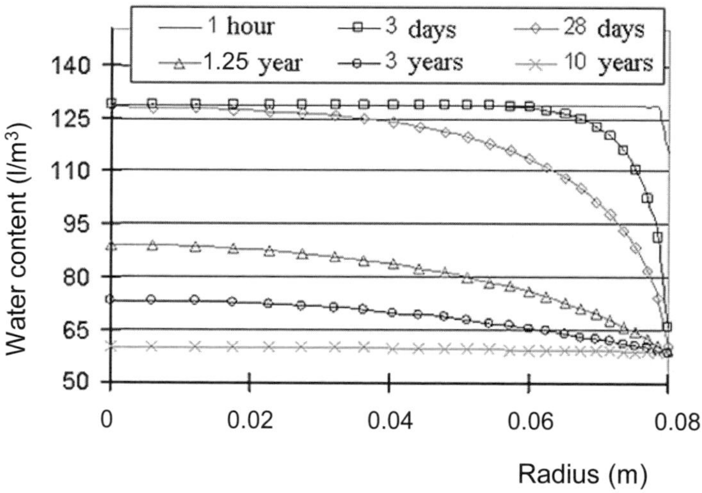

Figure 9.18. *Profile of the water content at different times*

9.4.2. *Modeling of the desiccation shrinkage*

The experimental results show that strain of intrinsic shrinkage is proportional to the variation of the intern relative humidity, for relative humidity values in-between 40 and 100% [VER 68, HAN 87, BAR 99]. Therefore, the existing modeling is generally based on this experimental observation [ALV 93, BAŽ 94]. Another approach consists of the expression of desiccation shrinkage as a linear function of the variation C of the water content in time [CAR 37, GRA 95, TOR 97]:

$$\dot{\varepsilon}_{ds} = k_{ds} \; \dot{C} \; \delta \qquad [9.12]$$

where $\dot{\varepsilon}_{ds}$ is the level tensor of the strain of intrinsic desiccation shrinkage, k_{ds} (m^3/l) the hydric compressibility coefficient, and δ the unite tensor. This relation considers desiccation shrinkage as isotropic.

Equation [9.12] means that local variation of the water content induces an instantaneous strain of desiccation shrinkage. This hypothesis is plausible. Indeed, liquid-vapor equilibrium can be regarded as quasi-instantaneous considering the characteristic time of humidity diffusion.

This modeling is still coherent considering the experimental results. Indeed, the desorption isotherm curve is quasi-linear for relative humidity values between 50 and 100% (Figure 9.16).

In the case where the material behavior is purely linear visco-elastic (crack-free), the constitutive law [9.12] leads to a simple relation between desiccation shrinkage strain and weight loss [BEN 00b]:

$$\dot{\varepsilon}_{ds} = K_{ds} \; \frac{\dot{P}}{P_0} \; \delta \text{ with } K_{ds} = 1000 \, k_{ds} \; \frac{\rho_c}{\rho_w} \qquad [9.13]$$

where ρ_c [kg.m^{-3}] is the concrete density, ρ_w [kg.m^{-3}] the water density, P [kg] the drying specimen mass, and P_0 [kg] its initial mass.

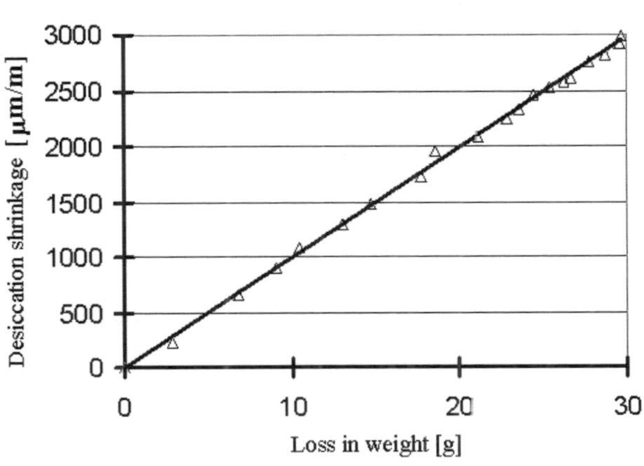

Figure 9.19. *Desiccation shrinkage as a function of weight loss for a cement paste [FOW 59]*

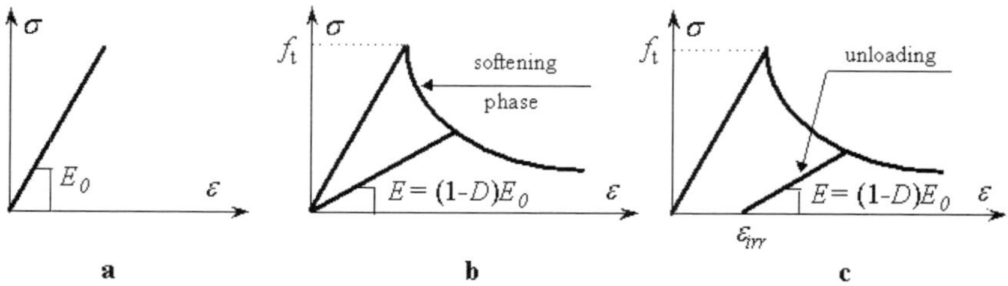

Figure 9.20. *Behavior in uniaxial tension: a) elastic model; b) elastic damage; c) coupled damage- plasticity model*

In equation [9.13], the coefficient of proportionality, K_{ds}, represents the slope of the linear part of the desiccation shrinkage-weight loss relation (Figure 9.7). Thus, parameter k_{ds} of equation [9.12] can easily be identified on the linear part of the desiccation-weight loss curve. Writing the behavior law of desiccation shrinkage in terms of water content and not in terms of relative humidity allows us to identify the model parameter.

This relation of proportionality [9.13] has also been experimentally highlighted [POW 59] on cement paste specimens of thin thickness (Figure 9.19) where development of superficial cracking is negligible.

Considering such a linear relation inevitably means that non-linear observed relations in the shrinkage-weight loss diagram (Figure 9.7) are exclusively attributed to cracking, and therefore, are a manifestation of the structural effect. Therefore, conjoint use of a cracking model with this linear model of desiccation shrinkage must allow this non-linearity to be found. In the contrary case where non-linearity is partially found, then the intrinsic behavior of the material must be non-linear.

That is why the simulations performed with two cracking models are compared to the experimental results along with the results of an elastic model (Figure 9.20a). They are performed on the same specimen of drying calculation. The first model is an elastic damage model [MAZ 84]: unloading does not reveal any irreversible strain (Figure 9.20b). The second model is a model coupling softening plasticity and damage [BEN 05]: some irreversible strains appear during unloading, showing the opening and closing of cracks (Figure 9.20c).

The mechanical behavior law is then given by:

$$\boldsymbol{\sigma} = (1-D)\cdot \boldsymbol{E_0} : \boldsymbol{\varepsilon}_e \qquad [9.14]$$

where σ is the stress tensor, D the damage variable which can be scalar or tensorial, and E_0 the tensor of initial elastic rigidity. The elastic strain ε_e is given by:

$$\varepsilon_e = \varepsilon - \varepsilon_{ds} \quad [9.15]$$

In the case of an elastic damage model, the elastic strain is given by:

$$\varepsilon_e = \varepsilon - \varepsilon_{ds} - \varepsilon_{irr} \quad [9.16]$$

for the elastoplatic-damage model. The desiccation shrinkage, ε_{ds}, in equations [9.15] and [9.16] is obtained from equation [9.12] once the drying issue has been resolved. The irreversible strain ε_{irr} is given by a plasticity formalism [BEN 05].

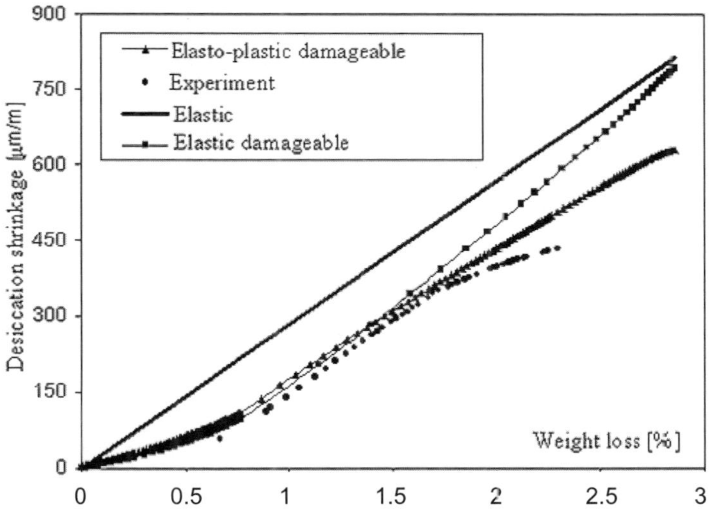

Figure 9.21. *Comparison between experimental evolution and simulated evolutions of desiccation shrinkage*

The comparison between the experimental results of desiccation shrinkage and the numerical results from modeling is represented in Figure 9.21. Both cracking models allow us to correctly reproduce the first two phases (Figure 9.7) of desiccation shrinkage. This enables us to check, *a posteriori*, that the first phase, where weight loss is not followed by apparent shrinkage, corresponds to the specimen surface level in time (Figure 9.22). Indeed, the obtained strain during the elastic calculation strongly exceeds the elastic limit, leading to skin cracking when a cracking model is used. Stress drops, due to softening, as soon as it has reached the elastic limit of 3 MPa (Figure 9.22).

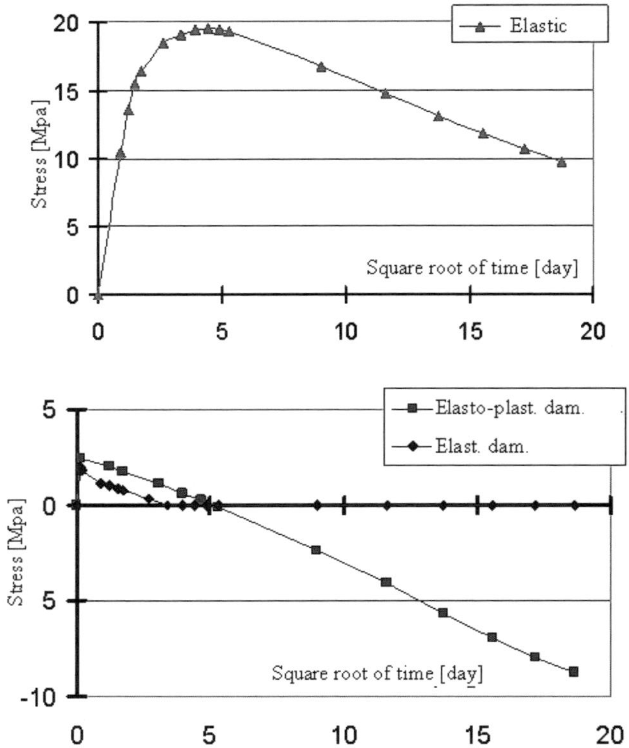

Figure 9.22. *Axial stress evolution on the specimen's skin for the three used models*

The observed non-linearity in this first phase of the shrinkage-weight loss diagram is then a structural effect, as the shrinkage linear model coupled to a cracking modeled indicates.

Later, the water content gradients decrease, cracking no longer prevails, and the observed shrinkage shows linear behavior.

Considering those two phases, the adopted linear hypothesis for the desiccation shrinkage model seems to be justified regarding the material's intrinsic behavior. Nevertheless, it becomes less obvious considering the final phase, which presents an asymptomatic behavior that cannot be described by any of the models. Whereas the response obtained with the damage model tends towards the elastic solution, the response given by the model, which takes into account the irreversible strains, is the model that comes the nearest. This observation allows us to propose some elements of discussion for this phase. The water content profile becomes then almost uniform (Figure 9.18) within the specimen and the self-stress system converts to an unloading mode. This closing does not generate any stress for the elastic damage model but produces skin compressive stresses in the case of the irreversible strain model (Figure 9.22).

Thus, a structural effect appears again due to crack closing, which almost explains the asymptotic behavior of the apparent desiccation shrinkage. The other part must certainly be explained by the material's intrinsic behavior. When the material has entirely dried, a consolidation of the solid skeleton might be at the origin of the delimited characteristics of desiccation shrinkage. It is then suitable to design an experimental procedure which will be able to confirm it.

Nevertheless, it should be mentioned that the identification process has enabled us to better understand the limit between the material's intrinsic behavior and the structural effects that affect experimental measurements.

A recent approach has been developed [BEN 07], where drying shrinkage strains are regarded as being a spherical elastic and creep response of the material under rising pore pressures during the drying process. The developed model is able to describe the main features of shrinkage behavior of cement-based materials. In this case, drying shrinkage is only a consequence of the stresses due to non-uniform drying.

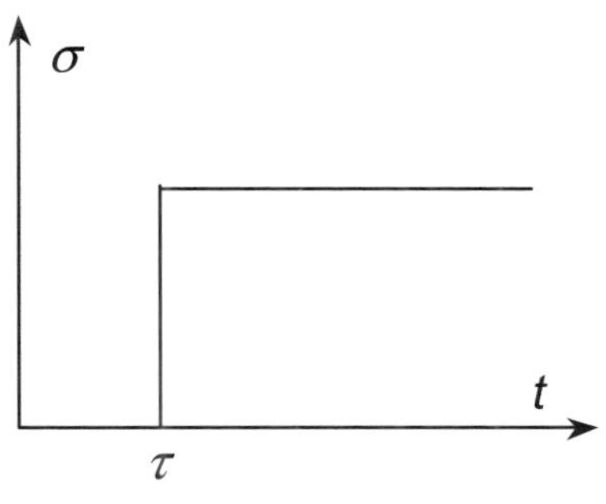

Figure 9.23. *Loading during a creep test in the laboratory*

9.4.3. Creep modeling

9.4.3.1. *Uniaxial modeling*

Most of the existing endorsed models split delayed strain into a shrinkage strain and a global creep strain. Recent codes like the Eurocode 2 [LER 96a] and ACI-B3

[BAŽ 98a] consider basic creep strain and desiccation creep strain separately. Those models introduce a creep compliance, which allows strain variation in time for the applied stress to be determined.

9.4.3.1.1. Creep compliance

Let us consider a creep experiment the loading history of which is presented in Figure 9.23.

For a stress level σ lower than 0.5 times the limit compressive stress *fc*, creep strain and stress are experimentally observed to be proportional. We can then write:

$$\varepsilon_m = \varepsilon(t) - \varepsilon_s(t, \tau) = J\ t, \tau) \,.\, \sigma(\tau) \qquad [9.17]$$

or:

$$\varepsilon_m(t, \tau) = \varepsilon_i(t) + \varepsilon_c(t, \tau) = \frac{\sigma(\tau)}{E(\tau)} + \varepsilon_c(t,\tau) \qquad [9.18]$$

where $\varepsilon_m(t, \tau)$ is the mechanical strain, $\varepsilon_s(t, \tau)$ the shrinkage strain or depending on the situation, the loading independent strain, $\varepsilon_i(\tau)$ the instantaneous strain during loading, $\varepsilon_c(t, \tau)$ the creep strain at time t for a loading age of τ, $\sigma(\tau)$ the stress at time τ, $J(t,\tau)$ the creep compliance at time t for a loading age of τ, and $E(\tau)$ the elasticity modulus at the loading time.

Moreover, we can also define a creep-kernel φ:

$$\varepsilon_c(t,\tau) = \frac{\sigma(\tau)}{E(\tau)} \varphi(t,\tau) \qquad [9.19]$$

which can be related to creep compliance by:

$$J(t,\tau) = \frac{1+\varphi(t,\tau)}{E(\tau)} \qquad [9.20]$$

Existing modeling consists of the definition of the analytical expression of $J(t,\tau)$ or $\varphi(t,\tau)$ as a function of a certain number of material parameters. Those aspects are discussed in more details in the section dealing with models used in the codes.

9.4.3.1.2. Superposition principle

Creep test then gives access to the compliance $J(t,\tau)$ or the creep kernel $\varphi(t,\tau)$. Nevertheless, loading history of the structures is not constant and therefore different from the one previously presented. With any stress history we have to be able to calculate creep strains. In order to do so, we generally use the Boltzmann superposition principle (linear visco-elasticity). We then have:

$$\varepsilon_m = \varepsilon(t) - \varepsilon_s(t, \tau) = \int_0^t J(t',\tau)\, d\sigma(t',\tau) \qquad [9.21]$$

where $d\sigma(t',\tau)$ is the strain increment at time τ with $t' \in [0, t]$.

No conceptual problem is then faced with the application of the superposition principle, which nevertheless has some limitations. The first one is a numerical issue: indeed, during calculation, the stress history has to be noted. This problem can be resolved by using a Dirichlet series, for compliance: an association of a series of Kelvin-Voight elements (a spring in parallel with a damper). We then have:

$$J(t,\tau) = J_1 + \sum_{i=2}^{n} J_i \left[1 - exp\left(-\lambda_i \left(t-\tau\right)\right)\right] \qquad [9.22]$$

where n is the number of Kelvin-Voigt elements and $\lambda_i = E_i/\eta_i$ with E_i and η_i the spring rigidity and the damper viscosity, respectively, of element i. In this case, the stress history does not have to be stored, as the strain state of every element of the series is enough to represent the history. To obtain more information about the application of this method and the recognition of temperature and hygrometry history, the reader is referred to [GRA 97].

9.4.3.1.3. Equivalent time method-incremental model

The application of the superposition principle during unloading leads to an overestimation of the creep recovery strain compared to experimental observations. Within concrete a small part of the creep strain is recovered after unloading [ACK 89]. French regulations of pre-stressed concrete [BPEL 83] have predicted an unloading compliance function different from the loading one. This has to be carefully dealt with as it obviously breaks the superposition principle and the application of this method to a loading-unloading succession causes inaccurate results.

This is why the equivalent time method [ACK 89] or the incremental model [EYM 91] have been developed. If we suppose that we have a loading made of two strain levels σ_1 applied at time τ_1 and σ_2 at time τ_2 with $\tau_2>\tau_1$. Creep strain, at time $t>\tau_2$ will be:

$$\varepsilon_c(t,\tau) = \frac{\sigma_2(\tau_2)}{E(\tau_{eq})}\varphi(t,\tau_{eq}) \qquad [9.23]$$

where τ_{eq} is the equivalent time such as:

$$\varepsilon_c(\tau_{eq}) = \frac{\sigma_2(\tau_2)}{E(\tau_{eq})}\varphi(\tau_2,\tau_{eq}) = \frac{\sigma_1(\tau_1)}{E(\tau_1)}\varphi(\tau_2,\tau_1) \qquad [9.24]$$

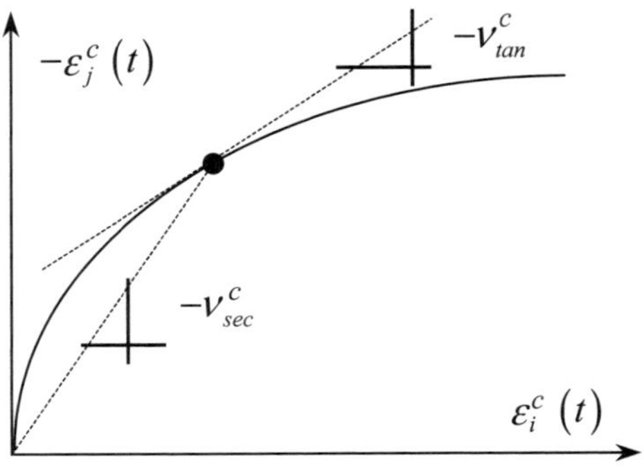

Figure 9.24. *Definition of tangential and secant creep Poisson coefficient*

The concrete creep strain is as it could have been if we had directly loaded concrete with stress σ_2 at τ_{eq}. The evolution of creep strain depends then on this single variable and on the strain and strain state at time τ_2.

Obviously, the equivalent time method presents some drawbacks too. During a total, or at least significant, drying, equation [9.24] cannot be solved. In that case, the initial method predicted a recovery creep of zero, which does not correspond to reality. An improvement of the method is possible using the results of the recovery tests to complete the solutions of equation [9.24] [EYM 91].

Nevertheless, the acquisition of recovery test results is a limit to the method except if the recovery laws being used are already known (for example, the laws proposed by Dreux in the French code BPEL). The extension of the method to a multi-axial loading is also not obvious (especially on a conceptual point of view) but has still been performed by using hypotheses applicable to pre-stressed concrete beams [EYM 94].

9.4.3.2. *Multi-axial modeling*

Generalization from uniaxial behavior to the multi-axial strain case is done by analogy with the concrete elastic behavior. The principal of creep strains, $\varepsilon_c = (\varepsilon_x, \varepsilon_y, \varepsilon_z)$

can be related to the principal stresses $\sigma_c = (\sigma_x, \sigma_y, \sigma_z)$ by a linear visco-elasticity law:

$$\varepsilon_c = \begin{bmatrix} \varepsilon_x(t) \\ \varepsilon_y(t) \\ \varepsilon_z(t) \end{bmatrix} = \begin{bmatrix} 1 & -\nu^c_{xy}(t) & -\nu^c_{xz}(t) \\ -\nu^c_{yx}(t) & 1 & -\nu^c_{yz}(t) \\ -\nu^c_{zx}(t) & -\nu^c_{zy}(t) & 1 \end{bmatrix} J(t,\tau) \begin{bmatrix} \sigma_x(\tau) \\ \sigma_y(\tau) \\ \sigma_z(\tau) \end{bmatrix} \quad [9.25]$$

where $J(t,\tau)$ is the creep compliance determined in the uniaxial configuration, $-\nu^c_{ij}(t)$ the creep Poisson coefficients, which have to be experimentally identified by the measurement of the three creep strain components. This coefficient can be taken as secant (Figure 9.24):

$$-\nu^c_{ij}(t) = -\frac{\varepsilon^c_i(t)}{\varepsilon^c_j(t)} \quad [9.26]$$

or tangential:

$$-\nu^c_{ij}(t) = -\frac{d\varepsilon^c_i(t)}{d\varepsilon^c_j(t)}. \quad [9.27]$$

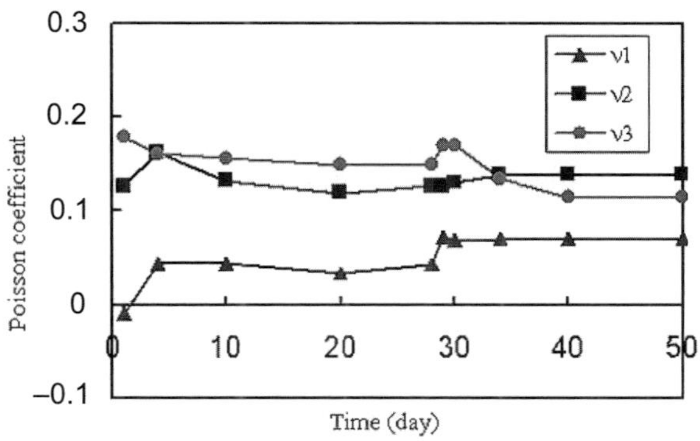

Figure 9.25. *Evolution of basic creep Poisson coefficient according to three loaded directions [GOP 69]*

Nevertheless, the few existing results on creep Poisson coefficient are often different from an author to another, and very often, those results do not specify whether the given evolutions concern basic creep or total creep (simultaneous mechanical loading and desiccation). Some authors find that it increases with time [NEV 83], that it is constant [PAR 73], and others that it decreases [GAU 82].

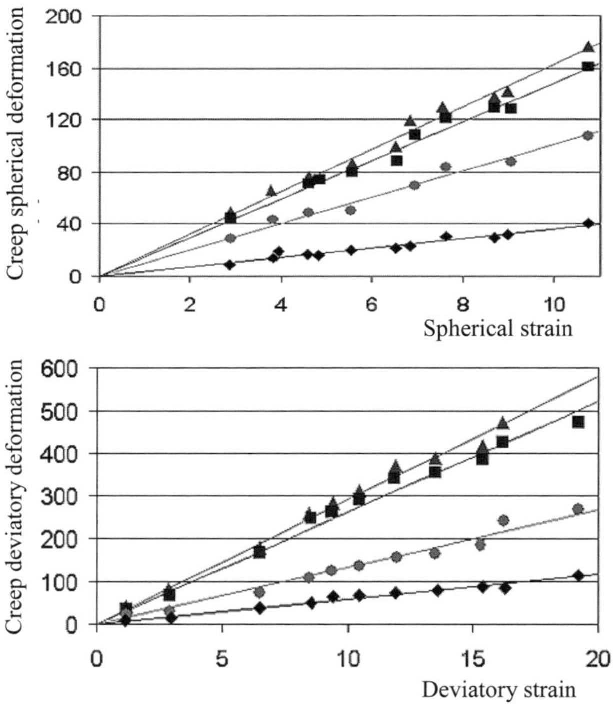

Figure 9.26. *Experimental highlight of the existence of spherical and deviatory creep mechanisms [BEN 00a]*

Therefore it has been suggested to use a basic creep Poisson coefficient equal to the elastic Poisson coefficient [BAŽ 98a, GRA 95]. However, the experimental results of Neville [NEV 83] show that it varies between 0.05 and 0.20 and that it is not related to either the elastic Poisson coefficient of the cement's paste, or the aggregates. Therefore Neville recommends that a basic creep Poisson coefficient of 0.16 (being 80% of elastic Poisson coefficient) should be considered. Ulm [ULM 98] suggests considering two evolutions of the basic creep Poisson coefficient regarding the distinct two mechanisms mentioned earlier, a short-term creep and a long-term one.

Desiccation creep also is also similarly contentious. Experimentally, we find a lower creep Poisson coefficient of desiccation when concrete dries. According to Neville's [NEV 83] results, the creep Poisson coefficient of desiccation would vary between –0.17 and –0.29. In his model, Bazant [BAŽ 97] proposes that a constant intrinsic creep Poisson coefficient value of 0.2 should be used. However, the creep Poisson coefficient of intrinsic desiccation is related to the mechanism of long-term basic creep [BAŽ 94, ULM 98], which is accelerated by desiccation. As long-term basic creep is supposed to occur with constant volume, the desiccation creep coefficient should be close to 0.5. The lack of experimental results does not allow the issue to be resolved.

Moreover, a certain number of authors find a creep Poisson coefficient lower for multi-axial compression (0.09 to 0.17) than for uniaxial compression. It would depend on the three applied main strains (Figure 9.25). Thus, it would be more important towards the less loaded direction in compression. This observation is significant. Indeed, the linear visco-elastic modeling and, therefore, superposition principle would no longer be valid. The use of equation [9.25] would then no longer be justified.

To overcome this, a range of basic creep modeling under multi-axial stresses has recently been proposed, based on splitting the basic creep process into a spherical element and a deviatory element. This method is valid according to experimental results analysis of basic creep under multi-axial conditions [BEN 00a] (Figure 9.26). The spherical element is associated with the movement of physisorbed water via capillary porosity. The deviatoric element is related to the C-S-H sheets' sliding in nanoporosity where water acts as a lubricant [BEN 01].

Starting from the proposed mechanisms, a constitutive law describing the evolution of basic creep has been obtained. Thus, spherical basic creep [BEN 01] can be written as:

$$\begin{cases} \dot{\varepsilon}_r^{sph}(t) + 2\cdot\dot{\varepsilon}_i^{sph}(t) = \dfrac{1}{\eta_r^{sph}}\cdot\left[h(t)\cdot\sigma^{sph}(t) - k_r^{sph}\cdot\varepsilon_r^{sph}(t)\right] \\ \dot{\varepsilon}_i^{sph}(t) = -\dfrac{1}{\eta_i^{sph}}\left\langle -2\cdot k_r^{sph}\cdot\varepsilon_r^{sph}(t) + k_i^{sph}\cdot\varepsilon_i^{sph}(t) + h(t)\cdot\sigma^{sph}(t)\right\rangle^{+} \end{cases} \quad [9.28]$$

where ε_r^{shp} and ε_i^{shp} are reversible and irreversible spherical basic creep strains, respectively. The first is related to the movement of water within macroporosity, whereas the second concerns micropores. Parameters k_r^{shp} and k_i^{shp} are the apparent incompressible volume property and that of the solid skeleton, respectively, η_r^{shp} and η_i^{shp} are the apparent viscosities that depend on the intrinsic viscosity of water and the geometry of the connected pores. The internal relative humidity, h, is introduced to consider the desiccation effect. Finally, σ^{shp} is the spherical stress and $\langle x\rangle^{-}$ represents x positive part.

The deviatoric strain process of basic creep is ruled by the following equations system [BEN 01]:

$$\begin{cases} \eta_r^{dev} \cdot \dot{\boldsymbol{\varepsilon}}_r^{dev}(t) + k_r^{dev} \cdot \boldsymbol{\varepsilon}_r^{dev}(t) = h(t) \cdot \boldsymbol{\sigma}^{dev}(t) \\ \eta_i^{dev} \cdot \dot{\boldsymbol{\varepsilon}}_i^{dev}(t) = h(t) \cdot \boldsymbol{\sigma}^{dev}(t) \end{cases} \qquad [9.29]$$

where σ^{dev} is the deviatoric stress tensor, $\boldsymbol{\varepsilon}_r^{dev}$ and $\boldsymbol{\varepsilon}_i^{dev}$ are the reversible and irreversible strain tensors, respectively, of deviatoric basic creep. Parameter k_r^{dev} stands for the rigidity and η_r^{dev} the viscosity associated to water layers strongly adsorbed in the hydrates gel porosity, and η_i^{dev} the viscosity associated to water layers within the same porosity but far from the solid's surface and therefore not strongly adsorbed.

It is worth noting that the identification material's parameters in equations [9.28] and [9.29] require the measurement of transversal strains in addition to longitudinal strain during the basic creep test.

Regarding desiccation creep, the structural part can be modeled thanks to the use of a proper mechanical model, correctly describing the softening behavior of concrete, as well as the irreversible characteristic of mechanical strains, related to microcracking [BEN 01]. This observation has already been done during the analysis of the structural element of desiccation shrinkage. Regarding the intrinsic part of desiccation creep, the most commonly used model is probably the strain-induced shrinkage model, proposed by Bazant and Xi [BAŽ 94]:

$$\dot{\varepsilon}_{dc} = \lambda \left|\dot{h}\right| \sigma \qquad [9.30]$$

where ε_{dc} is the intrinsic element of the desiccation creep, λ [Pa^{-1}] a material parameter to be identified, and h the relative humidity variation in time showing water microdiffusivity. This model is similar to a rheologic model of damper type where viscosity η_{dc} is equal to:

$$\eta_{dc} = 1/\left(\lambda \left|\dot{h}\right|\right) \qquad [9.31]$$

which means its response to a relative humidity change is purely viscous. Thus, for a constant stress, the strain kinetics of desiccation creep is controlled by the drying process kinetics and the solid skeleton has no influence on the microdiffusion of the water molecules. But at the scale on which microdiffusion is considered [BAŽ 94], the degree of adsorption of the water layers by the solid skeleton, must play a major role in the kinetics of creep intrinsic desiccation. Numerical simulations, presented later, actually show that it is not possible to identify the value of the λ parameter which corresponds to the experimentally observed kinetics.

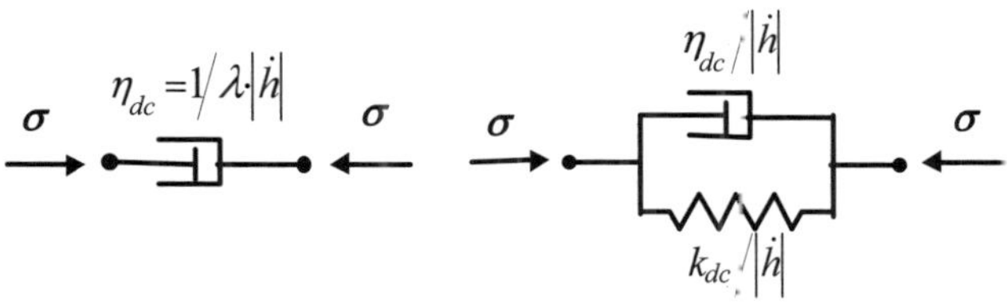

Figure 9.27. *Rheologic models associated to: (left) equation [9.31], (right) equation [9.32]*

To separate desiccation creep and drying kinetics, we usually use a model of Kelvin-Voigt type in which strongly adsorbed water layers can present a certain rigidity, k_{dc}, related to their ordered structure. The constitutive law of intrinsic desiccation creep strain becomes then [BEN 05]:

$$\eta_{dc}\,\dot{\varepsilon}_{dc} + k_{dc}\,\varepsilon_{dc} = \theta\left|\dot{h}\right|\sigma \qquad [9.32]$$

where θ (seconds) is a unit conversion parameter. Therefore, this model allows drying and intrinsic desiccation creep strain kinetics to be separated.

The constitutive relations giving basic creep strain, ε_{bc}, (equations [9.29] to [9.32]) and desiccation creep strain, ε_{dc}, have been introduced in the damage-plasticity model (equation [9.34]) along with the elastic strain given this time by:

$$\varepsilon_e = \varepsilon - \varepsilon_{irr} - \varepsilon_{ds} - \varepsilon_c \qquad [9.33]$$

with $\varepsilon_c = \varepsilon_{bc} + \varepsilon_{dc}$.

In those creep constitutive equations, stress σ is replaced by effective stress $\tilde{\sigma}$ [KAC 83]

$$\tilde{\sigma} = \frac{\sigma}{1-D} \qquad [9.34]$$

which considers that creep only occurs in the material's undamaged part.

Moreover, relative humidity strains need to be considered, but can vary over time. Therefore, we use a discretization of the strains and the relative humidity during the usual time step to calculate the creep strain, applying the superposition principle.

We chose here to approach the evolution of effective stress and relative humidity with a piecewise linear function [BEN 01]:

$$\begin{cases} h(t) = h_n + \dfrac{(t - t_n)}{\Delta t_n} \times \Delta h_n \\ \tilde{\boldsymbol{\sigma}}(t) = \tilde{\boldsymbol{\sigma}}_n + \dfrac{(t - t_n)}{\Delta t_n} \times \Delta \tilde{\boldsymbol{\sigma}}_n \end{cases} \quad \text{with} \quad \begin{cases} t \in [t_n, t_{n+1}] \\ \Delta h_n = h_{n+1} - h_n \\ \Delta \tilde{\boldsymbol{\sigma}}_n = \tilde{\boldsymbol{\sigma}}_{n+1} - \tilde{\boldsymbol{\sigma}}_n \end{cases} \qquad [9.35]$$

where h_r is the relative humidity and $\tilde{\boldsymbol{\sigma}}_n$ the stress with a time step of index n. This discretization allows numerical calculation precision to be improved [BAŽ 82], compared to a step approximation (Heaviside function), especially in the case of significant time steps.

Application to this discretization to creep models leads to the following equation:

$$\varepsilon_{n+1}^{c} = A_c(t_n) + B_c(t_n)\,\tilde{\sigma}_n + C_c(t_n)\,\tilde{\sigma}_{n+1} \qquad [9.36]$$

where terms $A_c(t_n)$, $B_c(t_n)$, and $C_c(t_n)$ only depend on calculated variables with a time step of index n. Actually this relation enables a strong coupling between creep and cracking during calculation of plastically acceptable stresses $\tilde{\boldsymbol{\sigma}}_{n+1}$. For further details, refer to [BEN 01b, BEN 05].

9.4.3.3. *Numerical simulations*

In this section, the results of numerical simulations performed on specimens tested for creep by Granger [GRA 95] are presented. It notably refers to the desiccation creep component, being either intrinsic or structural parts. To do so, we have to mention that basic creep parameters have been identified in order to be consistent with experimental results, which allows us to focus on the component analysis of desiccation. Regarding desiccation shrinkage, experimental values will be used.

In Figure 9.28, evolutions of numerically obtained delayed strains are compared to the those obtained experimentally for two models of intrinsic desiccation creep. Those simulations show that the modified version of Bazant model allows us to reproduce kinetics and the amplitude of desiccation creep strain. Regarding the initial version (equation [9.30]), a value of parameter λ which allows the amplitude and kinetics cannot be found on a short term. Indeed, simulation with the value λ_2 [BEN 05] enables us to precisely reproduce the experimental evolution of the delayed strain, for about 200 days. Using the same parameter for longer durations of delayed strain leads to an overestimation of about 45% after 900 days of simultaneous drying and loading.

Thus, regardless of the considerations related to the choice of one model over another, we have to consider the duration of the experimental tests (typically 1 to 3 months), as the drying process is a slow phenomenon (lasting more than 10 days for a studied specimen). The identification of the intrinsic desiccation parameter for this duration leads to inaccurate results.

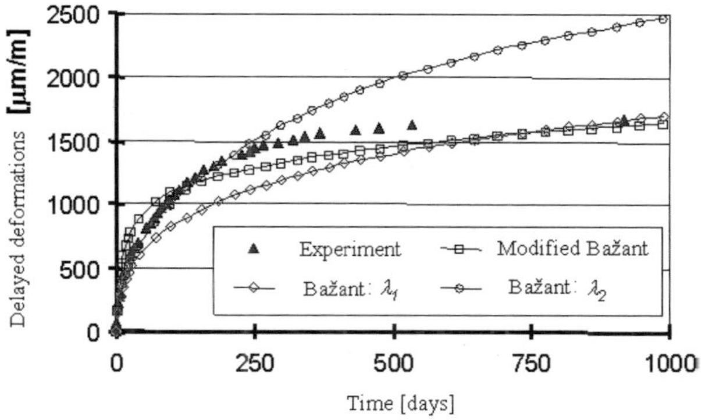

Figure 9.28. *Comparison between experimental evolution and simultaneous evolutions of total delayed strains*

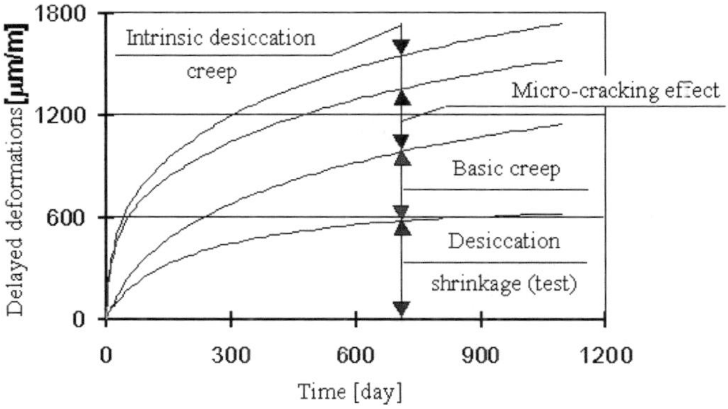

Figure 9.29. *Evolutions of deformations components: Bažant model (with λ_1)*

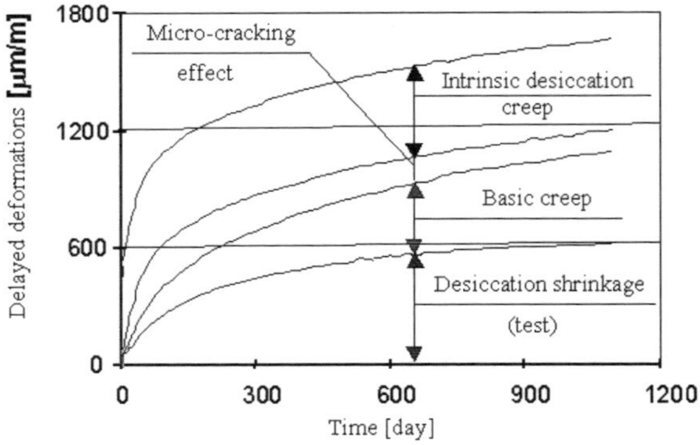

Figure 9.30. *Evolutions of strain components: modified Bažant model*

The contributions of every component of delayed strains to the total delayed strain are reported as a function of time in the case of the Bažant model (Figure 9.29) and its modified version (Figure 9.30). This comparison shows the variation of the strain of intrinsic desiccation creep depending on the model used, although predicted values of the total delayed strain are the same. Thus, this strain progressively and slowly increases when the original Bažant model is used, whereas it rapidly reaches its asymptotic value in the other case.

In addition, the contribution of the structural part of desiccation creep is still significant for the duration of test, whatever model of intrinsic desiccation creep is used. Nevertheless, Bažant and Xi [BAZ 94] reported that component contribution was only significant for a few days; after that, its contribution becomes equal to zero. The results obtained here can be explained by the admitted cracking model. It allows us to take into account a cracking closing characteristic. Thus, the structural part can originate in the two following processes:

– the occurrence of microcracking due to the water content gradient at the start of drying;

– the partial closing of the induced microcracking while the water content becomes distributed homogenously at the end of drying.

9.4.4. *Conclusion*

Throughout the performed simulations, structural effects can be seen to affect the delayed strains. These are well identified in the case of desiccation shrinkage, however, they remain hard to quantify in the case of desiccation creep requiring the use of increasingly sophisticated cracking models, which require numerical calculation. Nevertheless, their contribution strongly depends on the chosen model. This complicates the process, which consists of reverse identifying the (local) intrinsic behavior of the material at its commencement using experimental methods of the constitutive laws. An additional issue arises when predicting the structure's long-term behavior by the use of the constitutive laws whose parameters are identified by relatively short tests.

9.5. Codified models

The engineer in charge of building structures or in charge of the maintenance of these structures when operating is particularly interested in models based on a limited number of parameters that describe the physical and mechanical properties involved in the design of a structure. This has led to the development of codified models. Different models of this type are proposed regarding creep and shrinkage of concrete in Europe and the United States.

In a context of engineering market globalization, it is often difficult for an engineer to evaluate the consequences of the use of a particular model versus a different model on the conception or reparation requirements of a structure. This issue can often be due to the fact that the proposed models in different codifications are often based on different strain decompositions and that their main physical characteristics are often different. This observation has led to the design of different comparative studies [ESP 98; ALM 99; HEI 99] often initiated by codification bodies or by recommendations by such organization as RILEM or ACI with the aiming of helping potential users.

With the same aim in mind, the goals and limitations of codified models of concrete creep and shrinkage are presented in this section. In accordance with the recommendations issued by RILEM, we will move to the presentation of a comparative study of the different models currently proposed in different codes.

9.5.1. *Goals and limitations of codified models*

Due to the complexity of the physical phenomena and current knowledge, codified models of the predictions of delayed strains in concrete are necessarily simplified and adapted to the resolution of specific problems. Regarding creep, they generally consider the linear hypothesis regarding strain and superposition principle to deal with the influence of the variable strain history. Prediction of shrinkage strain and formulation of the creep function generally consider a mean behavior hypothesis towards desiccation. Codified model formulations are a combination of rational elements and experimental observations carried out on laboratory specimens, of relatively small dimensions compared to the real structures. In principle, they are only adapted to an analysis of one-dimensional elements that assume an average behavior of the sections compared with desiccation.

By assuming constant temperature and relative humidity conditions, they are mainly used in norms and codes to:

– evaluate deformed structures (deflections);

– calculate pre-stressed losses;

– estimate some stress redistributions between the parts of hyperstatic systems (in particular, to evaluate the influence of constructive phases);

– understand the influence the influence of creep on the stability limit states of slender structures.

9.5.2. *RILEM recommendations on codified models characteristics*

Principle limitations of codified models having been presented, RILEM recommendations suggest a series of characteristics that should be present in codified models of creep and shrinkage prediction to avoid contradicting the acquired knowledge of experimental observations and diffusion theory. In order to facilitate the analysis of models that are currently used or about to be part of the codes, we will consider the main recommendations.

9.5.2.1. *Total strain expression*

Total strain at age t of a concrete fiber, in a uni-axial stress state, submitted to a constant compressive stress in time $\sigma(t')$ applied in t' is given:

$$\varepsilon_c(t,t') = \varepsilon_{cs}(t,t_s) + \sigma(t') \cdot J(t,t',t_s) \qquad [9.37]$$

where J is the creep function, $\varepsilon_{cs}(t,t_s)$ the shrinkage strain, and t_s the age of the concrete when shrinkage began.

For historical reasons, several calculation algorithms use the non-dimensional concept of creep coefficient $\varphi(t,t',t_s)$:

$$J(t,t',t_s) = \frac{1}{E(t')}\left(1+\varphi(t,t',t_s)\right) \qquad [9.38]$$

It is better that codified models specify creep due to the creep function J, meaning total strain (minus shrinkage) divided by the stress rather than creep coefficient φ. Indeed, using Young's modulus, E, for structure calculation is highly risky if this modulus does not correspond to the one that should have been used to deduce φ from the experimental strains. If a creep coefficient has to be used in calculation, we can deduce it from the following equation whatever the values of $E(t')$ are:

$$\varphi(t,t',t_s) = J(t,t',t_s) \cdot E(t') - 1 \quad [9.39]$$

9.5.2.2. *Expression of shrinkage strain*

Shrinkage strain, ε_{cs}, is delimited and must tend towards a final value, $\varepsilon_{s\infty}$. Shrinkage strain has to be a function of the reduced variable $(t{-}t_s)/\tau_{sh}$ where τ_{sh} represents the time required to reach a given percentage of desiccation (half of the final shrinkage for instance). The diffusion theory shows, which is experimentally confirmed, that:

– τ_{sh} is proportional to the square of the effective thickness (mean radius) of the section;

– on the short term, shrinkage strain must increase as a function of $((t{-}t_s)/\tau_{sh})^{1/2}$;

– on the long term, final shrinkage strain, E, must be asymptotically treated as an exponential function raised to time power $e^{\left\{-k\left(t-t_s\right)^n\right\}}$.

9.5.2.3. *Expression of creep strain*

Creep must be split into basic creep and desiccation creep, preferably as the sum of those two contributions.

Basic creep is the one that occurs during the absence of any hydric exchange with the surroundings. It could have no limit. On the short term, basic creep measures are adequately represented by a power function of $(t{-}t')$ with an exponent equal to about 1/8. On the longer term, basic creep is adequately represented by a linear function of $\log(t{-}t')$, with a slope independent of t'. Basic creep is an intrinsic characteristic of the material: it does not depend on the mean radius of the section.

Desiccation creep presents a similarity with the evolution of shrinkage strain. It is then delimited and its mathematical description must have the same characteristics as the formulation used to predict shrinkage strain. Desiccation creep and shrinkage strain are not intrinsic characteristics of the material: they depend on the structure's transversal dimensions (mean radius).

At the present time, the influence of the composition parameters on the models can only be empirically considered. Determination by the exploitation of an international data bank [MUL 93; BAZ 08] would be advisable. This bank would contain a multitude of creep and shrinkage results obtained from experimental tests, which have been undertaken using standardized procedures if possible (a subgroup of the RILEM TC 107-CSP commission attempted the creation of a pre-project to define such a procedure [ACK 93]).

The RILEM recommendations suggest as a minimum that the norms should highlight the prediction variability of delayed strains by demanding that the calculation of creep-sensitive structures considers this statistical variability. Regarding the previously mentioned data bank, for the codified models there should be an indication of the variation of the model's predictive capabilities.

9.5.3. *Comparison of the performance of the principal models currently proposed in the case of different codes in Europe and in the United States*

The question of performance measurement of a prediction model of concrete delayed strains is only partially resolved as there is no unanimously accepted methodology. So far, the quantitative method almost exclusively used has been the comparison of the model's prediction regarding strain measurements performed on concrete specimens submitted to shrinkage or raw creep tests in compression in the laboratory. Those same results have been used to calibrate the model's composition parameters [ESP 98].

The performance comparison of different models means that those models are evaluated by the same quality measurement of their prediction and that the same set of experimental results is used to calculate this measure. It is advisable that this set of results is as large as possible so that the highest variance of tests conditions can be investigated: this is what makes a data bank universally acknowledged as valuable [MUL 93; BAZ 08]. A comparative study of this type has been carried out for different codified models [BAŽ 95]. It relies on the RILEM database. The authors of the study proposed a number of specifications in an attempt to create a complete process of performance evaluation for such a model [BAŽ 98b]. It is nevertheless worth noting that this type of evaluation does not provide any indication of the performances of the models when they are applied to the calculation of the influence of delayed effects on the behavior of real structures. Laboratory specimens differ from models in that they have a complex stress history, different geometries, and unpredictable environmental variables [ESP 98].

It can be stated that an important part of prediction variability of the models essentially results from the composition parameters of the concrete. Almost no model takes into account the nature of the aggregates and the influence of cement characteristics is reduced to a single parameter when it is considered. However, these two elements have a significant influence on the kinetics and the amplitude of delayed strains. Unfortunately, they are not usually manageable during calculation. To illustrate the present situation of codified models of creep and shrinkage prediction regarding this issue, we present a comparative analysis of several models proposed in the case of different codes in Europe or the United States, focusing on:

– the nature of the involved parameters;

– the consequences of the choice of basic parameters on the prediction of concrete delayed strain under different conditions.

9.5.3.1. *Description of the studied models*

In order to cover a wide range of proposed models in codes, five models are studied: Eurocode 2 – 1991 version [EUR 91], ACI 209 [ACI 92], AFREM [LER 96b], ACI B3 [BAŽ 98a] and ACI GZ [GAR 98]. Comparative studies of the theoretical and physical bases of some of those models are proposed in the literature [ESP 98; LER 98; ALM 99; HEI 99]. This aim of the present section is not to present the constitutive equations of every model in detail, but to focus on the comparison of material, geometrical, and environmental parameters required for their use.

With the exception of the AFREM model, which is based on a partition of the total strain of concrete into physical mechanisms, the models split total strain $\varepsilon(t)$ into a shrinkage strain independent of the strain $\varepsilon_{sh}(t,t_0)$ and a strain dependent of the applied strain by the intermediate compliance function $J(t, t',t_0)$:

$$\varepsilon(t) = \varepsilon_{sh}(t) + J(t,t',t_0)\cdot\sigma(t) \qquad [9.40]$$

where t is the current time, t' the age of the concrete at loading time, t_0 the age of the concrete when drying begins, and σ the applied stress. With what follows, we consider the time origin from the instant the concrete's striking occurs For historical reasons, some of the models [EUR 91; ACI 92; GAR 98] use the non-dimensional concept of creep coefficient $\varphi(t, t',t_0)$ which is related to the compliance function by:

$$J(t,t',t_0) = \frac{1}{E(t')}\left(1+\varphi(t,t',t_0)\right) \qquad [9.41]$$

where $E(t')$ is the Young's modulus value of concrete at loading time t'.

With the exception of ACI 209, the models specify the validity domains using different parameters, which are presented in Table 1, where f_{cj} is the characteristic compressive strength of concrete at j days and a/C is the ratio of aggregate mass to the cement mass for a given volume of concrete.

Parameter	ACI 209	EC 2	ACI B3	ACI GZ	AFREM
Minimum/maximum compressive strength (MPa)		20/90	17/70	20/70	40/80
Maximum stress level		$0.45.f_{cj}$	$0.4.f_{cj}$		$0.7.f_{cj}$
Minimum/maximum surrounding relative humidity (%)	40-100	40-100	40-100	40-100	
Minimum/maximum W/C			0.35/0.85	≤0.6	
Minimum/maximum a/C			2.5/13.5		
Minimum/maximum temperature (°C)		0/80		15/30	
Minimum age at loading time	7 days			2 days	
Minimum/maximum cement dosage (kg/m^3)			160/720		

Table 9.1. *Specified validity domains for the studied models*

The use of every model requires a given number of material, geometrical, and environmental parameters that we summarize in this section. Among those parameters, we can classify some as "fundamental parameters", which have to be directly identified, and some as "secondary parameters", which can be deduced form the "fundamental parameters" with the help of explicit equations or conversion tables. The "fundamental" and "secondary" terms must be understood in this context and not for their importance in terms of physical influence on creep and shrinkage., We will then focus on the identification of the fundamental parameters as these can be regarded as a crucial point for the engineer. We will develop some theories on the proposed identification of secondary parameters in every codification. To facilitate the comparison, we consider some homogenous notations for all the codes that can then be different to the original notations used in the corresponding reference documents.

The model proposed in EUROCODE 2 [EUR 91] is based on the CEB-FIP MC 90 model [CEB 90]. A creep coefficient is used in the expression of the compliance function:

$$J(t,t',t_0) = \frac{1}{E(t')} + \frac{\varphi(t,t',t_0)}{E_{c28}} \qquad [9.42]$$

This coefficient can be expressed as the following general form in which the global creep coefficient φ_0 is corrected by an ageing function φ_a:

$$\varphi(t,t',t_0) = \varphi_0(t',r_m,h,f'_{c28}) \cdot \varphi_a(t,t',h,r_m) \quad [9.43]$$

where r_m is a factor relative to the structure's transversal dimensions, h the surrounding relative humidity, f'_{c28} the characteristic compressive strength of concrete at 28 days, and E_{c28} the characteristic value of Young's modulus of concrete at 28 days. Shrinkage strain is given as:

$$\varepsilon_{sh}(t,t_0) = \varepsilon_{sh0}(h,i_c) \cdot \beta(t,t_0) \quad [9.44]$$

where ε_{sh0} is a function enabling the evaluation of infinite shrinkage and β a function describing the evolution of shrinkage as a function of time, and i_c a parameter depending of the cement type.

The used creep coefficient in the ACI 209 model [ACI 92] is generally given as:

$$\varphi(t,t') = \varphi_\infty(t',h,r_m,f'_{cm28},i_c,i_{cr},\rho) \cdot \beta(t,t') \quad [9.45]$$

where φ_∞ is the ultimate creep coefficient and β a function describing its evolution as a function of time, f'_{c28} the average compressive strength of the concrete at 28 days, ρ the density of concrete and i_{cr} a parameter dependant of the applied curing type. This code also proposes a refined version whose complementary parameters regarding mainly concrete's composition are required for the calculation of the ultimate creep coefficient:

$$\varphi_\infty = \varphi_\infty(t',h,r_m,f'_{cm28},i_c,i_{cr},\rho,i_s,i_{fa},i_a i_{cc}) \cdot \beta(t.t') \quad [9.46]$$

where i_s, i_{fa}, i_a, and i_{cc} are parameters depending, respectively, on the slump's test results, percentage of fine aggregates (D<4.8 mm), percentage of entrapped air, and cement dosage. Shrinkage strain is given in this code as:

$$\varepsilon_{sh}(t) = \varepsilon_{shu}(t,i_{cr},h,r_m) \cdot \beta_{sh}(t,t_0,i_{cr}) \quad [9.47]$$

where ε_{shu} is the ultimate shrinkage strain and β_{sh} a function describing its evolution in time. A refined version of the evaluation of shrinkage strain is also proposed by this code. This version uses the same complementary parameters than those proposed for the refined version of the calculation of creep strain.

The ACI GZ model [GAR 98] also uses a creep coefficient. It is given as:

$$\varphi(t,t',t_0) = \phi_1(t,t') \cdot \phi_2(t_0,t',\varepsilon_{sh}) \cdot \mu(t,t',r_m,h,f'_{cm28},f'_{cmt'}) \qquad [9.48]$$

where ϕ_1 is a function describing the general shape of the concrete's creep, μ a parameter relative to the structure's transversal dimensions, and ϕ_2 a function considering drying that occurred before concrete loading, which can reduce creep ultimate strain. $f'_{cmt'}$ is the concrete's average compressive strength at time t' and ε_{sh} the shrinkage strain given by the function:

$$\varepsilon_{sh} = \varepsilon_{shu}(i_{c,},f'_{cm28},f'_{cmt'}) \cdot \beta_1(h) \cdot \beta_2(t,t_0,r_m) \qquad [9.49]$$

where ε_{shu}, β_1 and β_2 are functions, respectively, describing the material shrinkage dependence, its dependence to surrounding relative humidity, and its dependence to the structure's transversal dimensions and i_c a parameter dependant of the cement type.

Although the usual notions of compliance function and shrinkage function are used in the ACI 93 model [BAŽ 98a], the evaluation of the compliance function is based, as for the AFREM model, on the dvision of concrete's total strain into separate physical mechanisms. The compliance function is then given in this model as:

$$J(t,t',t_0) = q_1(f'_{cm28}) + C_0(t,t',f'_{cm28},c,w/c,a/c) + C_d(t,t',t_0,h,r_m,k_s,f'_{cm28}) \quad [9.50]$$

where q_1 is the instantaneous strain due to a unit strain loading, C_0 the compliance function for basic creep, and C_d the compliance function of desiccation creep. k_s is a parameter dependant on the geometrical shape and the structure's transversal dimensions, c the amount of cement in the concrete, W/C and a/C are the mass ratios of water/cement and aggregates/cement, respectively. Shrinkage strain is given in this model as:

$$\varepsilon_{sh}(t) = \varepsilon_{shu}(i_c,w,t_0,f'_{cm28},i_{cr}) \cdot S(t,t_0,r_m,k_s) \cdot k_h(h) \qquad [9.51]$$

where ε_{shu} is the ultimate shrinkage strain, S a function describing shrinkage evolution as a function of time and the structure's transversal dimensions, and k_h a parameter dependant of the surrounding humidity, i_{cr} a parameter dependant of the applied curing type, W the initial water content of concrete.

As we have mentioned before, the AFREM model [LER 96b] is based on the decomposition of the total strain ε of concrete during an instantaneous elastic strain ε_e, a basic creep strain ε_{bc}, a desiccation creep strain ε_{dc}, an endogenous shrinkage strain ε_{as} and a desiccation shrinkage strain ε_{ds}:

$$\varepsilon = \varepsilon_e + \varepsilon_{as} + \varepsilon_{ds} + \varepsilon_{bc} + \varepsilon_{dc} \qquad [9.52]$$

The instantaneous elastic strain is obtained thanks to Hooke law. Explicit relations are given for every other strain component:

$$\varepsilon_{bc} = \varepsilon_{bc}\left(t, t', f_{c28}, i_{sf}\right) \qquad [9.53]$$

where i_{sf} is a parameter identifying the presence or not of silica fume in the composition of concrete.

$$\varepsilon_{dc} = \varepsilon_{dc}\left(t, t', t_0, f_{c28}, h, r_m\right) \qquad [9.54]$$

$$\varepsilon_{as} = \varepsilon_{as}\left(t, f_{c28}\right) \qquad [9.55]$$

$$\varepsilon_{ds} = \varepsilon_{ds}\left(t, t_0, f_{c28}, r_m, h\right) \qquad [9.56]$$

The sets of the “fundamental” parameters required for every model are presented in Table 9.2 where a cross “x” means knowledge of the corresponding parameter is necessary. Some of those models propose, as we previously introduced, refined versions. In that case, introduced additional parameters are identified in Table 9.2 by a “o”.

Previous descriptions of the different codified models shows that only models B3 and AFREM divide creep strain into a basic creep strain and a desiccation creep strain. We also notice that only the AFREM mdoel divides shrinkage strain into an endogenous shrinkage strain and a desiccation shrinkage strain. In Table 2 is can be seen that those divisions of the physical mechanisms do not increase the required number of parameters. Note also that the new version of EC2 also uses a division of shrinkage and creep into endogenous or basic and desiccation parts.

The type of cement is only considered by the AFREM model, which has been specially developed for high-strength or high-performance concretes, in which the presence or absence of silica fume in the composition of the concrete plays a major role. This information is required in models ACI and EC2. Nevertheless, no clear equivalence rule is given between the definition of cement type used in models ACI and the definition used in the EC2.

Although all models involve a geometrical dimensions factor, it is worth noting that this model is differently defined regarding the codes. In addition, only model ACI B3 involves a supplementary parameter that explicitly considers the geometrical shape of the structure

Parameter	ACI 209	EC 2	ACI B3	ACI GZ	AFREM
Strength of concrete 28 days	x	x	x	x	x
Transversal dimensions of the structure	x	x	x	x	x
Surrounding relative humidity	x	x	x	x	x
Cement type	x	x	x	x	
Silica fume present					x
Concrete density	x				
Curing type	x		x		
Curing time	x				
Total water dosage			x		
Aggregates dosage			x		
Geometrical shape of the structure			x		
Slump	o				
Percentage of fine aggregates (D<4.8 mm)	o				
Percentage of entrapped air	o				
Cement dosage	o		x		

Table 9.2. *"Fundamental" material, geometrical, and environmental required parameters of the models studied*

9.5.3.2. *Comparative analysis of the obtained results by laboratory test simulation*

Some tests carried out on cylindrical, 1 m long, and 0.16 m diameter specimens are simulated for four concrete compositions. In order to cover a representative range of concrete compositions usually used in different structure types, the models have been tested on four compositions, presented in Table 9.3. Except for B25 whose characteristics are out of the validity domain of the AFREM model, the compositions were chosen to allow the application of the five studied models.

Composition	B25	B40	B60 without SF	B60 with SF
Cement	CEM III/A 42.5 LA	CEM III/A 42.5 LA	CEM I 52.5 CP2	CEMII 52.5 R PM SF
Type of cement/ASTM	I	I	I	I
Type of cement/EC2	2	2	2	2
w (kg/m^3)	177	178	150	150
c (kg/m^3)	375	375	425	425
W/C	0,59	0,45	0,35	0,35
a (kg/m^3)	1954	1859	1825	1825
Silica fume dosage (kg/m^3)	0	0	0	35
f_{c28} (MPa)	30,3	40,6	82,7	70,1

Table 9.3. *Composition of the tested concretes*

For every simulated test, stress is quasi-instantaneously applied and maintained constant in time. Different curing conditions, age at loading time and surrounding relative humidity have been applied in order to show the effect of the models' basic parameters' choice on their predictions' sensitivity. Those conditions are presented in table 4. Although it does not correspond to any real application, those models have been tested to their limits (quasi-instantaneous response and long-term response) in order to show the sturdiness of their mathematical description and the effects of the physical basic hypotheses introduced in their constitutive equations on their asymptotic behavior.

Figures 9.31 to 9.38 present the evolutions of shrinkage strain ε_{sh} $(t-t_0)$ and compliance function $J(t,t_0,t')$ obtained from the five models for the four concrete compositions. For the models using the notion of creep coefficient, compliance function $J(t,t_0,t')$ has been obtained using equation [9.40].

General observation of those curves shows that the variation of the predicted strain by the different models is essentially due to shrinkage strain. In Figures 9.31 to 9.35 for usual drying conditions, the ACI 209 and ACI GZ models predict the highest shrinkage strains. This observation can be explained by the fact that those two models have been essentially set up on American experimental databases. Nevertheless, those databases mainly contain test results on concretes formulated from American standards, which authorize a higher aggregates/water content than the European ones.

Composition	B25	B40	B60 without FS	B60 with FS
Curing type	Norm., humid	Norm., humid	Norm., humid	Norm., humid
Age of concrete at the end of curing, t_0 (days)	5	10	7	7
Age of concrete at loading, t' (days)	20	45	30	30
Room temperature, T (°C)	20	20	20	20
Relative room temperature, h (%)	70	95	65	85

Table 9.4. *Loading, curing conditions, and applied environmental conditions for the tests carried out on the concrete's four compositions*

In Figures 9.31, 9.35 and 9.37, it is interesting to observe that because of the lack of ultimate delimitation of shrinkage, desiccation shrinkage predicted by the ACI GZ model indefinitely increases. Nevertheless, we can consider that under usual environmental conditions and for commonly used concrete compositions, those two effects should not have any practical consequences for a reasonable lifespan of the structure. However, the engineer in charge of the conception or maintenance should consider these observations in the choice of such a concrete or such a model, when particular conditions, such as maintaining a particularly dry environment, must be considered in the creation of the structure.

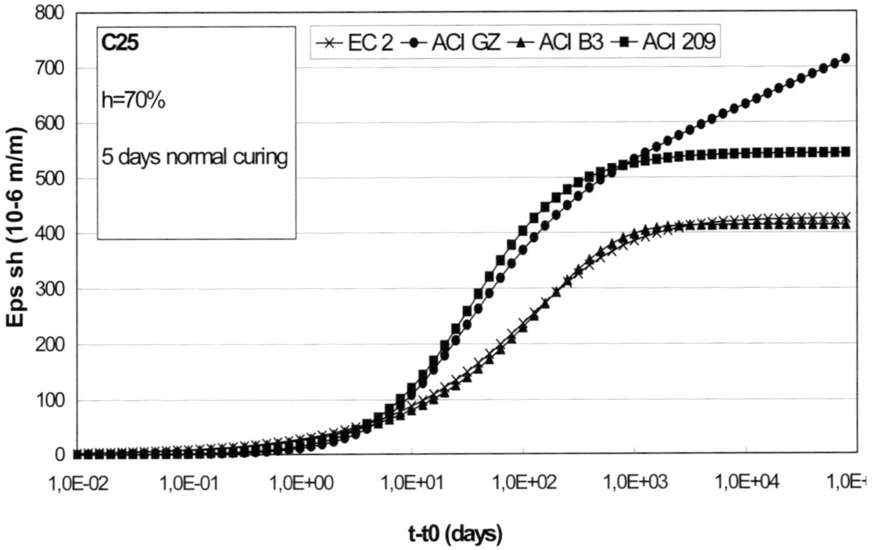

Figure 9.31. *B25 composition: evolution of the predicted shrinkage strain by the four models*

The observation of short-term predictions of different models show a few differences between quasi-instantaneous compliance functions provided by the ACI 209, Eurocode 2, and AFREM models, which result from the choice of the age of the concrete when Young's modulus is evaluated (at loading time t' or at 28 days). The B3 model, which introduces a reduction factor of the instantaneous strain, provides a value of this systematically lower function than those models at loading time. We can also observe that the adopted mathematical expression of creep compliance function in the ACI GZ model leads to a significant underestimation of the short-term creep strain. Finally, we can observe that only the AFREM model predicts a short-term shrinkage strain corresponding to endogenous shrinkage.

9.5.3.3. *Comparative analysis of the obtained results for a simple structure*

In order to complete the previous analyses, a comparative study of the results obtained by different models for a given structure are performed here on a simple structure. The studied structure is a pre-stressed beam. The composition of the concrete corresponds to B40, given in Table 9.3. This formulation is commonly used for building construction or the prefabrication of pre-stressed structural elements.

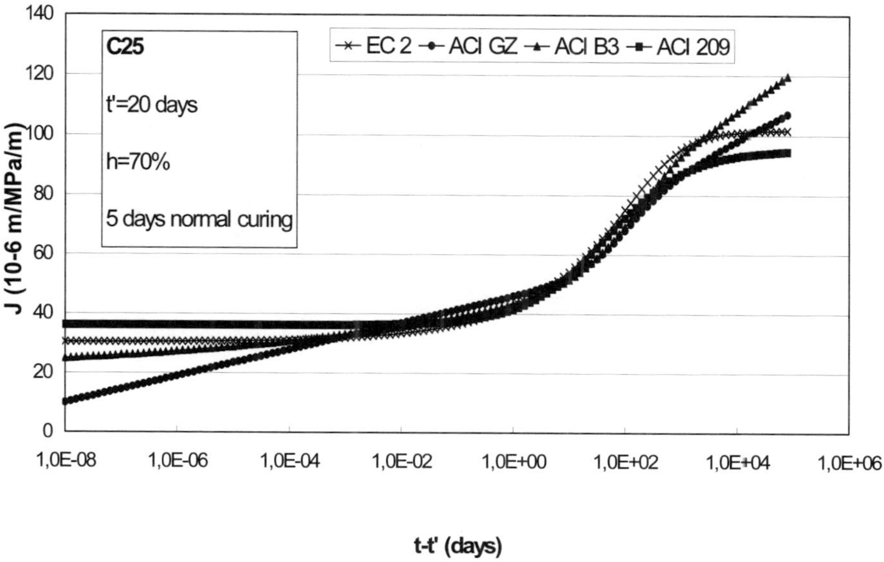

Figure 9.32. *B25 composition: evolution of the predicted compliance function by the four models*

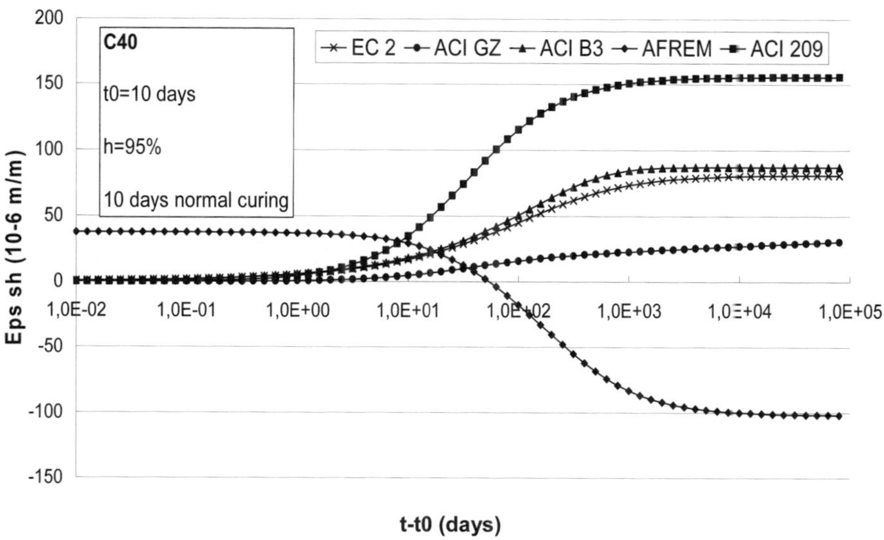

Figure 9.33. *B40 composition: evolution of the predicted shrinkage strain by the five models*

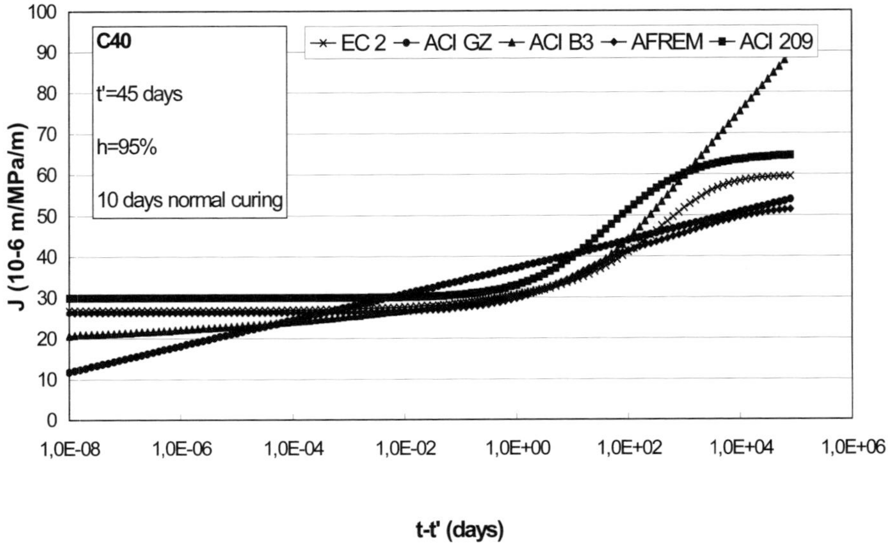

Figure 9.34. *B40 composition: evolution of the predicted compliance function by the five models*

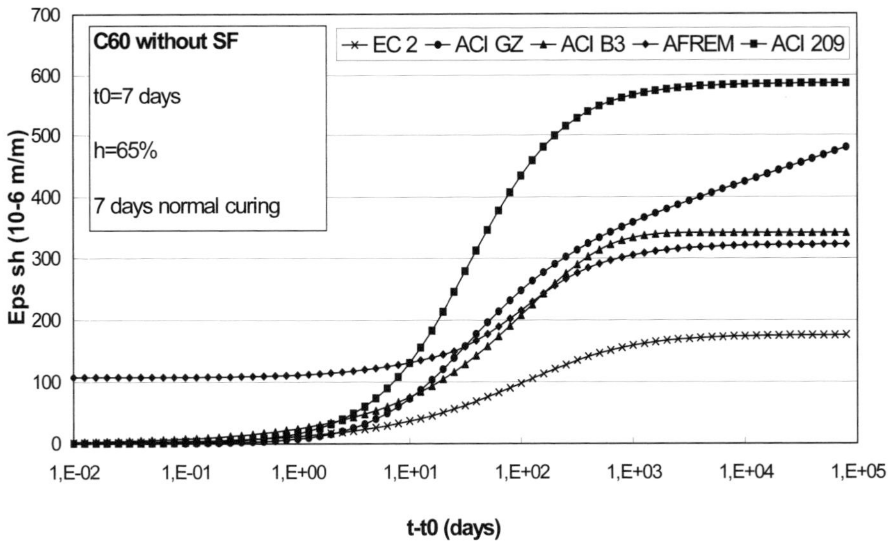

Figure 9.35. *B60 without SF composition: evolution of the predicted shrinkage strain by the five models*

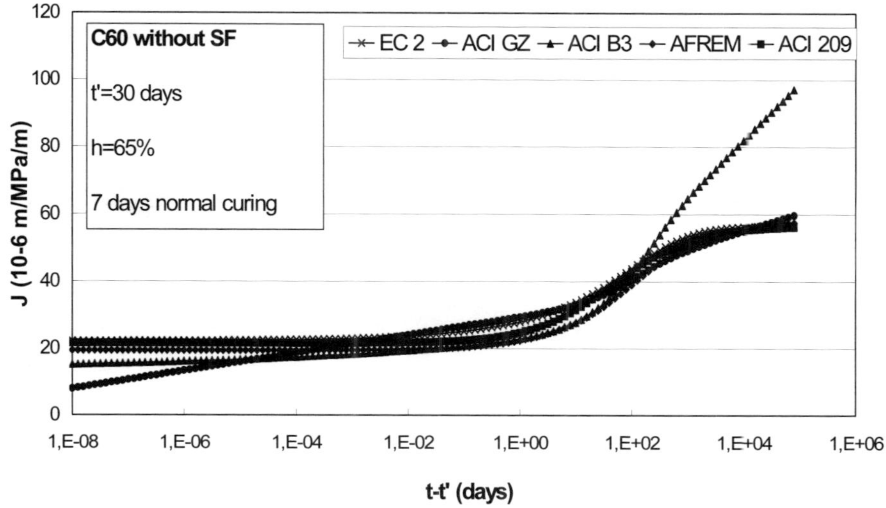

Figure 9.36. *B60 without SF composition: evolution of the predicted compliance function by the five models*

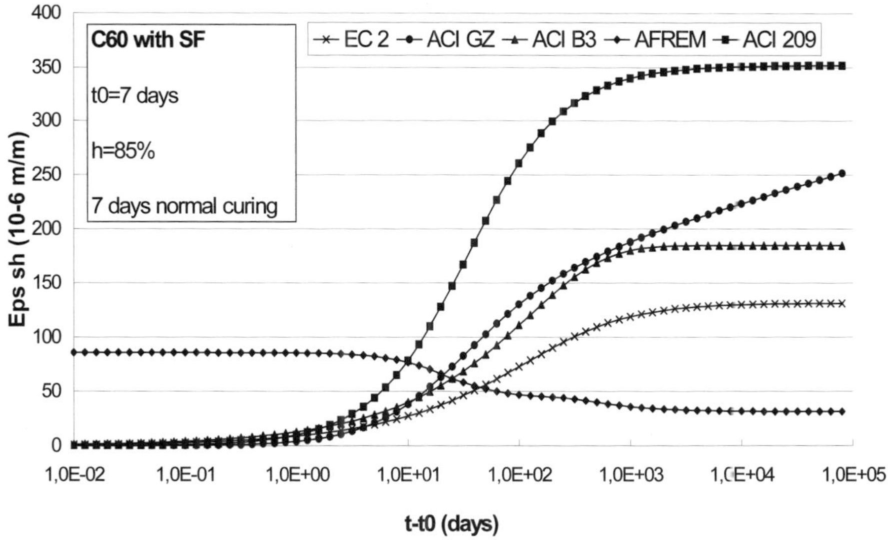

Figure 9.37. *B60 with SF composition: evolution of the predicted shrinkage strain by the five models*

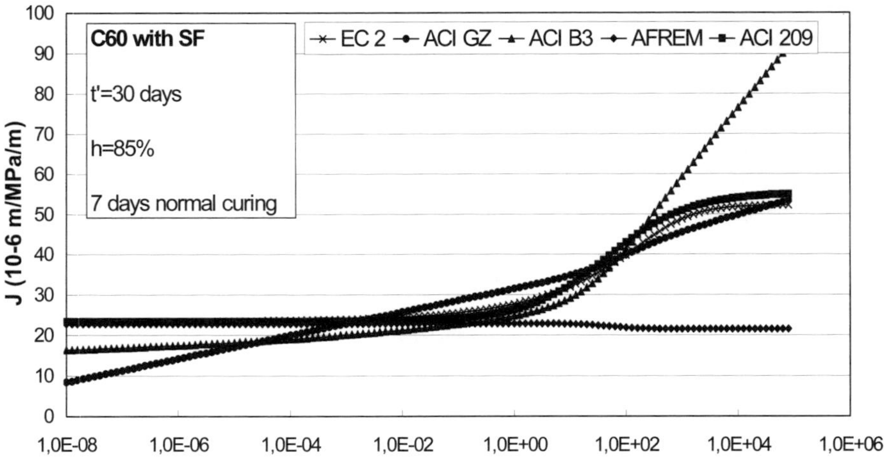

Figure 9.38. *B60 with SF composition: evolution of the predicted compliance function by the five models*

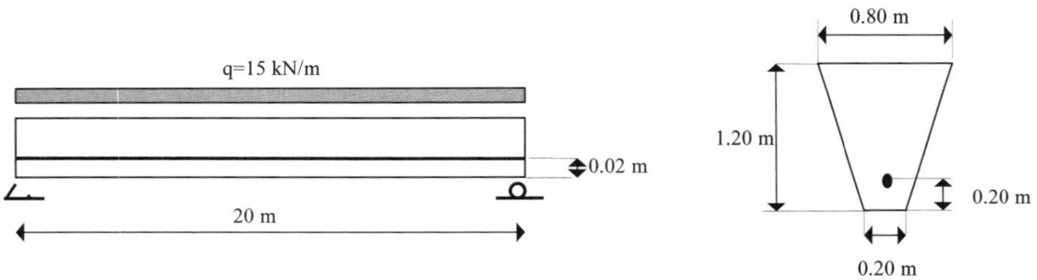

Figure 9.39. *Geometry of the studied beam*

The geometry of the studied beam is presented in Figure 9.39. The axial pre-stress is made of two 12T13 cables. An initial pre-stress of 2.4 MN is applied. A 3 day curing is carried out. A longitudinal load is applied at 25 days.

By ignoring steel relaxation, tension loss of the pre-stressed cables $\Delta\sigma s(t)$due to concrete delayed strains is calculated by the EUROCODE2 [EUR 91]:

$$\Delta\sigma_s(t) = \frac{\varepsilon_{sh}(t-t')\cdot E_s + \rho\cdot\varphi(t,t',t_0)\cdot(\sigma_{cl} + \sigma_{cp}(t'))}{1 + \rho\frac{A_s}{A_c}\left[\left(1 + \frac{A_c}{I_c}z_{pc}^2\right)(1 + \chi\cdot\varphi(t,t',t_0))\right]} \quad [9.57]$$

where E_s is the Young's modulus of steel, $\rho = E_s/E(t)$, $x=0.8$, A_s, and A_{cs} respectively, are the pre-stress and concrete steel sections. I_c is the section inertia, z_{pc} the distance between the center of gravity of the concrete section and the pre-stressed cables. σ_{cl} and σ_{cp} are the stresses within concrete induced by permanent loadings and pre-stress, respectively. Creep coefficient $J(t,t',t_0)$ is obtained for every model from equation [9.41].

Evolutions of the tension losses of the cables predicted by the different models are presented in Figure 9.40, in which a significant dispersion of the results obtained by the different models can be observed. This clearly shows that more significant estimations of pre-stress losses are obtained by the use of the ACI B3 and ACI GZ models, which is essentially due the more significant estimation of shrinkage in these models, as previously noted.

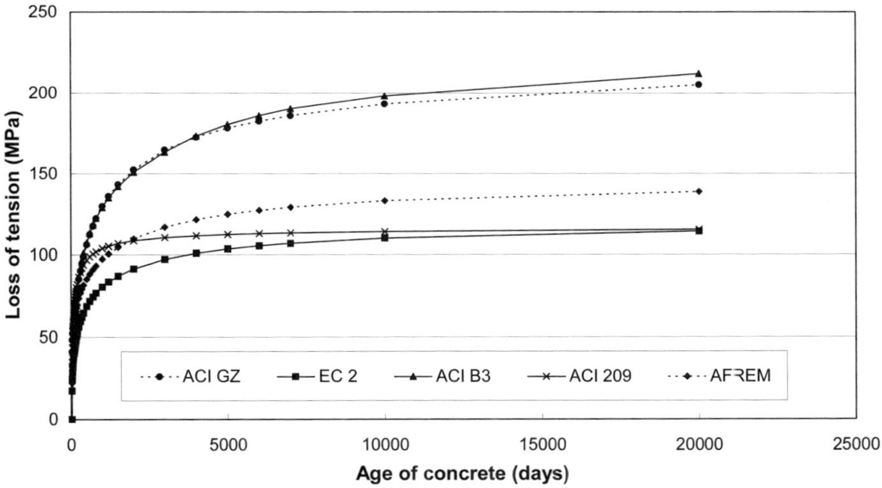

Figure 9.40. *Evolution in time of the cable's tension losses by the concrete's delayed strains*

Codified models of delayed strain in concrete are necessarily simplified and adapted to the resolution of specific problems because of the complexity of the physical phenomena and current limited knowledge. They rely on a certain number of hypotheses that systematically generate a prevision error. Knowing the existence of this error might still have some consequences on the design; therefore, the current normative trend consists of distinguishing the structures as a function of their sensitivity to delayed strains and of adapting for every structure of the model and the way in which delayed strains are studied.

An important part of prevision variability of the models mainly results from the composition parameters of the concrete. The current efforts of model development are moving towards the introduction and control of these parameters.

9.6. Conclusion

We have seen that the modeling of delayed strains is currently a complex problem, especially regarding creep. There are numerous physical mechanisms involved, which are inter-related or not, implicit or explicit, and conceded or contested. However, there are numerous experimental results that when exploitable, especially regarding protection against desiccation or measurement of transversal strains, provide access to the effects (measured delayed strains) of these mechanisms. Regarding modeling, shrinkage modeling seems to be unanimously accepted by the scientific community, whereas there is an important lack of consensus for creep modeling.

This situation is further complicated when the delayed behavior of concrete under multi-axial strains is studied. Something is still missing regarding structural effects induced by those delayed strains, and it is the identification of those structural effects that will enable us to quantify the intrinsic behavior of a material and then to better model it.

That is why this lack of knowledge is at the origin of the number of different models in codes, in addition to variability resulting from composition parameters. These differences between models are one of the first issues an engineer in charge of a project meets.

These issues highlight the questions that have to be resolved by research. The answers to these questions will enable the performances of present models to be improved, especially in terms of their parameter control.

9.7. Bibliography

[ACI 92] ACI, *Prediction of Creep, Shrinkage and Temperature Effects in Concrete Structures*, 209 R-9, ACI, 1992.

[ACK 83] P. ACKER, "De la controverse sur les lois de fluage du béton et des moyens d'en sortir", *Cement Concrete Res.*, vol. 13, 580-582, 1983.

[ACK 88] P. ACKER, *Comportement Mécanique du Béton: Apports de l'Approche Physico-Chimique*, Research Report of the L.C.P.C., no. 152, 1988.

[ACK 89] P. ACKER, M.Y. LAU, "Comportement mécanique du béton: validation expérimentale de la méthode du temps équivalent", *Bull. Liaison .LPC*, 163, 1989.

[ACK 93] P. ACKER, "Recommendation for peasurement of time-dependent strains of concrete loaded in compression", in *5th RILEM International Symposium on Creep and Shrinkage of Concrete*, Barcelona, pp. 849-858, 1993.

[ACK 98] P. ACKER, Z.P. BAŽANT, J.C. CHERN, C. HUET, F.H. WITTMANN, "RILEM recommendation on measurement of time-dependent strains of concrete. Rilem Draft Recommendation", *Mater. Struct.*, vol. 31, pp. 507-512, 1998.

[ACK 01] P. ACKER, "Micromechanical analysis of creep and shrinkage mechanisms", in F.J. ULM, Z.P. BAŽANT, F.H. WITTMANN, *Creep, Shrinkage and Durability Mechanics of Concrete and other Quasi-Brittle Materials, Concreep 6 Proceedings*, Boston, Elsevier, pp. 15-25, 2001.

[AIT 98] P. AÏTCIN, A. NEVILLE, P. ACKER, "Les différents types de retrait du béton", *Bull. Liaison LPC*, vol. 215, pp. 41-51, 1998.

[ALI 64] I. ALI, C. KESLER, "Mechanism of creep in concrete", in *Symposium on Creep of Concrete*, Detroit, ACI Publication SP-9, pp. 35-57, 1964.

[ALM 99] A. Al-MANASEER S. LAKSHMIKANTAN, "Comparison between current and future design code models for creep and shrinkage", *Rev. Française Génie Civil*, vol. 3, pp. 39-59, 1999.

[ALV 93] A.M. ALVAREDO, F.H. WITTMANN, "Shrinkage as influenced by strain-softening and crack formation", in Z.P. BAŽANT and I. CAROL, *Creep and Shrinkage of Concrete*, E&FN Spon, pp. 103-113, 1993.

[BAR 83] J.C. BARON, R. SAUTEREY, *Le Béton Hydraulique: Connaissance et Pratique*, Presse de l'ENPC, 1983.

[BAR 94] V. BAROGHEL-BOUNY, Caractérisation des pâtes de ciment et des bétons. Méthodes, analyse, interprétation, PhD thesis, ENPC Paris, 1994.

[BAR 97] J. BARON J.P. OLLIVIER, *Les Bétons: Bases et Données pour leur Formulation*, Edition Eyrolles, 1997.

[BAR 99] V. BAROGHEL-BOUNY, M. MAINGUY, T. LASSABATERE, O. COUSSY, "Characterization and identification of equilibrium and transefert moisture properties for ordinary and high-performance cementitious materials", *Cement Concrete Res.*, vol. 29, pp. 1225-1238, 1999.

[BAŽ 70] Z.P. BAŽANT, "Constitutive equation for concrete creep and shrinkage based on thermodynamics of multiphase systems", *Mater. Struct.*, vol. 3, pp. 3-36, 1970.

[BAŽ 72] Z.P. BAŽANT, "Thermodynamics of hindered adsorption and its implication for hardened cement paste and concrete", *Cement Concrete Res.*, vol. 2, pp. 1-16, 1972.

[BAŽ 82] Z. P. BAŽANT, F.H. WITTMAN, *Mathematical Modeling of Creep and Shrinkage of Concrete*, John Wiley & Sons Limited, 1982.

[BAŽ 85] Z. P. BAŽANT, J. C. CHERN, "Concrete creep at variable humidity: constitutive law and mechanism", *Mater. Struct.*, vol. 18, pp. 1-20, 1985.

[BAŽ 94] Z.P. BAŽANT, Y. XI, "Drying creep of concrete: constitutive model and new experiments separating its mechanisms", *Mater. Struct.*, vol. 27, pp. 3-14, 1994.

[BAŽ 97a] Z.P. BAŽANT, A.B. HAUGGAARD, S. BAWEJA, F.J. ULM, "Microprestress-solidification theory for concrete creep 1: aging and drying effects", *J. Eng. Mech.*, vol. 123, pp. 1188-1194, 1997.

[BAŽ 97b] Z.P. BAŽANT, A.B. HAUGGAARD, S. BAWEJA, "Microprestress-solidification theory for concrete creep 2: algorithm and verification", *J. Eng. Mech.*, vol. 123, pp. 1195-1201, 1997.

[BAŽ 97c] Z.P. BAŽANT, Y. XIANG, "Crack growth and life time of concrete under long time loading", *J. Eng. Mech.*, vol. 123, pp. 350-358, 1997.

[BAŽ 98a] Z.P. BAŽANT, S. BAWEJA, *Creep And Shrinkage Prediction Model For Analysis And Design Of Concrete Structures: Model B3*, Structural Engineering Report 96-3/ITIc, March, ACI, 1996 (revised May 1998).

[BAŽ 98b] Z.P. BAŽANT, *Criteria For Rational Prediction Of Creep And Shrinkage Of Concrete*, Structural Engineering Report 98-7/B675c to ACI Committee 209, July, ACI, 1998.

[BAŽ 01] Z. P. BAŽANT, "Prediction of concrete creep and shrinkage: past, present and future", *Nucl. Eng. Des.*, vol. 203, pp. 27-38, 2001.

[BAZ 08] Z.P. BAŽANT, G.-H. LI, "Comprehensive database on concrete creep and shrinkage", *ACI Mater. J.*, vol. 105, pp. 635-637, 2008.

[BEL 01] F. BELZUNG, F.H. WITTMANN, L. HOLZER, "Influence of composition of pore solution on drying shrinkage", in F.J. ULM, Z.P. BAZANT, F.H. WITTMANN, *Creep, Shrinkage and Durability Mechanics of Concrete and other Quasi-Brittle Materials, Concreep 6 Proceedings*, Boston, Elsevier, pp. 39-48, 2001.

[BEN 79] A. BENTUR, R.L. BERGER, F.V. LAWRENCE, N.B. MILESTONE, S. YOUNG, J.F. YOUNG, "Creep and drying shrinkage of calcium silicates pastes: a hypothesis of irreversible strains", *Cement Concrete Res.*, vol. 9, pp. 83-96, 1979.

[BEN 00a] F. BENBOUDJEMA, F. MEFTAH, G. HEINFLING, J.-M. TORRENTI, *Analyse des essais de fluage propre réalisés sur des pâtes de ciment, mortiers et bétons, avec mesures des déformations principales*, EDF, September 2000.

[BEN 00b] F. BENBOUDJEMA, F. MEFTAH, G. HEINFLING, J.-M. TORRENTI, *Étude hydro-mécanique des éprouvettes de fluage biaxial*, EDF, October 2000.

[BEN 01] F. BENBOUDJEMA, F. MEFTAH, A. SELLIER, J.-M. TORRENTI, G. HEINFLING, "On the prediction of delayed strains for concrete subjected to drying and loading simultaneously", in F.J. ULM, Z.P. BAZANT, F.H. WITTMANN, *Creep, Shrinkage and Durability Mechanics of Concrete and other Quasi-Brittle Materials, Concreep 6 Proceedings*, Boston, Elsevier, pp. 245-250, 2001.

[BEN 05] F. BENBOUDJEMA, F. MEFTAH, J.-M. TORRENTI, "Interaction between drying, shrinkage, creep and cracking phenomena in concrete", *Eng. Struct.* vol. 27, pp. 239–250, 2005.

[BEN 07] F. BENBOUDJEMA, F. MEFTAH, J.-M. TORRENTI, “A viscoelastic approach for the assessment of the drying shrinkage behaviour of cementitious materials”, *Mater. Struct.* vol. 40, pp. 163–174, 2007.

[BOU 07] M. BOUASKER, Etude numérique et expérimentale du retrait endogène au très jeune âge des pâtes de ciment avec et sans inclusions, PhD thesis, University of Nantes, 2007.

[BPEL 83] BPEL, “Règles techniques de conception et de calcul des ouvrages et constructions en béton précontraint par la méthode des états limites”, *Bulletin Officiel*, Fascicule Spécial 83-31 bis, 1983.

[BRO 76] N.H. BROWN, B.B. HOPE, “The creep of hydrated cement paste”, *Cement Concrete Res.*, vol. 6, pp. 475-486, 1976.

[BRU 69] S. BRUNAUER, J. SKALNY, E.E. BODOR, “Adsorption on nonporous solids”, *J. Colloid Interface Sci.*, vol. 30, pp. 546-552, 1969.

[CAR 37] R.W. CARLSON, “Drying shrinkage of large concrete members”, *J. Am. Concrete Inst.*, vol. 33, pp. 327-336, 1937.

[CEB 90] CEB, “Comité Euro-international du béton, CEB-FIP Model code 1990”, *CEB Bulletin d’information*, vol. 213/214, Thomas Telford, 1993.

[CLÉ 00] J.-L. CLEMENT, F. LE MAOU, “Etude de la répétabilité des essais de fluage sur éprouvettes de béton”, *Bull. Liaison LPC*, vol. 228, pp. 59-69, 2000.

[CON 04] G. CONSTANTINIDES, F.-J. ULM, “The effect of two types of C-S-H on the elasticity of cement-based materials: results from nanoindentation and micromechanical modeling”, *Cement Concrete Res.*, vol. 34, pp. 67–80, 2004.

[COU 01] O. COUSSY, V. BAROGHEL-BOUNY, P. DANGLA, M. MAINGUY, “Evaluation de la perméabilité à l’eau liquide des bétons à partir de leur perte de masse durant le séchage”, *Rev. Française Génie Civil*, vol. 5, pp. 269-284, 2001.

[DAI 01] J.-F. DAÏAN, “Evaluation des propriétés de transfert dans les matériaux cimentaires”, *Rev. Française Génie Civil*, vol. 5, pp. 179-202, 2001.

[DAY 83] R.L. DAY, B.R. GAMBLE, “The effect of changes in structure on the activation energy for the creep of concrete”, *Cement Concrete Res.*, vol. 13, pp. 529-540, 1983.

[DIA 87] W.P.S. DIAS, G.A. KHOURY, P.J.E. SULLIVAN, “An activation approach for the temperature dependence of basic creep of hardened cement paste”, *Mag. Concrete Res.*, vol. 39, pp. 141-147, 1987.

[ESP 98] B. ESPION, “Evaluation du retrait et du fluage du béton dans les codes de calcul actuels en Europe et aux Etats-Unis”, *Bull. Liaison LPC* special issue XX, 21-27, 1998.

[EUR 91] EUROCODE 2, “Règles générales et règles pour les bâtiments”, Part 1-1, in *Calcul des Structures en Béton*, CEN, 1991.

[EYM 91] R. EYMARD, P. ACKER, J.M. PIAU, “Ingénierie des ouvrages en béton: la prise en compte des effets thermiques et hydriques”, *Ann. Ponts Chaussées*, vol. 58, pp. 24-30, 1991.

[EYM 94] R. EYMARD, "Allowing for the creep of concrete in a finite element structural calculation", *Comp. Struct.*, vol. 53, pp. 921-928, 1994.

[FEL 68] R.F. FELDMAN, P.J. SEREDA, "A model for hydrated Portland cement paste as deduced from sorption length change and mechanical properties", *Mater. Struct.*, vol. 1, pp. 509-520, 1968.

[FEL 72] R.F. FELDMAN, "Mechanism of creep of hydrated Portland cement paste", *Cement Concrete Res.*, vol. 2, pp. 509-520, 1972.

[GAR 98] N.J. GARDNER, "Design provisions for shrinkage and creep of concrete", in *Paris ACI Chapter Symposium*, April 1998.

[GAU 82] C. GAUCHER, *Fluage du Béton: Tentative de Caractérisation du Comportement Rhéologique non Linéaire dans la Représentation par Intégrales Multiples*, LCPC, no. 108, 1982.

[GER 00] B. GÉRARD, J. MARCHAND, "Influence of cracking on the diffusion properties of cement-based materials part I: influence of continuous cracks on the steady-state regime", *Cement Concrete Res.*, vol. 30, pp. 37-43, 2000.

[GLU 62] J. GLUCKLICH, O. ISHAI, "Creep mechanism in cement mortar", *ACI J.*, vol. 59, pp. 923-948, 1962.

[GLU 72] J. GLUCKLICH, A. AMAR, "The volumetric creep of mortars subjected to triaxial compression", *La déformation et la rupture des solides soumis à sollicitations pluriaxiales*, *RILEM International Symposium*, Cannes, vol. 1, pp. 79-95, 1972.

[GOP 69] K.S. GOPALAKRISHNAN, A.M. NEVILLE, A. GHALI, "Creep Poisson's ratio of concrete under multiaxial compression", *ACI J.*, vol. 66, pp. 1008-1020, 1969.

[GRA 95] L. GRANGER, Comportement différé du béton dans les enceintes de centrales nucléaires : analyse et modélisation, PhD thesis, ENPC, 1995.

[GRA 97] L. GRANGER, M. PRAT, P. BISCH, A. MILLARD, P. MESTAT, P. PIJAUDIER-CABOT, "Ouvrages en béton sonnes aux effets du fluage", Chapter 14, in *Calcul des ouvrages généraux de construction*, Hermes, pp. 612-654, 1997.

[GUE 97] I. GUENOT-DELAHAIE, "Contribution à l'analyse physique et à la modélisation du fluage propre du béton", PhD thesis, ENPC, 1997.

[HAN 60] T.C. HANSEN, "Creep of Concrete. The influence of variations in the humidity of the ambient atmosphere", in *6th Congress of the International Association of Bridge and Structural Engineering*, Stockholm, 57-65, 1960.

[HAN 87] W. HANSEN, "Drying shrinkage mechanisms in Portland cement pastes", *J. Am. Ceramic Soc.*, vol. 70, pp. 323-328, 1987.

[HEI 99] G. HEINFLING, A. COURTOIS, B. ESPION, "A comparative study of code type models as an illustration of a guide for the engineer", *Rev. Française Génie Civil*, vol. 3, pp. 219-243, 1999.

[ISH 68] O. ISHAI, "The time dependent deformational behaviour of cement past, mortar and concrete", in *Proceedings of International Conference On the Structure of Concrete, Cement and Concrete Association*, London, 345-368, 1968.

[JEN 92] H.M. JENNINGS, Y. XI, "Relationships Between Microstructure and Creep and Shrinkage of Cement Paste", in J. SKALNY, *Material Science of Concrete*, vol. III, The American Ceramic Society, pp. 37-69, 1992.

[JOR 69] I.J. JORDAAN, J.M. ILLSTON, "The creep of sealed concrete under multiaxial compressive stresses", *Mag. Concrete Res.*, vol. 21, pp. 195-204, 1969.

[KAC 83] L. M. KACHANOV, *Introduction to Continuum Damage Mechanics*, Martinus Nijhoff Publishers, 1986.

[KOT 83] M.D. KOTSOVOS, "Effect of testing techniques on the post-ultimate behaviour of concrete in compression", *Mater. Struct.*, vol. 16, pp. 3-12, 1983.

[LAS 97] T. LASSABATERE, J.-M. TORRENTI, L. GRANGER, "Sur le couplage entre séchage du béton et contrainte appliqué", *Actes du Symposium Saint-Venant*, Paris, pp. 331-338, 1997.

[LER 96a] R. LE ROY, F. DE LARRARD, G. PONS, "Calculs des déformations instantanées et différées des bétons à hautes performances", in *Extension du domaine d'application des règlements de calcul BAEL-BPEL aux bétons à 80 Mpa, Bull. LPC,* special issue XIX, 1996.

[LER 96b] R. LE ROY, Déformations instantanées et différées des bétons à hautes performances, PhD thesis, ENPC, 1996.

[LER 98] R. LE ROY., R. CHAUSSIN, "Comparaison entre le modèle AFREM de retrait et fluage des bétons et les règles BPEL 91", *Bull. LPC*, special issue XX, 29-37, 1998.

[MAI 01] M. MAINGUY, O. COUSSY, V. BAROGHEL-BOUNY, "Role of air pressure in drying of weakly permeable materials", *J. Eng. Mech.*, vol. 127, pp. 582-592, 2001

[MAN 41] G.A. MANEY, "Concrete under sustained working loads: evidence that shrinkage dominates time yield", *Proceedings ASTM*, vol. 41, pp. 1021-1030, 1941.

[MAZ 84] J. MAZARS, Application de la mécanique de l'endommagement au comportement non linéaire et à la rupture de béton de structure, PhD Thesis, University of Paris VI, 1984.

[MEF 00] F. MEFTAH, J.-M. TORRENTI, W. N. NECHNECH, F. BENBOUDJEMA, C. DE SA, "An elasto-plastic damage approach for the modeling of concrete submitted to the mechanical induced effects of drying", in *Shrinkage 2000 Proceedings*, RILEM Publications, pp. 341-354, 2000.

[MEL 99] Règles BAEL91 modifiées 99, Règles techniques de construction et de calcul des ouvrages et construction en béton armé suivant la méthode des état-limites, no. 99-27, April 1999.

[MUL 64] W.G. MULLEN, W.L. DOLCH, "Creep of portland cement paste", in *Proceedings of the American Society of Testing Materials*, vol. 64, pp. 1146-1171, 1964.

[MUL 93] H.S. MÜLLER, "Considerations on the development of a database on creep and shrinkage tests", in *5th RILEM International Symposium on Creep and Shrinkage of Concrete*, Barcelona, pp. 859-872, 1993.

[NEV 57] A.M. NEVILLE, "The measurement of creep of mortar under fully controlled conditions", *Mag. Concrete Res.*, vol. 9, pp. 9-12, 1957.

[NEV 60] A.M. NEVILLE, "Recovery of creep and observations on the mechanism of creep of concrete", *Appl. Sci. Res.*, vol. 9, pp. 71-84, 1960.

[NEV 83] A.M. NEVILLE, W.H. DILGER, J.J. BROOKS, *Creep of plain and structural concrete*, Construction Press, Longman Group Ltd, 1983.

[PAR 73] L.J. PARROTT, "An examination of the effect of age at loading upon the creep of hardened cement pastes", *Mag. Concrete Res.*, vol. 25, pp. 197-200, 1973.

[PES 00] G. PESCHEL, "Disjoining forces in pore water", in M. SETZER, *Pore Solution in Hardened Cement Paste*, Aedificatio Publishers, pp. 3-26, 2000.

[PIH 82] S.E. PIHLAJAVAARA, "Estimation of drying of concrete at different relative humidities and temperatures of ambiant air with special discussion about fundamental features of drying and shrinkage", in Z.P. BAŽANT, F.H. WITTMAN, *Creep and Shrinkage in Concrete Structures*, John Wiley & Sons Ltd, 1982.

[POW 59] T.C. POWERS, "Causes and control of volume change", *J. Portland Cement Assoc. Res. Devel. Lab.*, vol. 1, pp. 29-39, 1959.

[POW 68] T.C. POWERS, "Mechanisms of shrinkage and reversible creep of hardened Portland cement paste", in *Proceedings of International Conference On the Structure of Concrete*, Cement and Concrete Association, pp. 319-344, 1968.

[RUE 68] W. RUETZ, "A hypothesis for creep of hardened cement paste and the influence of simultaneous shrinkage", Proceedings of International Conference On the Structure of Concrete, Cement and Concrete Association, pp. 365-387, 1968.

[SER 76] Service Du Materiel Du LCPC., "Appareil de mesure de fluage en compression des bétons de ciment", *Bull. Liaison LPC*, vol. 84, 1976.

[TOR 96] J.-M. TORRENTI, "Comportement mécanique du béton: bilan de six années de recherche, *Etudes et recherches des Laboratoires des Ponts et Chaussées*, "*Ouvrages d'Art*", OA 23, LCPC, 1996.

[TOR 97] J.-M. TORRENTI, L. GRANGER, M. DIRUY, P. GENIN, "Modélisation du retrait du béton en ambiance variable", *Rev. Française Génie Civil*, vol. 1, pp. 687-698, 1997.

[TOR 05] J.-M. TORRENTI, F. BENBOUDJEMA, "Mechanical threshold of concrete at an early age", *Mater. Struct.*, vol. 38, pp. 299-304.

[TOU 96] F. TOUTLEMONDE, F. LE MAOU, "Protection des éprouvettes vis-à-vis de la dessiccation-Le point sur quelques techniques de laboratoire", *Bull. LPC*, vol. 203, pp. 105-119, 1996.

[ULM 98] F.-J. ULM, P. ACKER, "Le point sur le fluage et la recouvrance des bétons", *Bull. LPC*, special issue XX, 73-82, 1998.

[ULM 07] F.-J. ULM, M. VANDAMME, C. BOBKO, J. A. ORTEGA, "Statistical indentation techniques for hydrated nanocomposites: concrete, bone, and shale", *J. Am. Ceramics Soc.*, vol. 90, pp. 2677–2692, 2007.

[VAN 09] M. VANDAMME, F.J. ULM, "Nanogranular origin of concrete creep", *Proc. Natl Acad. Sci. USA,* doi:10.1073/pnas.0901033106, 2009.

[VER 68] G.J. VERBECK, R.A. HELMUTH, "Structures and physical properties of cement paste", in *5th International Symposium on Chemistry of Cement*, Cachan, vol. 3, pp. 1-36, 1968.

[VLI 00] M.R.A. VAN VLET, Size effect in tensile fracture of concrete and rock, PhD thesis, Delft University of Technology, 2000.

[WIR 00] L. WIRTANEN, V. PENTTALA, "Influence of temperature and relative humidity on the drying of concrete", *Concrete Sci. Eng.*, vol. 2, pp. 39-47, 2000.

[WIT 70] F.H. WITTMANN, "Influence of moisture content on the creep of hardened cement", *Reol. Acta*, vol. 9, pp. 282-287, 1970.

[WIT 76] F.H. WITTMANN, "The structure of hardened cement paste – a basis for better understanding of materials properties", in *Proceedings Conference On Hydraulic Cement Pastes: Their Structure and Properties,* Sheffield, Cement and Concrete Association, pp. 96-117, 1976.

[WIT 82] F.H. WITTMANN, "Creep and shrinkage mechanisms", in Z.P. BAZANT, F.H. WITTMANN, *Creep and Shrinkage of Concrete Structures*, J Wiley and Sons, 1982.

[WIT 01] F.H. WITTMANN, "Mechanisms and mechanics of shrinkage", in F.J. ULM, Z.P. BAŽANT, F.H. WITTMANN, *Creep, Shrinkage and Durability Mechanics of Concrete and other Quasi-Brittle Materials*, *Concreep 6 Proceedings*, Boston, Elsevier, pp. 3-12, 2001.

[XI 94] Y. XI, Z.P. BAŽANT, L. MOLINA, H.M. JENNINGS, "Moisture diffusion in cementitious materials: moisture capacity and diffusivity", *Adv. Cement Based Mater.*, vol. 1, pp. 258-266, 1994.

[YOU 74] J.F. YOUNG, "Assessment of the influence on microstructure on time dependent deformations of hardened cement past", in *Proceedings of First Australian Conference On Engineering Materials*, Sidney, pp. 3-28, 1974.

Closing Remarks: New Concretes, New Techniques, and Future Models

From concrete to concrete materials

Concrete has been and still is a construction material that is changing following the advances of sciences and engineering. Over the past century, it has been considered as a "simple construction material", probably because it remains very simple to prepare and to cast, but concrete is not a "single" material. There are many concretes today, with different colors, and different mechanical or durability properties. Concrete engineering has departed from the old idea that this material should resist up to 30 MPa in compression, with a water to cement ratio equal to 0.5. This revolution of cement-based materials followed several steps that we are going to revisit in these concluding remarks inspired by a contribution by Prof. Y. Malier in the French edition of this book [MAL 05].

Decrease the amount of water

Water in concrete plays several roles. The first is to ensure the workability of the fresh mix And the second, of course, to promote hydration. Compared to the amount of water that was usually chosen in classical concrete formulations, the amount of water needed for hydration is much lower that what should be used in practice. This additional water stays in the material upon hydration. It is not trapped, however, and exchange with the outside environment is possible. This exchange generates microcracking and delayed strains. The aim of modern approaches to concrete is to

---

Written by Jean-Michel TORRENTI, Gilles PIJAUDIER-CABOT and Jean-Marie REYNOUARD.

decrease the amount of water, along with the inherent negative consequences of free water in concrete. The extensive use of admixtures (superplasticizers) for this purpose is advocated.

Control the workability

In some instances, the workability of concrete may play a pivotal economic role. Self-placing mortar and self-compacting concretes were designed with this aim. They were a revolution in the concrete casting process, decreasing the manpower needed for casting, decreasing the noise level, and decreasing the difficulties due to the vibration during casting, and at the same time accelerating the placing of concrete. Self-placing concretes were first used in Japan more than a decade ago. This is still a new material whose properties are thoroughly investigated. Among these, such materials with specific rheological properties when freshly mixed have heralded research directed at describing the flow of concrete inside moulds and around reinforcements. These approaches, that one might call *ab initio structural concrete models*, are currently the focus of many promising interdisciplinary works.

Use an optimal granular structure

Over the past 20 years, modern theories on granular packing devised in physics have had a strong impact on concrete technology. Optimization of granular packing is one of the techniques available for designing high-strength concrete, along with the reduction of water. In order to optimize the granular structure of concrete, it is necessary to cover at least four decades of granular sizes. Unhydrated cement, along with sand, coarse and fine gravel provided this missing size in the range of 0.1 μm. Mineral admixtures, such as silica fume or filler, also provided a useful complement, and with the optimization of granular packing in concrete, high-performance cementitious materials were born.

Today there are numerous examples of such materials, with compressive and tensile strength improved by an order of magnitude, and with durability properties that were even more enhanced than the mechanical properties. Today's concretes are yesterday's high-performance concretes. New materials were developed with applications reaching outside the civil engineering industry. Visiting cards can be made of macro-defect-free cementitious materials and UHPC (ultra high-performance concrete) has been used in architectural ornaments in several buildings in Paris and Villeneuve d'Ascq (France).

New standards for new concretes

Decreasing the water content and increasing the granular packing are two methods with the same aim: to produce a concrete with designed "*ad-hoc* properties". This is the point of departure from the recent European standards; however, tailoring concrete for each application is still an endeavor.

There are at least four categories of properties that should be considered:

– the workability of fresh concrete, with applications on site and in the precast concrete industry;

– the mechanical properties of concrete, including the time-dependent response of the material and the aging phase;

– the durability of concrete, which mainly deals with mass exchange and the chemical stability of the material;

– the esthetic aspect of cast concrete, which has become a true architectural asset in the construction industry, but also surface properties, such as rugosity and abrasion.

Producing "the" concrete that corresponds to specific properties required by a specific usage is an inverse problem, which is a great challenge for research and for the engineering practice. This is a key problem, without which the progress of standards might not follow the pace of innovation.

There is always a time lag between the development of new concrete materials and their regulation and consideration in our various construction codes and regulations. Any progress in developing robust techniques for designing concrete materials with selected properties and specifications will add confidence in this rather new approach of concrete technology. In return, it will help to close the gap between research and standardization to promote faster innovation in the construction industry.

The progress of standards cannot follow a "blue sky" approach. It has to be driven by the appeal of new, environmentally and economically effective materials, in conjunction with appropriate construction techniques and innovative design. This is a "global approach to new concrete materials" which reaches towards innovation in architecture, engineering, maintenance, safety, and economy. The construction industry has entered the era of system engineering, following many other industries. Concrete technology is one pillar upon which advances should rest, because it is the source of many innovations and also because it is the most widely used construction materials.

Future challenges and future materials

Although concrete is a rather old material, many properties remain to be understood and explained. Creep and shrinkage, for instance, are among these issues dealt with in this book, on which consensus in the engineering community has not been reached yet.

Reaching for advances in nanosciences

Unresolved questioning calls for new investigation techniques and new approaches. In this respect, concrete is probably one of the oldest nano-organized manufactured material. The investigation of its properties should benefit from the advances in imaging microstructures and measuring on-site mechanical properties. At the same time, the complexity involved in theoretical upscaling techniques should benefit from research progress in nanosciences, whether inspired from biological materials or solid-state chemistry. The advent of extraordinary computer-simulation capabilities and computer grids developed for particle physics, also heralds new *ab initio materials simulations* aimed at predicting various properties involved in mass transfer or mechanical loads, the counterparts of future abovementioned *ab initio structural concrete computations*. Despite their wide implementation, concrete materials have many properties that remain unexplained. This is also the case for many materials whose properties are currently being revisited, such as metal alloys, glass, or polymers.

Aiming for new sustainable materials

Manufacturing concrete has a rather important impact on our environment. We are not speaking about our architectural environment, on which societal consequences of construction plays indeed a very important role, but we will deal in this section with the consequences induced by the production of aggregate and clinker.

Regulations will very soon prevent the usage of natural aggregate and at the same time it will be delicate to find adequate sites for aggregate extraction. Modern sustainable approaches call for recycling as much as possible construction materials. Again, new concretes will be born from this process, new concretes for which the characterization of their properties and new tailoring techniques should be investigated.

Recycling is the latest issue for concrete, with long-term visions entailing concrete production with less Portland-type cement, which is not obtained through environmentally friendly processes. The consumption of carbon dioxide is rather high (around 800 kg/t of cement), cement is not a "green material" and soon its production will be regulated. As a consequence it will become more and more expensive. Developing a new material to replace Portland cement is quite a challenge for which collaborations between engineering, physics, and chemistry is certainly required.

This is a challenge that newly educated students in concrete technology, civil engineering, and building construction should face. They should be tempted to follow the steps presented in this book – going from the simple mechanical models to advanced non-linear modeling and advanced computer methods, and then accounting for time-dependent effects. Multiphysics and multiscale modern approaches are expected to enrich this process, but a global approach should always be undertaken, and at the same time inspiration should be drawn from other fields of materials science to promote new ways of thinking.

Bibliography

[MAL 05] MALIER Y., "L'apport de nouveaux bétons", in J.-M. REYNOUARD, G. PIJAUDIER-CABOT, *Comportement mécanique du béton*, Traité MIM, Hermès, 2005.

List of Authors

Farid BENBOUDJEMA
LMT
ENS Cachan
France

Benoît BISSONETTE
Centre de recherche sur les infrastructures en béton
University of Laval
Quebec
Canada

Nicolas BURLION
LML
Polytech Lille
Villeneuve d'Ascq
France

Jean-François DESTREBECQ
LaMI
Blaise Pascal University
Clermont-Ferrand
France

Jean-François DUBÉ
LMGC
University of Montpellier II
France

Fabrice GATUINGT
LMT
ENS Cachan
France

Jean-François GEORGIN
LGCIE, INSA de Lyon
Villeurbanne
France

Khalil HAIDAR
R&DO – GeM
Ecole Centrale de Nantes
France

Grégory HEINFLING
EDF – SEPTEN
Villeurbanne
France

Fabrice LEMAOU
LCPC
Paris
France

Jacques MARCHAND
Centre de recherche sur les infrastructures en béton
University of Laval
Quebec
Canada

Fekri MEFTAH
Laboratoire de Mécanique et
Matériaux du Génie Civil (L2MGC)
University of Cergy Pontoise
France

Buquan MIAO
University of Moncton
Canada

Gilles PIJAUDIER-CABOT
Laboratoire des fluides complexes
University of Pau et des Pays de l'Adour
Anglet
France

Jean-Marie REYNOUARD
LGCIE, INSA de Lyon
Villeurbanne
France

Jean-Michel TORRENTI
LCPC
Paris
France

Franz-Josef ULM
MIT
Cambridge
USA

Vincent WALLER
Unibéton
Guerville
France

Index

E

F

H, I

L, M, N

P

R

S

T

V, W